AAPG TREATISE OF PETROLEUM GEOLOGY

The American Association of Petroleum Geologists
gratefully acknowledges and appreciates
the leadership and support of the
AAPG Foundation in the development of the
Treatise of Petroleum Geology

DIAMOND JUBILEE
AAPG
1917-1991

SOURCE AND MIGRATION PROCESSES AND EVALUATION TECHNIQUES

EDITED BY
ROBERT K. MERRILL

TREATISE OF PETROLEUM GEOLOGY
HANDBOOK OF PETROLEUM GEOLOGY

TREATISE EDITORS:
NORMAN H. FOSTER
AND
EDWARD A. BEAUMONT

DIAMOND JUBILEE
AAPG
1917-1991

Published by
The American Association of Petroleum Geologists
Tulsa, Oklahoma, U.S.A. 74101

Published November 1991

ISBN: 0-89181-600-3

Association Editor: Susan A. Longacre
Science Director: Gary D. Howell
Publications Manager: Cathleen P. Williams
Special Projects Editor: Anne H. Thomas
Production: Kathy and Dana Walker, Editorial Technologies

Contents

TREATISE OF PETROLEUM GEOLOGY

ADVISORY BOARD

American Association of Petroleum Geologists Foundation
Treatise of Petroleum Geology Fund*

Major Corporate Contributors
($25,000 or more)

Amoco Production Company
BP Exploration Company Limited
Chevron Corporation
Exxon Company, U.S.A.
Mobil Oil Corporation
Oryx Energy Company
Pennzoil Exploration and Production Company
Shell Oil Company
Texaco Foundation
Union Pacific Foundation
Unocal Corporation

Other Corporate Contributors
($5,000 to $25,000)

ARCO Oil & Gas Company
Ashland Oil, Inc.
Cabot Oil & Gas Coporation
Canadian Hunter Exploration Ltd.
Conoco Inc.
Marathon Oil Company
The McGee Foundation, Inc.
Phillips Petroleum Company
Transco Energy Company
Union Texas Petroleum Corporation

Major Individual Contributors
($1,000 or more)

John J. Amoruso
Thornton E. Anderson
C. Hayden Atchison
Richard A. Baile
Richard R. Bloomer
A. S. Bonner, Jr.
David G. Campbell
Herbert G. Davis
George A. Donnelly, Jr.
Paul H. Dudley, Jr.
Lewis G. Fearing
Lawrence W. Funkhouser
James A. Gibbs
George R. Gibson
William E. Gipson
Mrs. Vito A. (Mary Jane) Gotautas
Robert D. Gunn
Merrill W. Haas
Cecil V. Hagen
Frank W. Harrison
William A. Heck
Roy M. Huffington
J. R. Jackson, Jr.
Harrison C. Jamison
Thomas N. Jordan, Jr.
Hugh M. Looney
Jack P. Martin
John W. Mason
George B. McBride
Dean A. McGee
John R. McMillan
Lee Wayne Moore
Gover E. Murray
Rudolf B. Siegert
Robert M. Sneider
Estate of Mrs. John (Elizabeth) Teagle
Jack C. Threet
Charles Weiner
Harry Westmoreland
James E. Wilson, Jr.
P. W. J. Wood

The Foundation also gratefully acknowledges the many who have supported this endeavor with additional contributions.

*Based on contributions received as of July 1, 1991.

PREFACE TO THE *HANDBOOK OF PETROLEUM GEOLOGY*

Source and Migration Processes and Evaluation Techniques is one of four books of the *Handbook of Petroleum Geology*, which is part of the *Treatise of Petroleum Geology*. The Treatise comprises three different publication sets: the *Reprint Series*, the *Atlas of Oil and Gas Fields*, and the *Handbook of Petroleum Geology*. The Treatise is AAPG's Diamond Jubilee project, commemorating AAPG's 75th Anniversary in 1991.

With input from an advisory board of more than 250 geologists and geophysicists from around the world, we designed this entire effort so that the set of publications will represent the cutting edge in petroleum exploration knowledge and application. The Reprint Series provides useful literature from various geological, geophysical, and engineering publications. In some cases, reprinted articles are from obscure sources. The Atlas is a collection of detailed field studies that illustrate the myriad ways oil and gas are trapped. It is also a guide to the petroleum geology of the basins where these fields are found. Field studies like those published in the Atlas are the documentation of petroleum geology. They form the basis for all of our assumptions regarding petroleum geology. From the standpoint of the explorationist who is building and selling prospects, details from field studies can be stored as memories and used to build stronger prospects and convince doubters of the validity of any unique features of a prospect—nothing is more convincing than a close analogy. The third part of the Treatise, the *Handbook of Petroleum Geology*, is a professional explorationist's guide to the methodology and technology used to find fields similar to those described in the Atlas.

The Handbook set is divided into four volumes, each of which addresses one of the four steps of oil and gas prospecting—evaluation of source rocks and migration, evaluation of reservoir quality and properties, evaluation of trapping conditions, and evaluation of economic opportunity. Accordingly, the four volumes of the Handbook are *Source and Migration Processes and Evaluation Techniques*; *Reservoirs*; *Traps*; and *The Business of Petroleum Exploration*. These publications should be kept close at hand so that when a question arises, an answer can be found quickly and easily.

This volume of the *Handbook of Petroleum Geology*, titled *Source and Migration Processes and Evaluation Techniques*, discusses source rock deposition and occurrence, expulsion of hydrocarbons from the source rock, migration of hydrocarbons in carrier beds to a trap, and techniques for evaluating each of these processes. Each chapter is a brief review of the topic addressed. Readers who are interested in more detail should read papers that are referred to in each chapter.

We, the Treatise Editors, are grateful to Robert K. Merrill for his hard work in serving as editor of this volume. The high quality of this book is a product of his efforts. We also want to thank all of the authors who contributed to the volume. Their participation in the project is what AAPG is all about.

Norman H. Foster and Edward A. Beaumont
Editors of the *Treatise of Petroleum Geology*

About the Editor

Robert K. Merrill was born in Lima, Peru, in 1945. He received his B.A. degree in geology from Colby College, Waterville, Maine, and his M.S. degree and Ph.D. in geology from Arizona State University, Tempe. His early work in geology included mineral exploration in Maine, petrography and stratigraphy studies for the American Stratigraphic Company in Alaska and the Rocky Mountains, and structural geology and environmental geology instruction at Arizona State University.

In 1974, he joined Cities Service Company as a staff geologist in the Basin Studies Group in Tulsa, Oklahoma. Subsequently, he worked both on the technical staff and in various exploration positions for Cities Service and its successor OXY USA, Inc. Through his work in numerous North American and South American hydrocarbon provinces, he developed a continuing interest in hydrocarbon exploration using a multidisciplinary approach. His interests include geochemistry, stratigraphy, subsurface fluid systems, petrography, structural geology, and geomorphology. In 1989 he began work with Unocal and is currently exploring for hydrocarbons in the Rocky Mountains.

Preface to this Volume

The science of petroleum exploration geology continues to become more technically sophisticated. As geologists, we have the wonderful opportunity to use more information than ever before in the continued effort to locate commercial accumulations of hydrocarbons. In their exploration efforts, geologists are interested in answering specific questions about source rock and oil relationships. These include

- Are source rocks present in the basin? If so, where?
- Are the potential source rocks oil or gas prone, or both?
- Has oil or gas been generated?
- Where was the oil and/or gas generated?
- When was the oil and/or gas generated?
- Where are the migration pathways for the oil and/or gas?
- Was the timing of migration optimum for entrapment, considering the basin's structural and depositional history?
- What was the depositional environment of the source rock (e.g., deltaic or marine)?
- Under what environmental conditions (salinity and water temperature) were these rocks deposited?

Organic geochemistry provides the geologist with insight into source rock characteristics (age, lithology, depositional environment, thermal history, and organic matter type) and oil history (relative migration, postsourcing alteration, mixed sources, and production contamination).

This volume was conceived as a reference for petroleum geologists to facilitate their review of geochemical concepts and analytical techniques. The purpose of this effort is not to make every geologist an organic geochemist, but to help geologists realize the power of understanding source rock, migration, and entrapment relationships and provide an understanding of what data could help them understand their problems. For more detailed discussions of geochemical concepts and interpretation, I refer the reader to these works:

Hunt, J. M, 1979, *Petroleum Geochemistry and Geology*: W. H. Freeman, New York, 617pp. (A new edition is forthcoming.)

Tissot, B. P., and D. H. Welte, 1984, *Petroleum Formation and Occurrence*, 2nd ed.: Springer–Verlag, New York, 699pp.

The volume has been organized into two parts. Part I deals with the importance of understanding what constitutes a source rock, how oil is generated, and how it migrates. Jacobson discusses the significance of organic facies and their relationship to source rocks. Having established the concept of organic facies, Palciauskas discusses the conversion of organic matter into hydrocarbons and the mechanisms for expulsion of oil from the indigenous organic matter into the rock, which may be accompanied by overpressuring the rocks containing the organic matter. From the generation and expulsion of hydrocarbons, the hydrocarbons are transferred to conduits and into reservoirs through secondary migration, as discussed by England. These hydrocarbons can be traced geochemically to potential source material, but migration processes can alter their signature. Palmer discusses how the effects of biodegradation and water washing can be recognized.

Modeling allows the geologist to understand what has happened to the rocks through geological time. Geochemical results can be related to geological situations through basin modeling. Burnham and Sweeney review the theoretical basis for modeling the maturation and migration of petroleum. Finally, the first part of the volume is completed by Dahl and Yükler who review their work in the Oseberg area of the Norwegian North Sea. A realistic model reconstructed from stratigraphy, tectonics, and geochemistry is presented, and the reader can see how geochemistry can be used to understand hydrocarbon entrapment and to direct exploration geologists to the most reasonable sites for hydrocarbon accumulation.

Part II of the volume reveiws geochemical methods used in exploration. As part of this preface, Table 1 summarizes many of the commonly used geochemical techniques used for evaluating source rocks and oils. Anders introduces this subject with an overview, followed by Noble's discussion of gas chromatography (GC) and gas chromatography–mass spectrometry (GCMS), two analytical techniques most commonly used for liquids analysis. Sofer then reviews stable isotopes in petroleum exploration. These tools are useful not only for liquids and gases but also for potential hydrocarbons generated from potential source rock. Philp and Galvez-Sinibaldi review the characterization of organic matter by various pyrolysis techniques to fingerprint the pyrolysis products and relate these back to the original source material or to simulate natural maturation processes. Jarvie brings us back to the original source material with his discussion of total organic carbon (TOC) analysis, followed by Senftle and Landis's discussion of vitrinite reflectance as a tool to assess thermal maturity. As exploration geologists, we need tools to evaluate source potential rapidly from possibly many wells in a basin. Herron shows us that this is possible through density, sonic, and resistivity relationships with source material and generated hydrocarbons.

To complete the review of geochemistry, I have included two papers to review surface and near-surface geochemical techniques. Saenz, Pannell, and Pingitore review some of these techniques and conclude that one of the keys to differentiating producing from barren structures is to identify anomalies that continue at depth. This is followed by Richers' article in which he presents some case studies of soilgas geochemistry in petroleum exploration where geochemsitry data are integrated with other geological information to gain maximum benefit. Geochemical terminology can be mysterious to the uninitiated, so I have concluded the volume with a glossary. This reference provides summaries of available techniques and some of the more commonly used indicators of source potential and maturity.

Geochemical conclusions must be made from a synthesis of geological and geochemical information. Some interpretive guidelines that I find useful accompany this preface as Tables 2, 3, and 4. The use of interpretive cutoff values is an approximation. The reader is cautioned to plan a geochemical program to consider geochemical data in relation to as many factors as possible. For instance, a commonly used parameter for source potential is total organic carbon. This information may be misleading unless considered with respect to lithology and maturity, both of which affect total organic carbon deposited or remaining in the rock. In planning a geochemical program, be sure to use the contractor's recommended sampling procedure to avoid contamination.

I would like to thank Ted Beaumont and Norm Foster who conceived this volume and who have given me significant support throughout this project. The authors deserve most of the credit for their willingness to assist me in this project, and I am grateful for their contributions. Many individuals, though unable to contribute, provided ideas and potential contributors to this volume and deserve recognition. Finally, I would like to thank Union Oil of California under whose employ I completed this project.

Robert K. Merrill

Table 1. ***(Facing page)*** **Geochemical techniques in relation to organic matter. (From Vlierboom, 1985.)**

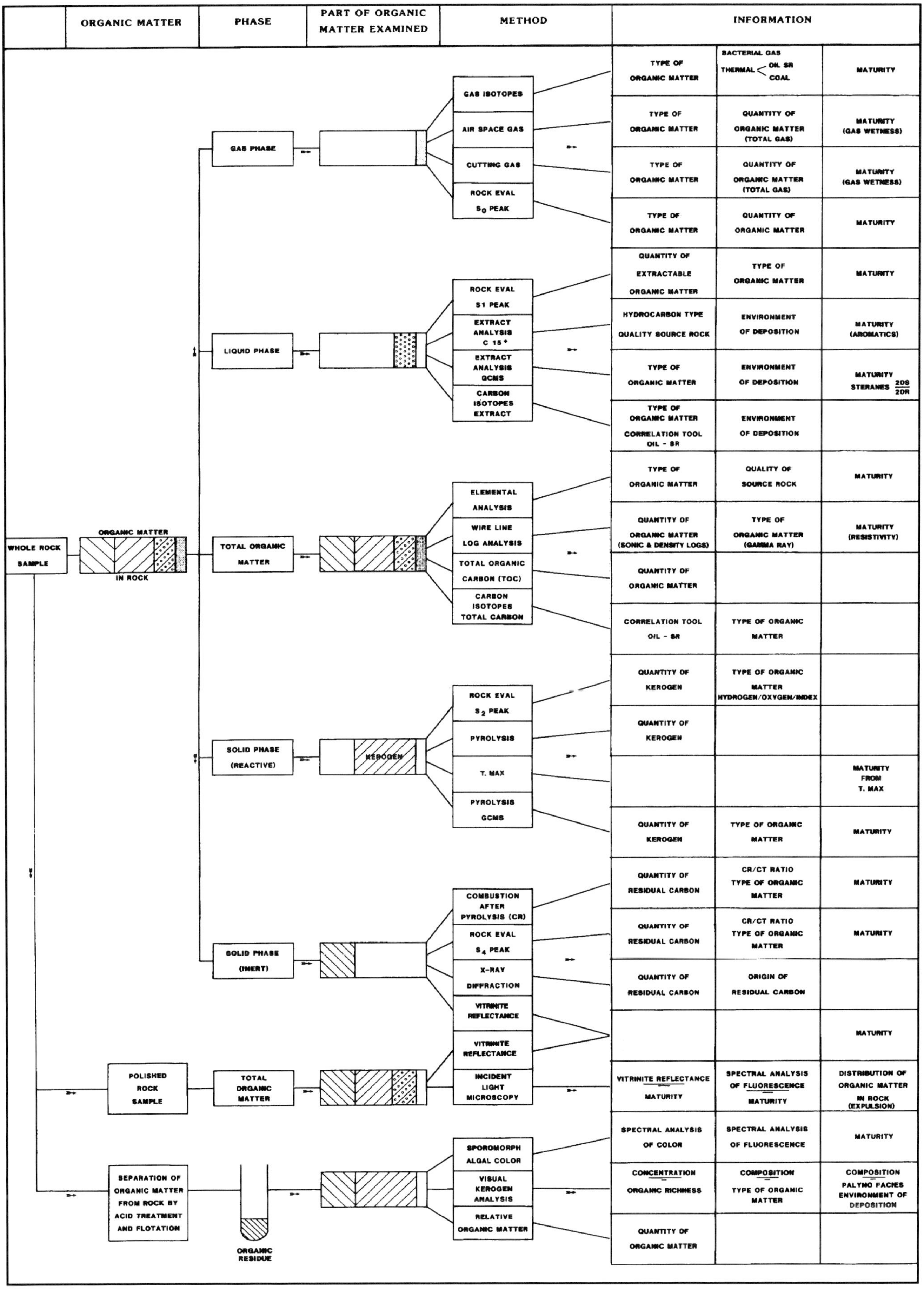
ORGANIC MATTER
PHASE
PART OF ORGANIC MATTER EXAMINED
METHOD
INFORMATION
WHOLE ROCK SAMPLE
ORGANIC MATTER IN ROCK
GAS PHASE
LIQUID PHASE
TOTAL ORGANIC MATTER
SOLID PHASE (REACTIVE)
KEROGEN
SOLID PHASE (INERT)
POLISHED ROCK SAMPLE
TOTAL ORGANIC MATTER
SEPARATION OF ORGANIC MATTER FROM ROCK BY ACID TREATMENT AND FLOTATION
ORGANIC RESIDUE
GAS ISOTOPES
AIR SPACE GAS
CUTTING GAS
ROCK EVAL S_0 PEAK
ROCK EVAL S1 PEAK
EXTRACT ANALYSIS C 15+
EXTRACT ANALYSIS GCMS
CARBON ISOTOPES EXTRACT
ELEMENTAL ANALYSIS
WIRE LINE LOG ANALYSIS
TOTAL ORGANIC CARBON (TOC)
CARBON ISOTOPES TOTAL CARBON
ROCK EVAL S_2 PEAK
PYROLYSIS
T. MAX
PYROLYSIS GCMS
COMBUSTION AFTER PYROLYSIS (CR)
ROCK EVAL S_4 PEAK
X-RAY DIFFRACTION
VITRINITE REFLECTANCE
VITRINITE REFLECTANCE
INCIDENT LIGHT MICROSCOPY
SPOROMORPH ALGAL COLOR
VISUAL KEROGEN ANALYSIS
RELATIVE ORGANIC MATTER
TYPE OF ORGANIC MATTER
BACTERIAL GAS
THERMAL < OIL SR / COAL
MATURITY
TYPE OF ORGANIC MATTER
QUANTITY OF ORGANIC MATTER (TOTAL GAS)
MATURITY (GAS WETNESS)
TYPE OF ORGANIC MATTER
QUANTITY OF ORGANIC MATTER (TOTAL GAS)
MATURITY (GAS WETNESS)
TYPE OF ORGANIC MATTER
QUANTITY OF ORGANIC MATTER
MATURITY
QUANTITY OF EXTRACTABLE ORGANIC MATTER
TYPE OF ORGANIC MATTER
MATURITY
HYDROCARBON TYPE
QUALITY SOURCE ROCK
ENVIRONMENT OF DEPOSITION
MATURITY (AROMATICS)
TYPE OF ORGANIC MATTER
ENVIRONMENT OF DEPOSITION
MATURITY STERANES 20S/20R
TYPE OF ORGANIC MATTER
CORRELATION TOOL OIL - SR
ENVIRONMENT OF DEPOSITION
TYPE OF ORGANIC MATTER
QUALITY OF SOURCE ROCK
MATURITY
QUANTITY OF ORGANIC MATTER (SONIC & DENSITY LOGS)
TYPE OF ORGANIC MATTER (GAMMA RAY)
MATURITY (RESISTIVITY)
QUANTITY OF ORGANIC MATTER
CORRELATION TOOL OIL - SR
TYPE OF ORGANIC MATTER
QUANTITY OF KEROGEN
TYPE OF ORGANIC MATTER HYDROGEN/OXYGEN/INDEX
QUANTITY OF KEROGEN
MATURITY FROM T. MAX
QUANTITY OF KEROGEN
TYPE OF ORGANIC MATTER
MATURITY
QUANTITY OF RESIDUAL CARBON
CR/CT RATIO TYPE OF ORGANIC MATTER
MATURITY
QUANTITY OF RESIDUAL CARBON
CR/CT RATIO TYPE OF ORGANIC MATTER
MATURITY
QUANTITY OF RESIDUAL CARBON
ORIGIN OF RESIDUAL CARBON
MATURITY
VITRINITE REFLECTANCE MATURITY
SPECTRAL ANALYSIS OF FLUORESCENCE MATURITY
DISTRIBUTION OF ORGANIC MATTER IN ROCK (EXPULSION)
SPECTRAL ANALYSIS OF COLOR
SPECTRAL ANALYSIS OF FLUORESCENCE
MATURITY
CONCENTRATION ORGANIC RICHNESS
COMPOSITION TYPE OF ORGANIC MATTER
COMPOSITION PALYNO FACIES ENVIRONMENT OF DEPOSITION
QUANTITY OF ORGANIC MATTER

Table 2. Kerogen types, their origin, and hydrocarbon potential.

ENVIRONMENT	KEROGEN TYPE	KEROGEN FORM / MACERAL	ORIGIN	HYDROCARBON POTENTIAL
Aquatic	I	Alginite	Algal bodies	Oil
		Amorphous Kerogen	Structureless debris of algal origin	
	II		Structureless, planktonic material, primarily of marine origin	
Terrestrial		Exinite	Skins of spores and pollen, Cuticle of leaves and herbaceous plants	
	III	Vitrinite	Fibrous and woody plant fragments and structureless, colloidal humic matter	Gas and some Oil Mainly Gas
	IV	Inertinite	Oxidized, recycled woody debris	None

Table 3. Zones of petroleum generating and destruction. (Modified from Dow, 1977, 1978, 1980.)

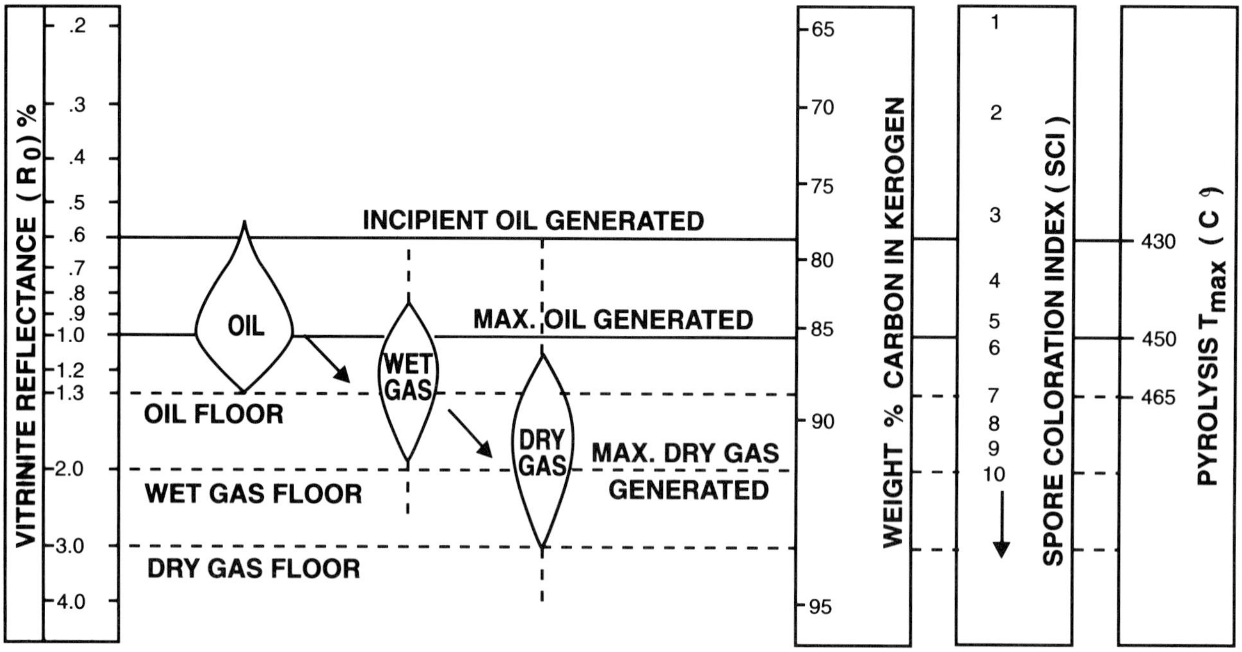

COMPARISON OF SEVERAL COMMONLY USED MATURITY TECHNIQUES & THEIR CORRELATION TO THE OIL & GAS GENERATION LIMITS.

Table 4. Rock-Eval pyrolysis parameters.

PYROLYSIS:

S_1 = Amount of free hydrocarbons in rock (mg HC/g); soluble organic matter in source rock
≥1 mg/g = 11) large amounts of kerogen-derived bitumen, or 2) presence of migrated hydrocarbons.

S_2 = Amount of hydrocarbons formed by the breakdown of kerogen (mg HC/g); amount of hydrogen-rich kerogen in source rock
≤ 1 mg/g = low hydrogen source potential due to 1) low quantities of organic matter, 2) hydrogen-poor organic matter, or 3) "cooked" organic matter.
≥ 5 mg/g = 1) excellent source potential, 2) high molecular weight soluble organic matter (biodegrated oil), or 3) coal.

S_3 = Amount of CO_2 formed by thermal breakdown of kerogen (mg HC/g); amount of Oxygen rich kerogen in source rock.
≥ 1.5 mg/g = relatively large amounts of terrigenous organic matter.

T_{max} - Heating temperature at which the top peak of S_2 occurs and is related to the degree of thermal maturation (°C)

	Notation	Definition	Equation	Rating
PRODUCTION INDEX (TRANSFORMATION RATIO)	P. I.	PRODUCTION INDEX	$\frac{S_1}{S_1+S_2}$	< 0.1 = immature 0.1 - 0.4 = Oil > 0.4 Gas
HYDROGEN & OXYGEN INDICES	H. I.	HYDROGEN INDEX	$\frac{(S_2 \times 100)\ mg/g}{\%\ TOC}$	< 150: GAS 150 – 300: MIXED >300: OIL (RO = 0.6%)
	O. I.	OXYGEN INDEX (S_3 or S_4)	$\frac{(S_3 \times 100)\ mg/g}{\%\ TOC}$	<40 mg/g low HI = land-derived organic matter, and/or mature high HI = good to excellent source potential >40 mg/g low HI =gas- prone organic matter; generally immature; high HI = good oil source; generally immature
POTENTIAL YIELD OR HYDROCARBON SOURCE POTENTIAL	P. Y.	POTENTIAL YIELD	(S_1+S_2) mg / g	**mg / g** <2 POOR 2-6 FAIR >6 VERY GOOD
KEROGEN TYPE		KEROGEN TYPE	$\frac{S_2}{S_3}$	<3: GAS PRONE 3 - 5: MIXED >5: OIL PRONE
REACTIVE CARBON INDEX	R. C. I.	REACTIVE CARBON INDEX	$\frac{(S_1+S_2)\ mg/g \times 10}{\%\ TOC}$	
	The Reactive Carbon Index is a function of thermal maturity. At thermal maturities greater than 1.5 % R (or Tmax > 475° C, and generation index > 0.6), the Reactive Carbon Index becomes a less useful.			
THERMAL MATURITY	T_{max}	MAXIMUM TEMPERATURE IN °C		<430: Immature 430 - 460: Oil generation >460: Gas generation or destruction
	G. I.	GENERATION INDEX	$\frac{S_1}{S_1+S_2}$	<0.1: Immature 0.15 - 0.30: Oil generation >0.30: Gas generation or destruction
INDICATIONS OF MIGRATED HYDROCARBONS	(1) HIGH S_1 (2) LOW T_{max} (3) HIGH S_1 / % TOC (4) HIGH S_1/ (S_1 + S_2)			

Part I

Petroleum Generation and Migration

Chapter 1

Petroleum Source Rocks and Organic Facies

S. R. Jacobson
Chevron Oil Field Research
La Habra, California, U.S.A.

INTRODUCTION

Geologists who map organic facies can project the geographical distribution of petroleum source rocks. These maps provide a foundation for calculating the timing of petroleum generation and for determining realistic migration pathways to potential reservoir traps. In some geographical areas, geologists may find organic facies useful for reconstructing stratigraphic sequences. Organic facies are the criteria for evaluating the physical, chemical, and biological processes that control the distributions of organic sediments in geological time and space.

Petroleum source rocks represent uncommon paleoenvironments. Their organic content is a function of biological productivity, oxygenation in the water column and in near-bottom sediments (preservation), and mineral supply (dilution). Biological input varies from hydrogen-rich algal/bacterial constituents to hydrogen-poor woody fragments. Physical and chemical processes alter biological material before its entombment in rocks. The single most important process affecting the preservation of oil-prone, hydrogen-rich, and therefore oil-generative sedimentary organic matter is oxygenation. Oxygen in the water column, soils, and subaqueous sediments encourages both biological and abiological destruction of hydrogen-rich molecules. Demaison and Moore (1980) discuss oxygen-restricted paleoenvironments where hydrogen-rich, petroleum-generative organic matter is best preserved.

Transported and precipitated minerals dilute the relative concentration of organic matter in sediments, and this initial organic matter concentration may ultimately determine whether the rock expels petroleum. Petroleum expulsion requires, among other factors, enough generated petroleum to saturate and exceed the available source rock (usually shale) porosity, especially where available porosity decreases significantly with burial-related compaction. Based on experience, Peters (1986) suggests that this threshold generally occurs at total organic carbon (TOC) values of approximately 0.5 wt. %. Palciauskas (this volume) describes the factors that drive petroleum expulsion.

Source rock candidates can be evaluated and mapped using organic geochemical data and geologically projected organic facies. The combined effort of organic geochemists, who provide sound interpretations of analytical results, and geologists, who position organic facies in realistic time and space, adds insights in both heavily explored and sparsely drilled basins. The value of organic facies grows commensurately with increasing knowledge of depositional systems and stratigraphic sequences. Geologists can map organic facies in sequence stratigraphic frameworks, even where small data sets are available. After determining source rock candidates and mapping their paleogeographical distributions, geologists can reconstruct burial histories and simulate thermal maturation (e.g., Lopatin, 1971; Waples, 1985; Ungerer and Pelet, 1987) to calculate the timing of petroleum generation. A senior oil company employee once stated that "timing was everything" (B. M. Wood, oral comm., 1979). Knowledge of the timing of certain geological events is the key to proficient petroleum prospecting. Well-grounded interpretations of source rock formation, thermal maturation, petroleum expulsion and migration, reservoir position and trapping all depend upon timing. Armed with an accurate chronostratigraphy, geoscientists can choose the most reasonable geological models for stalking illusive petroleum reservoirs.

What are petroleum source rocks? How are they formed? Where are they found in stratigraphic sequences? What are organic facies and how can they be mapped and projected in areas having little data? When are measured values reliable and what phenomena alter these values? In this chapter, definitions and discussion of the processes that relate petroleum source rocks to sequence stratigraphy preface detailed descriptions of kerogen types and organic facies. This introduction provides a foundation for appreciating some geochemical tools for interpreting the formation and occurrence of petroleum.

PETROLEUM SOURCE ROCKS

Petroleum source rocks must contain sufficient quantities of sedimentary organic matter with a requisite ratio of hydrogen to carbon. In thermally immature rocks, sedimentary organic matter is dominated by kerogen (organic matter insoluble in common organic solvents) with small amounts of bitumen. Bitumen is organic matter soluble in organic solvents, and the term is used here in a restricted sense for oil generated in petroleum source rocks before it is expelled.

Living organisms and their metabolic processes form sedimentary organic matter. Durand (1980) recognized that organic matter decomposes in the water column and in sediments, reducing the quantity and quality (hydrogen-richness) of the organic matter preserved in rocks. Organisms attack sedimentary organic matter for its carbon and hydrogen and metabolically convert some of it to simple molecules (e.g., CO_2, H_2O, CH_4, NH_3, N_2, and H_2S). Abiological oxidation also forms CO_2 and H_2O. These compounds are nonhydrocarbons (except for methane, CH_4) and often escape early in the depositional and burial process. Therefore, the organic residue preserved in rocks and available for thermal conversion to fossil fuels represents only a small proportion of the original biological input. An even lower percentage of this buried fraction is converted to petroleum (oil and gas), expelled, migrated, concentrated, and trapped in reservoirs.

Petroleum source rocks are evaluated by their carbon richness as total organic carbon (TOC) in weight percent (wt. %). They are also evaluated by their hydrogen richness, or quality, as atomic H/C or, using Rock-Eval pyrolysis, as hydrogen index (HI), which is the ratio of pyrolyzable hydrocarbon mass to rock mass. Thermally immature hydrogen-rich organic matter (H/C > 1 or HI > 300) has oil-generative potential, whereas lower hydrogen content usually denotes gas-generative potential.

Heat transforms thermally immature sedimentary organic matter to oil and gas. This thermal maturation process is a function of time and temperature. The time–temperature factor may be insufficient to convert even the most oil-prone organic matter to petroleum (thermal immaturity), or it may generate petroleum (thermal maturity) or generate, expel, and overheat the residual organic matter, leaving only charred carbon (thermal postmaturity) in the source rock. Heating is related to burial depth, crustal tectonics, and proximity to igneous bodies.

Lopatin (1971) first described maturation modeling by calculating time–temperature indices for coals, basing his calculations on conversion kinetics of vitrinite, a gas-generative organic rock constituent (maceral). Waples (1980) brought the Lopatin method to English-reading geologists not fluent in Russian and applied "vitrinite kinetics" to petroleum generation. Subsequently, Waples (1985) provided adjustments that accommodate some of the variations in oil-prone organic matter not considered in Lopatin's coal studies. Ungerer and Pelet (1987) (see also Tissot et al., 1987) revolutionized the technique by introducing kinetics based on pyrolytically determined activation energies for oil-generative organic matter assemblages. An activation energy is the amount of energy required for a chemical reaction to proceed. In sedimentary organic matter conversion to petroleum, many reactions take place. However, because of chemical complications inherent when many different reactions occur simultaneously and sequentially, as in sedimentary organic matter conversion to petroleum, the use of experimentally derived pyrolysis activation energies is a practical way to calculate realistic conversion rates. Burnham et al. (1988) have further developed the method.

At any designated location such as a well site, the Lopatin method enhanced by the input of pyrolysis activation energies requires (1) burial history for strata including source rock candidates, (2) measured or estimated geothermal gradient(s), (3) sediment compaction, and (4) thermal conductivities for the lithostratigraphic units. The simulated maturation process integrates temperatures through time for stratigraphic units of interest.

Thermal maturation indicators (Heroux et al., 1979) for calibrating simulation models include vitrinite reflectance (R_o), thermal alteration index (TAI) from palynomorphs, clay crystallinity, and hydrocarbon molecular ratios. Other methods for calibration include the conodont alteration index (CAI) (Epstein et al., 1977; Rejebian et al., 1987), Rock-Eval T_{max} (Tissot and Welte, 1984; Peters, 1986), the porphyrin maturity parameter (Sundararaman et al., 1988), and biomarker ratios (Mackenzie, 1984). Each technique has limits in sample requirements, thermal resolution, and the thermal range in which it can be applied with accuracy. Sensitivity studies demonstrate that reconstruction of the local paleoheat flow (geothermal gradients) through the geological time of interest is often the source of the greatest discrepancies between calculated maturities and thermal maturation indicators. Oils frequently carry gas chromatographic (GC) signatures useful for correlation to source rock extracts (thermal or solvent). These signatures are detectable with capillary column GC or may require the higher resolution available with mass spectrometry (GCMS). Reservoir oils and extracted bitumens contain source rock information. This "molecular stratigraphy" can reveal depositional features of petroleum source rocks, such as marine or lacustrine, clay rich or clay poor, redox potential of waters, and in some cases, the age of the source rock (using age-diagnostic biomarkers).

The petroleum yield of source rocks is related to the richness and quality of its incorporated sedimentary organic matter. Several methods have been proposed for determining potential volumetric yields and percentage of conversion of source rocks (e.g., Tissot et al., 1980; Tissot and Welte, 1984; Peters, 1986; Cooles et al., 1986; Tissot et al., 1987; Baskin, 1990).

Bitumen generated in source rocks saturates the source rock porosity. In source rocks with sufficient quantities of generated bitumen, the primarily physical process of expulsion may follow depending upon (1) the porosity, strength, and composition (e.g., clay, carbonate, or evaporite) of the source rock and its adjoining strata; (2) the concentration and distribution of kerogen (e.g., layered or disseminated); (3) whether the bitumen or mineral matrix is load bearing; (4) the properties of generated products (e.g., gas pressure and viscosity); and (5) the heating rate.

Recently, an evaluation of depositional controls, distribution, and effectiveness of the world's source rocks has been provided by Ulmishek and Klemme (1990). New schemes are emerging for evaluating petroleum generative potential in basins (Tissot et al., 1980; Demaison and Huizinga, in press).

PETROLEUM SOURCE ROCK FORMATION

Petroleum source rock formation requires both (1) sufficient biological productivity to create large quantities of organic matter and (2) suitable depositional conditions for its concentration and preservation. Creation of most hydrogen-rich, oil prone, sedimentary organic matter is ultimately attributed to photosynthesis and nutrient supply. Under appropriate conditions, phytoplankton and/or algal/bacterial mats form in lacustrine, estuarine, and marine paleoenvironments, and land plants that comprise coals form in terrestrial environments. Nutrient replenishment, sunlight, temperature, pH and Eh of waters, sediments, and soils influence biological productivity.

Source rocks form where depositional conditions shelter organic matter from destruction by oxidation and from extreme dilution by mineral input. Organic matter is preserved in oxygen-restricted sediments at depths below wave base, in stagnant waters where density or temperature-stratified water columns form, and in other areas where oxygen replenishment is low (Demaison and Moore, 1980). The exposure time of hydrogen-rich organic matter to chemically destructive agents is crucial. Therefore, water column conditions may be as important as those in the upper few centimeters of sediment. Organic-rich sediments often retain millimeter-thick depositional laminae, which reflect no sediment-churning bioturbation. Bioturbation records sediment invasion mainly by animals grazing on hydrogen-rich sedimentary organic matter.

SEQUENCE STRATIGRAPHIC ANALYSIS

Sequence stratigraphic analysis places strata in genetic frameworks. The method has developed significantly since the seminal work of Vail et al. (1977). An elaborate nomenclature of surfaces and rock packages is evolving to serve this revolutionary paradigm. Geologists always experience word wars in periods of innovation (e.g., plate tectonics, petroleum origin, and Darwinian evolution). To monitor the seemingly impenetrable vocabulary of sequence stratigraphy, the reader is directed to seek modern treatments, such as Van Wagoner et al. (1988).

Currently used sequence stratigraphic terminology is illustrated in Figure 1. In this framework, the best potential source rocks often occur in condensed sections where sedimentation is restricted to pelagic and hemipelagic particles (Loutit et al., 1988). This environment of rising relative sea level is characterized by limited detrital input. In these relatively deeper water environments, biological productivity may be increased by nutrient upwelling and sedimentary organic matter may be protected from the ravages of oxygen introduced by wave action and active sediment transport. Source rocks do not form in the high-energy depositional regimes where coarse-grained clastics are deposited (e.g., siliceous sands or calcareous öolites) or where reefs grow. Source rocks usually occur as fine-grained sediments (clays and marls) in low-energy environments where low-density organic sediments are deposited and can be preserved. Organic-rich rocks are colored dark gray, black, or chocolate brown, reflecting their relatively high volumetric ratio of organic matter to minerals.

Sequence stratigraphy is usually applied to marine rocks; however, agents forming lacustrine strata essentially respond to many of the same biological, chemical, and physical factors. Runoff and evaporation in lakes substitute for relative sea level changes in marine environments. Organic-rich, lacustrine condensed sections occur where biological productivity is sufficient, oxygenation is low, and mineral input is constrained.

Deltaic deposition of coals may also reflect position in stratigraphic sequences. Most woody coals generate gas, although some bear the algal or bacterial components that can generate oil. Gas-prone terrestrial

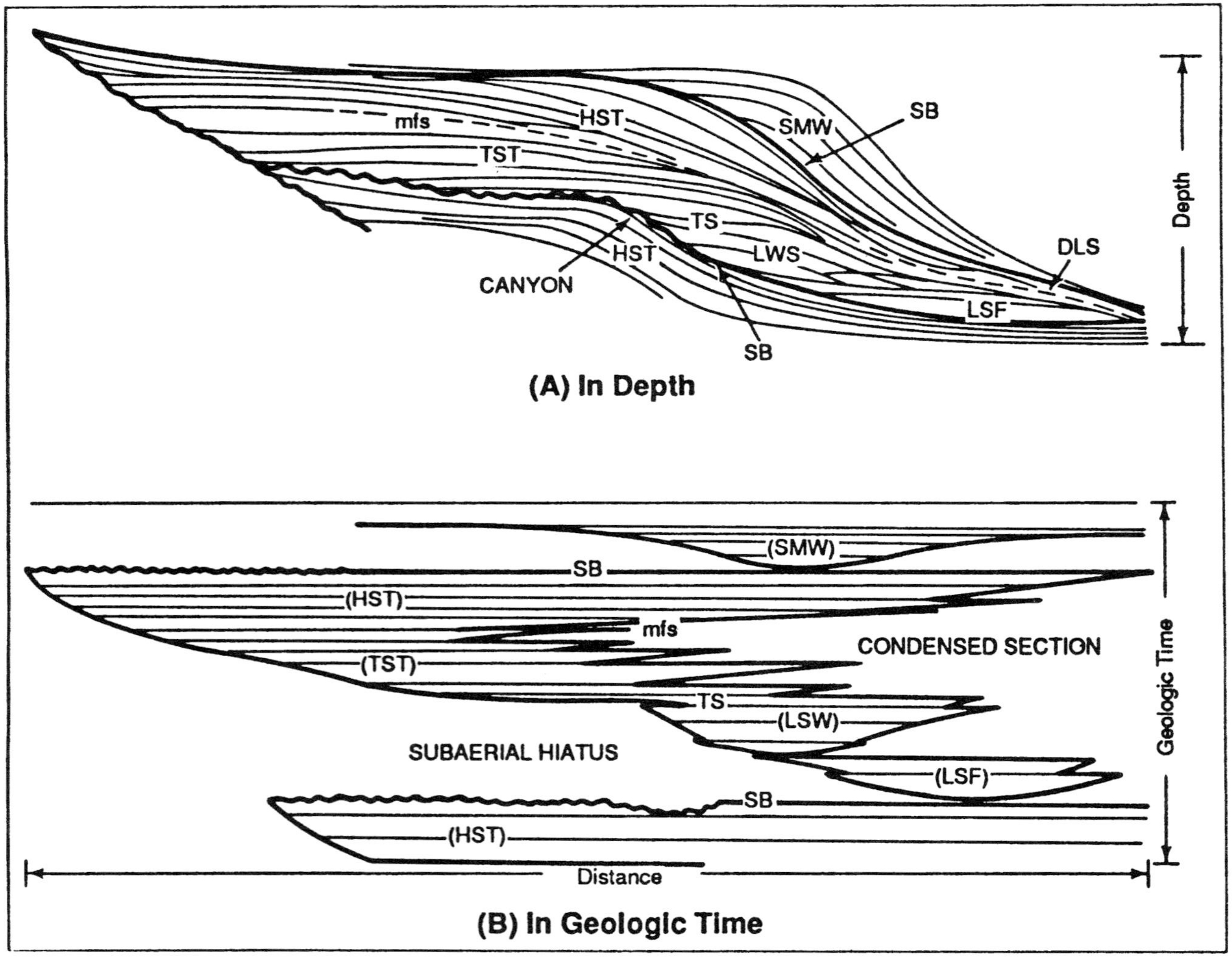

LEGEND

SURFACES

(SB) Sequence boundaries
(DLS) Downlap surfaces
(mfs) = maximum flooding surface
(TS) Transgressive surface
(First flooding surface above maximum regression)

SYSTEMS TRACTS

HST = Highstand systems tract
TST = Transgressive systems tract
LST = Lowstand systems tract
LSF = Lowstand fan
LSW = Lowstand wedge
SMW = Shelf margin wedge systems tract

Figure 1. Schematic diagram of stratal surfaces and systems tracts used in the sequence stratigraphy depositional model.

organic matter composes most coals formed in paralic and limnic swamps and abandoned river channels. These facies are often associated with prograding facies of highstand systems tracts.

DEFINITION OF ORGANIC FACIES

Organic facies have been described from rocks of diverse ages and from various areas (Jones and Demaison, 1982; Jones, 1987). The primarily chemical criteria used to characterize organic facies reflect numerous complex biological, physical, and chemical processes. The term *organic facies* has had an evolving etymology (see Jones, 1987, p. 2). Organic facies are not palynomorph facies, palynofacies, pollen facies, kerogen facies, organic matter facies, or maceral facies. They cannot be determined by microscopy alone. The definition of the term as used here is from Jones and Demaison (1982): "An organic facies is a mappable subdivision of a designated stratigraphic unit, distinguished from adjacent subdivisions on the basis of its organic constituents, without regard to the inorganic aspects of the sediments." Organic facies are determined by their elemental (C, H, and O) composition and, in ambiguous circumstances, are further differentiated by their microscopically determined maceral (organic particle) abundances. The concept of mappability presupposes that organic facies represent sufficiently large stratigraphic thicknesses and areal extent and have similar organic geochemical properties. This homogeneous character is rarely found on a fine scale, and the larger "mappable" scale requires a geological interpretation. To apply any facies concept, a geologist must (1) determine the scale at which variations are important and (2) carefully examine and interpret perturbations in data trends.

METHOD OF ORGANIC FACIES DETERMINATION

In thermally immature rocks, organic facies can be determined by kerogen typing. According to Jones (1987), this technique mainly relies on (1) kerogen identification adequately reflecting petroleum generative potential, (2) kerogen distribution being "neither random nor capricious," and (3) the distribution of similar kerogen extending over tens to hundreds of meters vertically and hundreds to thousands of square kilometers areally. The latter criterion is sometimes recognizable in massive dark shale units, but must be evaluated with care. There has to be enough stratigraphic sampling to recognize volumetrically significant variations in organic facies within lithostratigraphic units. Petroleum source rocks have organic facies variations both vertically and horizontally.

Table 1. Generalized geochemical characteristics of organic facies A through D.[a]

Organic Facies	H/C (at % R_o = 0.5)	H (mg HC/g TOC)	OI (mg CO_2/g TOC)
A	>1.45	>850	<30
AB	1.35–1.45	650–850	20–50
B	1.15–1.35	400–650	30–80
BC	0.95–1.15	250–400	40–80
C	0.75–0.95	125–250	50–150
CD	0.60–0.75	50–125	40–150+
D	>0.60	>50	20–200+

[a]The H/C values define the organic facies. Hydrogen index (HI) and oxygen index (OI) values from Rock-Eval may vary slightly from indicated values for specific organic facies. (After Jones, 1987.)

Kerogen typing is based on elemental analyses for hydrogen, carbon, and oxygen. By plotting atomic H/C versus O/C of coal constituents (macerals), Van Krevelen (1961) recognized that each kind of coal maceral had consistent ratios. These ratios reflect the hydrocarbon generative potential of the maceral type. Van Krevelen's work was extended to potential petroleum source rocks by Tissot and his co-workers (e.g., Tissot and Welte, 1984). The somewhat time-consuming elemental analyses of kerogen were replaced by Rock-Eval data for whole rock samples. The Rock-Eval derived hydrogen index (HI) and oxygen index (OI) were substituted for H/C and O/C.

From data published by many workers, main pathways emerged in Van Krevelen-type plots that reflected the dominance of four main organic matter components. These four types have evolved to kerogen types I, II, III (Tissot and Welte, 1984), and IV. These narrowly defined kerogen types are found with surprising frequency, although intermediate values are common. Jones (1987) presents a scheme (Table 1) for separating this range of data into seven organic facies, which he designates with letters to distinguish them from kerogen types (Figures 2 and 3). He suggests that organic facies reflect both the original biological input and the preservational processes. Jones supports his criteria with numerous published examples.

ALTERATION OF PETROLEUM SOURCE ROCKS AND ORGANIC FACIES

Alteration affects organic material before it is buried. This can include biodegradation, physical abrasion, and chemical changes related to Eh and pH in the water column and in sediments. Individual organic matter constituents react differently to each of these factors. Also, selective alteration of particular constituents can occur after burial and kerogen formation. The measurable properties of petroleum source rocks and

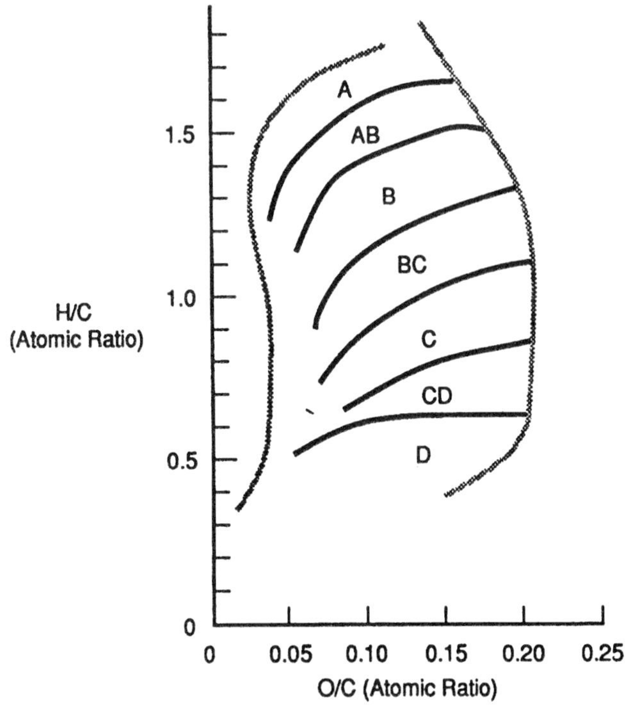

Figure 2. Modified Van Krevelen diagram for organic facies A through D. (After Jones, 1987.)

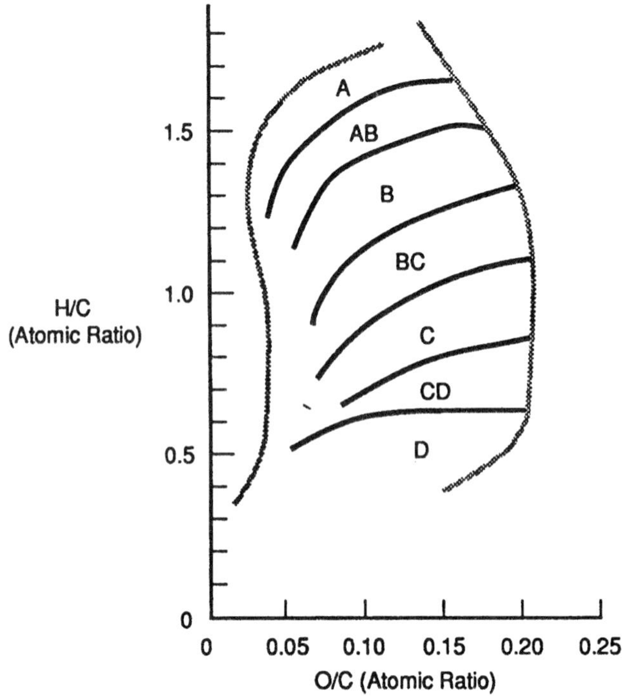

Figure 3. Van Krevelen-type diagram for organic facies A through D, using Rock-Eval parameters on whole-rock samples. Discrimination of organic facies CD and D is analytically less definitive than others, and thus these boundaries are represented by gray shading.

their organic facies change after weathering and thermal maturation. In both cases, organic matter quantity and quality are diminished.

An example of extreme weathering occurs in numerous thin (millimeters to a few centimeters thick), thermally immature to marginally mature chocolate brown oil shales of the Galena Group (Middle Ordovician) of the North American Mid-Continent. In shallow cores (<1000 ft, or ~300 m, deep) from east-central Iowa (Hatch et al., 1987; Jacobson et al., 1988), these thin brown shales have TOC values occasionally exceeding 40 wt. % and Rock-Eval hydrogen indices (HI) near 1000. These thin layers contain more than 85% organic matter (mainly carbon, hydrogen, oxygen, sulfur and nitrogen) by volume. Along the Mississippi River at Guttenberg and Specht's Ferry, Iowa, these chocolate brown shales have selectively weathered away completely, leaving recesses between the more resistant carbonate layers. Preferential weathering of most fine-grained sedimentary rocks occurs whether or not rocks are organic rich and whether or not the fine-grained sediments are well cemented. Weathering often penetrates beyond the observable changes in color and induration and reduces TOC and HI values. It appears that the quantity and quality of all are diminished when exposed.

Generating and expelling petroleum reduces organic matter quantity and quality of source rocks (e.g., Daly and Edman, 1987). Therefore, organic facies maps must reflect the original organic constituents and changes caused by weathering and thermal maturation. Experienced organic geochemists can often reconstruct organic facies for thermally mature samples by tracking the kerogen types back to thermally immature values along pathways of Van Krevelen diagrams.

KEROGEN TYPES

The following descriptions of kerogen types indicate their biological input, stratigraphy, and depositional processes that control their oil-generative properties. Kerogen types are defined on H/C and O/C values (or HI and OI from Rock-Eval). In thermally immature samples, the chemically extreme kerogen types I and IV (and therefore the equivalent organic facies A and D) contain macerals having relatively uniform chemical properties. These end-members are dominated by the most and least hydrogen-rich constituents. Other kerogen types (and therefore their equivalent organic facies) are frequently mixtures of macerals. Microscopy is the method of choice for distinguishing the constituents of mixed organic matter assemblages.

Before enumerating the criteria for discriminating

kerogen types, it is important to consider the "mineral matrix effect." Some mineral (polar clay) constituents retard the release of hydrocarbons from powdered whole rock samples during Rock-Eval pyrolysis, under-evaluating the quantity, quality, and thermal maturation data. Although this factor, the mineral matrix effect, is well known to organic geochemists, it is frequently overlooked when interpreting Rock-Eval-dependent values used to determine kerogen type and organic facies. The mineral matrix effect occurs when polar clays react with polar organic molecules during the nonhydrous Rock-Eval procedure (Espitalié et al., 1980; Horsfield and Douglas, 1980; Orr, 1983, Dembicki, et al., 1983; Katz, 1983; Peters, 1986; Crossey et al., 1986; Langford and Blanc-Valleron, 1990).

Pioneers of pyrolysis found that some minerals inhibit hydrocarbon expulsion during whole-rock pyrolysis and not during kerogen pyrolysis (Espitalié et al., 1980; Horsfield and Douglas, 1980; Dembicki et al., 1983). The effect of different matrix constituents (Espitalié et al., 1980; Horsfield and Douglas, 1980; Dembicki et al., 1983; Katz, 1983) varies from strongest to weakest: illite > Ca-bentonite > kaolinite > Na-bentonite > calcium carbonate > gypsum (Espitalié et al., 1980). Variations in the mineral matrix effect related to organic richness occur in whole-rock samples with TOC values less than 10% (Espitalié, et al., 1980; Horsfield and Douglas, 1980; Dembicki et al., 1983).

Geological thermal maturation processes differ from those of Rock-Eval pyrolysis. Whole-rock Rock-Eval samples are heated rapidly in an anhydrous environment. Geological burial processes cause clays to undergo physical and chemical alteration usually preceding the slow and systematic thermal conversion (generation) of kerogen to petroleum. These changes occur in hydrous environments, which probably reduce the reactive capabilities of clays, usually before significant hydrocarbon generation has occurred. Nevertheless, some degree of mineral matrix effect probably does persist under geological conditions.

Kerogen Type I

Kerogen type I is predominantly composed of the most hydrogen-rich organic matter preserved in the rock record. Often the organic matter is structureless (amorphous) alginite and, when immature, fluoresces golden yellow in ultraviolet (UV) light. A large proportion of type I kerogen can be thermally converted to petroleum and therefore is rarely recognizable in thermally mature or postmature rocks. Sometimes in thermally immature rocks, morphologically distinct alginite is structurally or chemically assignable to specific algal or bacterial genera. These organic-walled microfossils have high H/C values because they formed hydrocarbons biologically. Some examples of pure assemblages with type I kerogen properties include the following: (1) the lacustrine alga *Botryococcus braunii,* which sometimes retains its diagnostic cup-and-stalk colonial morphology and/or its unique chemical compound, botryococcane (Moldowan and Seifert, 1980); (2) *Tasmanites* spp., which are low-salinity, cool water, marine algal phytoplankton with unique physical features (Prauss and Reigel, 1989); and (3) the Ordovician marine organic-walled colonial microfossil *Gloeocapsomorpha prisca,* with its diagnostic physical appearance and unique chemical signature (Reed et al., 1986). Where kerogen type I is widespread, it is mapped as organic facies A. It usually forms in stratified water columns of lakes, estuaries, and lagoons.

Kerogen type I is concentrated in condensed sections where detrital sediment transport is low and primarily pelagic. Condensed sections occur in offshore facies of transgressive systems tracts in marine and lacustrine settings. Although this extention of terminology from marine to lacustrine environments may be unfamiliar at first, lacustrine rocks are formed by the same dynamic processes that form marine rocks (i.e., sediment supply, climate, tectonics, and subsidence), although changes in lake levels often reflect local changes in runoff, evaporation, and sediment basin filling rather than the global and relative sea level changes postulated for marine sediments (Haq et al., 1988).

Kerogen Type II

Kerogen type II in its pure (monomaceral) form is characterized by the relatively hydrogen-rich maceral exinite. Examples include spores and pollen of land plants, primarily marine phytoplankton cysts (acritarchs and dinoflagellates), and some land plant components such as leaf and stem cuticles. As with kerogen type I, the occurrence of kerogen type II depends on high biological productivity, ow mineralic dilution, and restricted oxygenation. The pure exinitic kerogen type II is preserved in condensed sections and represents macerals that are slightly less hydrogen rich than kerogen type I.

Kerogen type II can also be formed from partial degradation of type I kerogen or from varying mixtures of type I and types II, III, and IV. For example, organic matter formed in different provenances can be combined, such as when planktonic algal material falls into sediments containing transported woody macerals (kerogen type III). Kerogen type II is recorded in transgressive systems tracts, sometimes landward of type I kerogen deposition.

Kerogen Type III

Kerogen type III contains sufficient hydrogen to be gas generative but not enough hydrogen to be oil prone. In its pure form, it is composed of vitrinite, a maceral formed from land plant wood. As with other intermediate kerogen types, however, various maceral mixtures or degradational processes can contribute to kerogen type III formation. Coal-forming environments represent several different kerogen types. Most coals form in paralic swamps and abandoned river channels. Vail et al. (in press) find that in regions where sediment supply is low, incised valleys contain these sediments as estuarine or coastal plain deposits.

Kerogen Type IV

Kerogen type IV is a term not universally employed by organic geochemists because it is difficult to distinguish type IV from type III using only Rock-Eval pyrolysis. It is an inert (does not generate hydrocarbons) end-member on the hydrocarbon generative spectrum. Kerogen type IV is composed of hydrogen-poor constituents such as inertinite, which is detrital organic matter oxidized directly by thermal maturation including fire (charcoal) or by biological or sedimentological recycling.

ORGANIC FACIES

In the following brief review of organic facies, their geochemical and microscopic characteristics are taken from the conclusions of Jones (1987, p. 78–80). Some new examples are also presented. Organic facies A through D represent the range of sedimentary histories from no oxidation to severe oxidation.

In the definition of organic facies by Jones and Demaison (1982) (already given), the organic "constituents" that distinguish organic facies might be macerals (organic particles) or molecular components. Jones (1987), however, only employs the elemental atomic ratios of C, H, and O and to some extent equivalent Rock-Eval parameters (HI and OI). Macerals are used to identify the components of mixtures and this has a bearing, for example, on whether relatively low-quality organic facies have oil- or gas-generative potential. Thermally immature mixed maceral assemblages with H/C values of 0.8 might contain some oil-generative type I or II kerogen with predominantly inert type IV kerogen or might contain all gas-generative type III kerogen.

Organic Facies A

Criteria for this relatively rare facies are $H/C > 1.45$ and $HI > 850$. The rocks are usually laminated and organic rich and are found in alkaline lakes and marine paleoenvironments. The organic matter, often brightly fluorescent, is derived primarily from a single type of algae or bacteria. Organic facies A commonly occurs in carbonate settings. It is found in condensed sections of lakes and marine margins that are protected from oxygenated waters. Examples include the lacustrine Green River oil shale of Eocene age in Wyoming, Colorado, and Utah; and the marine Galena oil rock of Ordovician age in Illinois, Iowa, Minnesota, and Wisconsin.

Organic Facies AB

Criteria for this facies are $H/C = 1.45–1.35$ and $HI = 850–650$. Rocks forming this facies are often laminated and organic rich. The organic matter is similar to that in organic facies A, except that it is diluted either with organic input of lesser quality or by partial degradation. Organic facies AB is found in both carbonates and shales. One example is the Upper Jurassic marine source rocks of Saudi Arabia.

Organic Facies B

Criteria for this facies are $H/C = 1.35–1.15$ and $HI = 650–400$. This facies is the source of petroleum for the majority of the world's oil fields, although organic facies AB may have contributed more of the world's oil (Jones, 1987). It is often laminated and may contain some terrestrial organic matter. It can be interbedded with less oil-prone facies, reflecting either fluctuations in bottom water anoxia or introduction of sediments with associated oxygen or poorer quality organic matter. Organic facies B and its systematic neighboring facies AB and BC can be mixtures representing biological source variation, some transported organic matter, or variations in preservation. Organic facies B encompasses most of the earth's best petroleum source rocks and is predominantly found in marine rocks, especially in deep water paleoenvironments associated with upwelling. Examples include the marine Kimmeridgian Clay source rocks of Jurassic age of the North Sea; the marine Monterey Formation of Miocene age in California; and the marine Phosphoria Formation of Permian age in Wyoming and Montana.

Organic Facies BC

Criteria for this facies are $H/C = 1.15–0.95$ and $HI = 400–250$. This facies is found in both marine and lacustrine paleoenvironments. It is often deposited in fine-grained siliciclastics where rapid deposition captures small oxygen volumes in the sediments. This "sedimentary oxygen" encourages biological activity in the sediments. Terrestrial organic matter can be a significant contributor, but bioturbation of bottom sediments may be sufficient to degrade marine organic matter to this quality. Examples include the marine Mowry Formation of Cretaceous age in Wyoming and prodelta muds from many deltaic and lacustrine deposits.

Organic Facies C

Criteria for this facies are H/C = 0.95–0.75 and HI = 250–125. This facies is predominantly gas prone. The organic matter is primarily woody and terrestrial and makes up most coals. Organic facies C is found in marine environments on Tertiary and Mesozoic shelf margins where it includes mixtures of hydrogen-rich and hydrogen-poor macerals or degraded hydrogen-rich macerals. Organic facies C and neighboring facies BC and CD are found in coal-forming swamp deposits, deltaic deposits, and bioturbated marine mudstones. The types of environments in which this facies occurs often correspond to the transgressive and early highstand systems tracts where some oxidation occurs and where different kerogen components can be deposited together (Vail et al., in press). One example is the lower Tertiary of Labrador.

Organic Facies CD

For this facies, the criteria are H/C = 0.75–0.60 and HI = 125–50. It is heavily oxidized and frequently represents terrestrial organic matter that has been transported through oxidizing environments. This facies may represent recycled organic matter that has been eroded one or more times from sediments. An example is in Cretaceous sediments lying on the Atlantic shelf of North America.

Organic Facies D

Criteria for this final facies are H/C < 0.60 and HI < 50. This facies contains highly oxidized organic matter, which may represent burnt wood (charcoal), recycled terrestrial material, and thermally postmature constituents. It may include fragments of larger woody components recycled from porous sandy units where oxidation prevailed. This facies is usually encountered in small concentrations and has no hydrocarbon generative capacity. It occurs in prograding sediments associated with sea level highstands and redeposited sediments of lowstands. Organic facies D is the regionally distributed type IV kerogen. It is frequently found in fluvial paleoenvironments, on offshore toes of deltas, and where organic carbon has been recycled. Sediments containing this kerogen can be found in all systems tracts, including the fine-grained parts of turbidites in lowstand systems tracts and the silts and muds of highstand prograding sediments.

SUMMARY

Organic facies are mappable units. Hydrogen-rich organic facies reflect hydrocarbon generative potential that helps exploration geologists map the distributions of possible petroleum source rock candidates. The criteria for assigning rocks to organic facies are hydrogen richness combined with maceral analysis of sedimentary organic matter. The method is based on elemental analyses (C, H, and O) of the organic matter, but may be effectively applied using hydrogen and oxygen indices from Rock-Eval pyrolysis.

The distribution of organic facies reflects varying influences of biological productivity, preservation, and geological processes. These processes are complexly woven into a fabric of water (oceanic and lacustrine) and atmospheric paleocirculation, nutrient and mineral source and supply, water depth, water temperature and chemistry, wave and storm activity, and bioturbation related to oxygen availability in bottom sediments.

Petroleum source rocks form where appropriate conditions for organic productivity and preservation of organic matter occur. Condensed sections often record these conditions because they reflect conditions of rising sea level and reduced detrital input. Thus, oil-prone source rock candidates often occur where condensed sections formed in poorly oxygenated sedimentary environments. These condensed sections are often associated with transgressive system tracts. Gas-generative organic facies such as coals may occur in the prograding deltaic deposits of some highstand system tracts.

Organic facies can be mapped with relatively small amounts of data in sequence stratigraphic frameworks. Furthermore, by increasing their familiarity with organic facies and source rock potential in stratigraphic sequences, geologists can more accurately predict petroleum source rock distributions in geological time and space.

Chapter 2

Primary Migration of Petroleum

V. V. Palciauskas
Chevron Oil Field Research
La Habra, California, U.S.A.

EXPULSION MECHANISMS

The expulsion of oil and gas from low-permeability source rocks (primary migration) has been a poorly understood phenomenon as compared to fluid movement in permeable rocks from source to trap (secondary migration). It has been known for a long time that secondary migration is driven primarily by buoyancy and hydrodynamic forces, which are most effective in highly permeable strata. In contrast, the most important driving forces for primary migration have not been clearly documented, and these will be the central focus of this chapter.

Many mechanisms have been proposed for the primary migration of petroleum, with the majority falling into two classes. One class of mechanisms is associated with the existence and/or movement of water through the source rock. For example, oil/gas solubility in water increases with increasing temperature, and hydrocarbons will be carried with the water as it flows out of source rocks due to compaction. Another example is hydrocarbon diffusion through the water-saturated pore space. In this case, water need not move relative to the rock and is simply the medium through which the hydrocarbons diffuse. Mass balance considerations generally show that these water-controlled mechanisms are not the dominant migration mechanisms operating in most basins. For a historical review of water-related mechanisms, see Tissot (1987).

It is now generally accepted that the bulk of petroleum is expelled as a separate oil phase. Some of the observations leading to this conclusion have been summarized by Jones (1978):

1. Most of the major oil accumulations of the world originated in source rocks with total organic carbon (TOC) content of ≥2.5 wt. %. There was insufficient water from compaction, clay dehydration, or any other source to transfer the oil from source to reservoir.

2. During peak generation and related primary migration, most bitumens reside in the pore space (including fractures) of the inorganic matrix and not in solution, attached to clays, or in the kerogen.

3. Primary migration accompanies peak generation. Observations supporting this conclusion include (a) bitumen dikes, (b) fractured shale production, (c) maximum similarity of oils in reservoirs and bitumens in source rocks at peak generation, and (d) the undersaturation with gas of most oils migrated from organic-rich rocks.

These general observations have been recently supported by detailed chemical analyses by MacKenzie et al.(1988) of shale source rocks from the North Sea which are currently expelling hydrocarbons. Their analyses of the Kimmeridgian Clay shales and the spatial distribution of hydrocarbons could be explained "if most expulsion occurs by pressure-driven flow of a discrete hydrocarbon phase through the pores of shales and into the pores of adjacent sandstones" (MacKenzie et al., 1988, p. 63). They also inferred that capillary forces can play a minor role in expelling hydrocarbons close to shale–sand boundaries and that diffusion of the lighter hydrocarbons can play a role in altering hydrocarbon fluid composition.

The movement of a separate petroleum phase can occur by the displacement of pore fluids in the surrounding inorganic matrix (immiscible displacement) or by fracture generation whereby the petroleum completely bypasses the matrix porewater. Most major oil accumulations originated in source rocks with high total organic content (TOC > 2.5 wt. %), in some cases exceeding 10 wt. % (e.g. the Kimmeridgian of the northern North Sea, the Silurian of Algeria, the upper Miocene of California, the Cretaceous "black shale" of the Prudhoe Bay area, and the Bakken Formation of Williston basin). In these source rocks, migration of a single petroleum phase by immiscible displacement and/or fracture generation is volumetrically the most important process and will be the primary focus of this chapter. For lean source rocks (TOC ≤ 0.5 wt. %), there might not be sufficient liquid hydrocarbons generated to permit single phase migration. In such cases, cracking to gas, and less efficient migration processes such as those associated with water, would become important.

CONCEPTUAL MODEL OF THE SOURCE ROCK

To quantify the expulsion process, a conceptual model for the physical state of a source rock is necessary. For the purpose of this paper, a source rock can be described as any organic-rich layer of prescribed boundaries with a volume V_{source}. Organic matter is assumed to be distributed throughout the source rock and occupies a volume V_{om}. This volume of organic matter does not have to form a connected "network." The remaining volume, V_{matrix}, is occupied by inorganic solids such as sands, carbonates, and clays and their associated porosity ϕ. Thus, V_{matrix} is the sum of mineral volume V_{im}, plus its associated pore space $V_{im}\phi/(1 - \phi)$. The porosity of the inorganic matrix depends on the mineralogy, texture, and effective stress to which it has been subjected. We can summarize the defined quantities as follows:

V_{om}:	Volume of organic solids
V_{im}:	Volume of inorganic solids
ϕ:	Porosity of inorganic matrix
V_{matrix}:	Volume of inorganic matrix (solids + pore space)
V_{source}:	Volume of the source rock

The precise relation between the volumes V_{source} and V_{matrix} depends on the distribution and thus the state of stress of the organic matter at the pore size scale. When organic matter particles or particle accumulations are larger than the grain size of the surrounding matrix, all or at least a large fraction of the organic matter supports the overburden stress (a nonzero effective stress). In this case, stress-bearing organic matter is not part of the intergranular pore space ϕ and thus not part of V_{matrix}. This implies that

$$V_{source} = V_{matrix} + V_{om} = \frac{V_{im}}{1 - \phi} + V_{om} \quad (1)$$

Effective stress on organic matter $\neq 0$

A common example of stress-bearing organic matter is where organic matter occurs as thin laminae between the inorganic layers.

Where organic matter resides wholly in the pore space of the mineral matrix, it experiences only pore fluid pressure. In this case the effective stress on the organic solids is zero and the volume of the matrix equals the volume of the source rock.

$$V_{source} = V_{matrix} = \frac{V_{im}}{1 - \phi} \quad (2)$$

Effective stress on organic matter = 0

These two cases are illustrated schematically in Figure 1. For the case of zero effective stress, organic matter volume must be less than the matrix pore volume. In most situations, the organic matter will not all be in stress- or nonstress-bearing locations, but could have fractions in both stress conditions. Petrographic evidence and theoretical considerations indicate that in most "good" source rocks, the organic matter is predominantly at nonzero effective stress (load bearing) and that this state of stress is the main driving force for primary migration. Therefore, for the remainder of this chapter, it will be assumed that all of the organic matter is load bearing.

HYDROCARBON CONVERSION

In this discussion hydrocarbon conversion kinetics is viewed as a temperature-induced transformation of organic matter (solid) into a single oil phase (liquid) of constant density ρ_o and viscosity μ_o. It is assumed that the generated oil will be expelled rapidly, so that no further alteration of the viscosity and density due to cracking kinetics and gas formation occurs.

The potential mass of oil, M_o (g), that can be generated from the source rock can be computed from the thermally immature (initial) hydrogen index HI (mg HC/g organic carbon), the percent organic carbon per unit rock mass TOC (g organic carbon/g rock), and the total rock mass $\rho_{im} V_{im}$ (g rock), as follows:

$$M_o = \text{HI} \times \text{TOC} \times \rho_{im} V_{im} \times 10^{-5} \quad (3)$$

where ρ_{im} is the density of the inorganic solids. From the viewpoint of expulsion, it is useful to convert this mass to a volume of oil by dividing it by the oil density ρ_o and then expressing the volume in terms of the number of pore volumes of the inorganic matrix. This ratio, denoted by ψ, represents the "relative richness" of the source rock

$$\psi \equiv \frac{M_o}{\rho_o V_{pore}} = \text{HI} \times \text{TOC} \times 10^{-5} \times \left(\frac{\rho_{im}}{\rho_o}\right) \times \frac{1 - \phi}{\phi} \quad (4)$$

For HI = 400 mg HC/g organic carbon , TOC = 10 wt. %, and ρ_{im}/ρ_o = 2.65/0.9, the relative richness of a source rock would be $\psi \approx (0.1/\phi)$, which implies that for a porosity of 0.10, the relative richness would have a value of about 1.

Petroleum formation kinetics depend on temperature and somewhat less on heating rate (e.g., Burnham

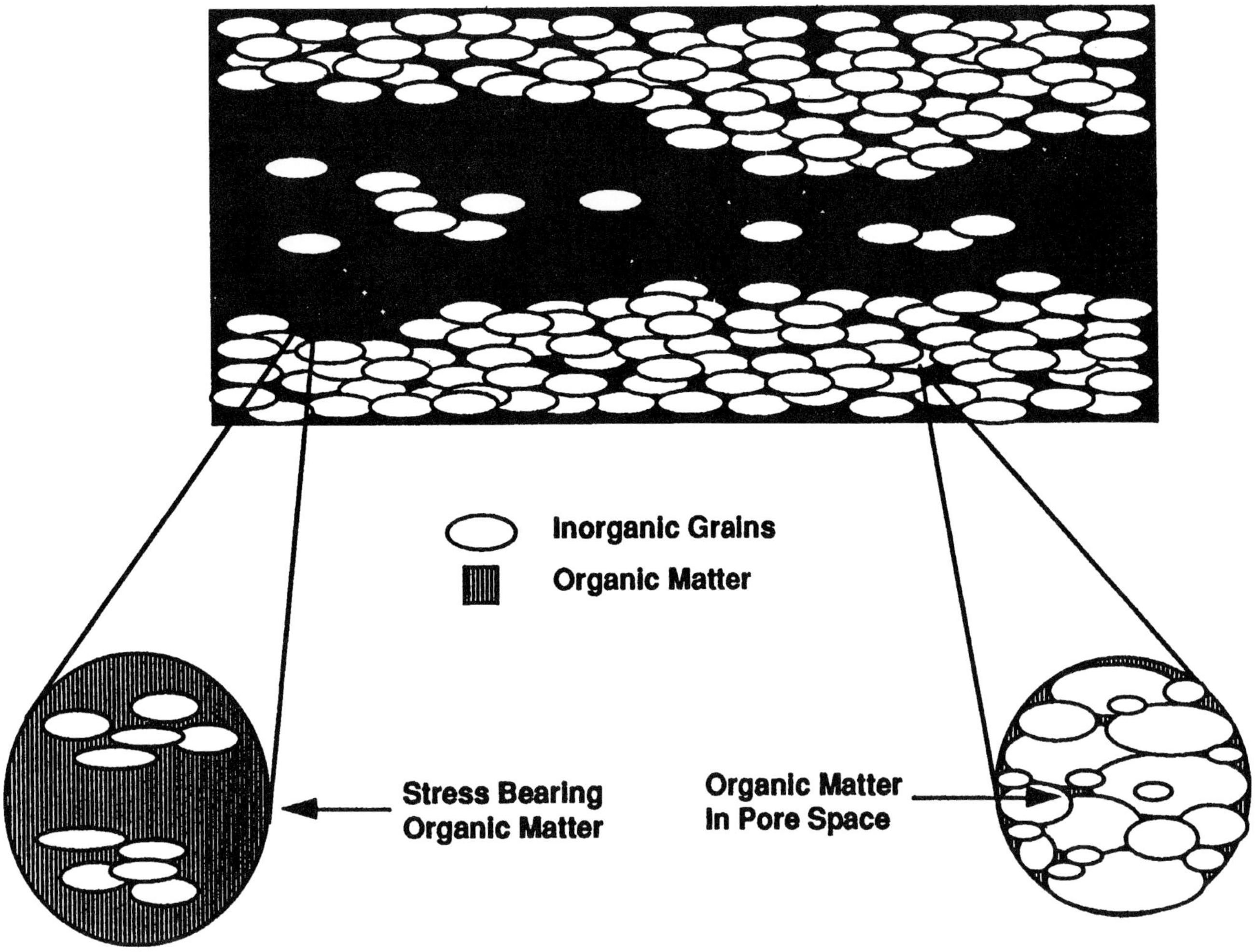

Figure 1. Schematic diagram illustrating how distribution of organic matter determines its state of effective stress.

and Sweeney, this volume). The results presented here are independent of time and thus of burial history, implying that the rate of conversion need not be specified but only how much conversion has taken place. For this reason, the conversion index (CI) will be defined as the volume fraction of the total oil potential that has been generated at time t

$$\text{Conversion index} \equiv \text{CI}(t) \equiv \frac{\text{Volume of oil generated}}{\text{Potential volume of oil}} \tag{5}$$

The rate of oil generation per unit rock volume is then the product of the total volume that can be generated, $M_o/\rho_o V_{im}$, and the rate of change of conversion index with time, $d(\text{CI})/dt$.

ROLE OF INORGANIC MATRIX

Rouchet (1981) pointed out that the local stress field can play an important role in primary migration. Because most of the organic matter is stress bearing as it is slowly broken up by thermal agitation, the liquid products can no longer maintain the stress differential $\sigma_1 - \sigma_3$ between the vertical (σ_1) and horizontal (σ_3) directions and will flow to achieve stress equilibrium. Subjected to a differential stress, the liquid fraction will form thin continuous films between the clusters of aromatic stacks and will be squeezed out of the kerogen into the surrounding inorganic matrix, preferentially in the direction of the least principal stress. At this stage, two pathways for the oil are possible: (1) oil can be injected into the pore space of the inorganic matrix displacing the water and resulting in two-phase immiscible flow, or (2) it can open a fracture through the matrix, thus bypassing the water-saturated pore space. For both cases, there must be sufficient oil conversion (sufficient TOC in the source rock) so that a continuous oil phase can be "squeezed" out of the kerogen and percolate through the inorganic matrix. This process does not require a continuous kerogen network through the source rock.

The particular pathway taken (fracture generation or immiscible displacement) depends on the ambient pore pressure (P_w) of the water, the capillary pressure (P_c), the magnitude of the least principal stress (σ_3), and the tensile strength (T_3) of the matrix. For oil to open a tensile fracture, the oil pressure (P_o) must exceed the sum of the least compressive stress (σ_3) plus the tensile strength of the matrix at its weakest point of cohesion, $P_o > \sigma_3 + T_3$. For oil to displace water in a pore, the oil pressure must exceed the sum of pore fluid pressure and capillary pressure (pore entry pressure), $P_o > P_w + P_c$. Thus, the relative magnitude of $\sigma_3 + T_3$ and $P_w + P_c$ determines which mechanism is operative:

Fracture formation: $$P_w + P_c > \sigma_3 + T_3 \quad (6)$$

Immiscible displacement: $$\sigma_3 + T_3 > P_w + P_c \quad (7)$$

The physical properties of the inorganic matrix greatly influence pore and capillary pressures and thus play a vital role in determining which mechanism is operative. For example, grain size is an important parameter directly affecting capillary pressure and, indirectly, pore pressure (P_w) through the matrix permeability. Because permeability varies as the square of the grain size diameter, fine- grained sediments have much lower permeabilities and are more likely to be overpressured. Typical examples would be clay-rich sequences in active depositional areas such as the U.S. Gulf Coast or the Niger delta. Thus, where all other factors are identical, P_w and P_c will increase as the grain size of the matrix decreases and will eventually satisfy condition (6), resulting in oil expulsion by fracture formation. To evaluate the pore pressure, a more detailed reconstruction of the burial and compaction history of the source rock is required. But because the horizontal stress is usually not well known, it would still be difficult to determine precisely when fracturing would begin. Sample computations indicate that a 100-m-thick source rock whose permeability is 0.1 microdarcy (μd) or less would be inclined to expel by fracture generation and propagation rather than by immiscible displacement. Figure 2 indicates in an approximate way how the permeability of the inorganic matrix affects the migration mechanism.

Note from Figure 2 that when permeability becomes larger than 100 μd or so, it is sufficiently large that buoyancy effects start becoming important. For still larger permeabilities, one would be in the secondary migration region. Long-range migration processes due to buoyancy and/or meteoric water from source to trap are discussed in another chapter (see England, this volume).

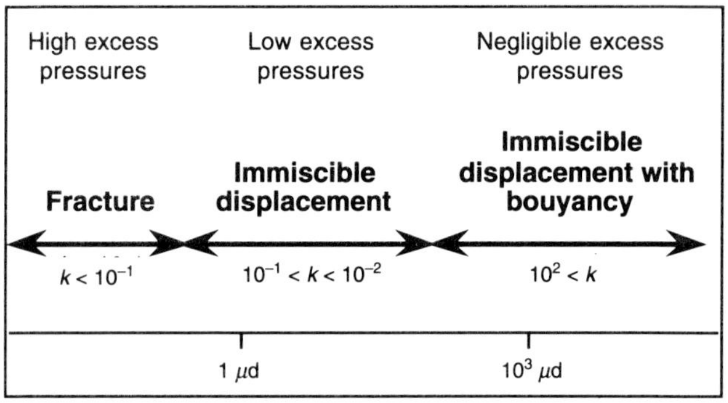

Figure 2. Relationship between source rock permeability *k* (μd), excess pressures generated, and petroleum migration mechanism.

RECOGNITION OF MIGRATION: SOURCE ROCK COMPACTION

As the organic matter is slowly converted to oil, the oil is squeezed out of the organic matter and into the pore space, displacing water or fracturing the rock. During this process, the volume of oil and/or water expelled from the source rock is approximately equal to the net decrease in volume of the organic matter, i.e., ΔV_{om}. Thus, a decrease in source rock volume occurs not only because of normal compaction of inorganic pore space, $\Delta\phi$, but also because of the decrease in volume of the organic matter, ΔV_{om}, which has been converted and expelled. Decrease in matrix porosity is caused by inelastic grain deformation in response to the continually increasing effective stress with burial, while the volume decrease due to conversion of organic matter is caused by the temperature-induced solid fluid transformation and is sharply peaked around 120°C. This additional volume decrease due to organic matter conversion is localized in the oil window and applies only to the organic matter in load-bearing positions, i.e., effective stress $\neq 0$.

It is instructive to compare the magnitude of these two independent processes. ΔV_{om} is equal to the change in the conversion index times the total volume of organic matter that can be converted. The change in the volume of the source rock in response to passing through the oil window can be written as

$$\Delta V_{source} = \Delta V_{matrix} + \Delta V_{om}$$

$$\frac{\Delta V_{source}}{V_{im}} = -\frac{\Delta\phi}{(1-\phi)^2} - \Delta(\mathrm{CI})\left(\frac{M_o/\rho_{om}}{V_{im}}\right) \quad (8)$$

Typically, the matrix porosity (ϕ) is about 0.05–0.10 around the oil window, and the change in porosity ($\Delta\phi$) is at most 0.02 through the oil window. This implies that the change in volume $\Delta V_{source}/V_{im}$ due to a decrease in matrix pore space is approximately 0.012–0.025 or equivalently 1.2–2.5%. For a source rock of initial HI = 400, TOC = 10 wt. %, and ρ_{im}/ρ_o = 2.65, we find the second term in equation (8) to be approximately 0.10 (i.e., 10%) or 4 to 8 times larger than the first. In localized regions of very high TOC values, the decrease in source rock volume due to conversion can be more than 10 times larger than that due to normal compaction. This suggests that mature source rocks should exhibit "collapse" microstructures of the inorganic matrix, where a high percentage of organic matter was converted and expelled.

EXPULSION OF OIL BY FRACTURE GENERATION

For very low permeability source rocks, the pore water pressures can be quite high ($P_w \approx \sigma_3 + T_3$) and close to where incipient microfracturing will occur (Palciauskas and Domenico, 1980). In this state, fracture generation will be the primary mechanism of expulsion. Once the fracture network is established, the rate of expulsion is approximately equal to the rate of generation, i.e., peak expulsion occurs during peak generation. Mass balance implies that the total amount of expelled petroleum is equal to the total amount generated, minus the liquid petroleum within the kerogen and in the fracture network.

Oil generated = Oil expelled + Oil retained in source rock (kerogen + fractures)

The amount of liquid petroleum in the fracture network can be estimated from fracture mechanics and generally leads to values that are not greater than 2% of the inorganic matrix volume. The amount of liquid petroleum retained within the kerogen is more difficult to estimate, but a physical model based on compaction processes (McKenzie, 1984) leads to values of 1–2% of the kerogen volume. Using the upper value of 2%, an estimate of how much oil is retained within the source rock fractures and kerogen would be 2% of the source rock volume, i.e., $(0.02)V_{source}$. Thus, a very rich source rock with a very fine grained, impermeable inorganic matrix could be expected to yield a very high expulsion efficiency, which is defined as follows:

$$\text{Expulsion efficiency} \equiv \frac{\text{Oil expelled}}{\text{Oil generated}} = 1 - \frac{\text{Oil retained}}{\text{Oil generated}} \tag{9}$$

As an illustration, let's express the total volume of oil that can be generated (M_o/ρ_o) as a fraction x of the initial source rock volume, that is, $(M_o/\rho_o) = xV_{source}$, where M_o can be computed through equation (3). Using $(0.02)V_{source}$ as the estimate of total retained oil, equation 9 leads to an expulsion efficiency equal to 1 - $(0.02/x)$. Assuming that x is 0.10 (i.e., 10% of the initial source rock volume), we get an expulsion efficiency of 0.80, or 80%. If x is 0.20 (i.e., 20% of the initial source rock volume), then the expulsion efficiency is 0.9, or 90%. Thus, the expulsion efficiency of very rich, impermeable source rocks can be very high.

EXPULSION OF OIL BY IMMISCIBLE DISPLACEMENT

For intermediate values of matrix permeability, the pore space can accommodate the rate of oil generation without significant increase in pore pressure and development of fractures. In this case, equation (7) is satisfied, and primary migration of a single oil phase through the inorganic matrix takes place by the process of immiscible displacement. Primary migration occurs as a displacement process whereby a nonwetting fluid (oil) is slowly injected into the pores of an inorganic matrix from the organic matter, displacing the wetting fluid (water). Overburden pressure again acts as the main driving force, squeezing the converted oil out of the load-bearing organic matter. Because the pore space of the matrix can be a significant volume fraction of the total volume of oil generated, it is important to estimate how much of the pore network will have to be saturated with oil before expulsion begins and the factors that control this saturation.

The model computations that follow are based on several assumptions: (1) only a single oil phase is formed with constant density ρ_o and viscosity μ_o, (2) the pore volume of the inorganic matrix is constant during the period when the source rock is generating oil, (3) the distribution of kerogen and oil saturation are spatially uniform (on a coarse scale), and (4) the flow of oil and water can be described by Darcy's law for each phase with the relative permeability concept. Relative permeabilities should reflect the pore structure and capillarity properties of the inorganic matrix. Buoyancy forces are minimal compared to the main driving force of overburden pressure, so that the influence of only three variables need be considered. These variables are

1. ψ, the relative richness of the source rock (i.e.,

potential volume of oil normalized by the constant pore volume),
2. μ_w/μ_o, the viscosity ratio of water to oil, and
3. the relative permeabilities of oil, $k_{ro}(S)$, and water, $k_{rw}(S)$, both of which depend on oil saturation S within the matrix.

Mass balance considerations in conjunction with equations governing the flow of immiscible fluids can then be solved for the average oil saturation and the amount of oil expelled. This solution is presented as equation 10:

$$S + \frac{\mu_w}{\mu_o}\int_0^S \frac{k_{ro}(S')}{k_{rw}(S')}dS' = \psi \times \text{CI} \qquad (10)$$

Saturation + Oil expelled = Oil generated

$$S + \text{Ex}(S) = \psi \times \text{CI}$$

The integral, denoted by Ex(S), is equal to the number of pore volumes of expelled oil for a given amount of conversion CI. For the given assumptions, equation 10 is a solution of the mass balance and flow equations for immiscible displacement. It shows how, given a physical model for the inorganic matrix $k_{ro}(S)$ and $k_{rw}(S)$, relative richness of source rock ψ, amount of conversion CI, and type of hydrocarbons generated μ_w/μ_o, the saturation S and volume expelled Ex(S) can be computed. This result does not dependent explicitly on time, but only implicitly through the conversion index CI. This means that when richness ψ and conversion index CI are specified, the saturation and volume expelled are uniquely determined and the results are independent of the details of the burial history. That is, the value of CI is important, but not how you reached it.

Relative Permeabilities of Oil and Water

The relative permeability model shown in Figure 3 is meant to reflect a wide distribution in pore throat sizes within the inorganic matrix. When oil is generated, it preferentially enters pores accessible through the largest pore throats, implying that during this initial saturation, the relative permeability of the oil increases rapidly. At the same time, the relative permeability to water decreases rapidly because it is a wetting fluid and preferentially occupies the microporosity of the matrix. This behavior is consistent with percolation theory and experimental data on the conductivity of random porous media and fracture systems. The principal idea is that in porous media with a wide distribution of pore throat sizes or fractures, the electrical and hydraulic conductivities are primarily determined by a small

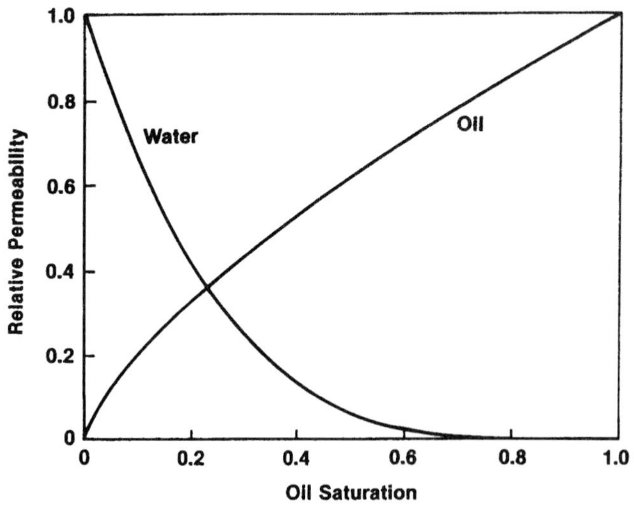

Figure 3. Model of matrix relative permeability to oil and water as a function of oil saturation.

fraction of the total pore space, i.e., the connected cluster of the most permeable pores and fractures that traverse the media. This model will be used for the remaining computations, but it should be emphasized that other source rocks can have pore structures and relative permeabilities that deviate significantly from those shown in Figure 3.

Fractional Flow of Oil

Because the inorganic matrix porosity is assumed to be constant, the volumetric outflow of oil plus water from the source rock must be approximately equal to the volumetric rate of oil conversion. The oil fraction of the total outflow depends on the relative permeabilities of oil and water, oil saturation, and the viscosity ratio:

$$\text{Outflow of oil} = \left[\frac{k_{ro}}{k_{ro} + \left(\frac{\mu_o}{\mu_w}\right)k_{rw}}\right] \times \text{Rate of oil conversion} \qquad (11)$$

Outflow of oil = Fractional oil flow
x Rate of oil conversion

Figure 4 shows the fractional oil flow as a function of oil saturation for viscosity ratios (μ_w/μ_o) ranging from 0.01 to 100. The most important feature to note is that as oils become more viscous, the same fractional flow of oil can only be achieved by higher saturations. Less viscous oils (API gravity ≈ 30°, viscosity ratio ($\mu_w/\mu_o \approx$ 1) and especially gas condensates (API gravity ≈ 50°, viscosity ratio $\mu_w/\mu_o \approx 10$) can have large fractional oil

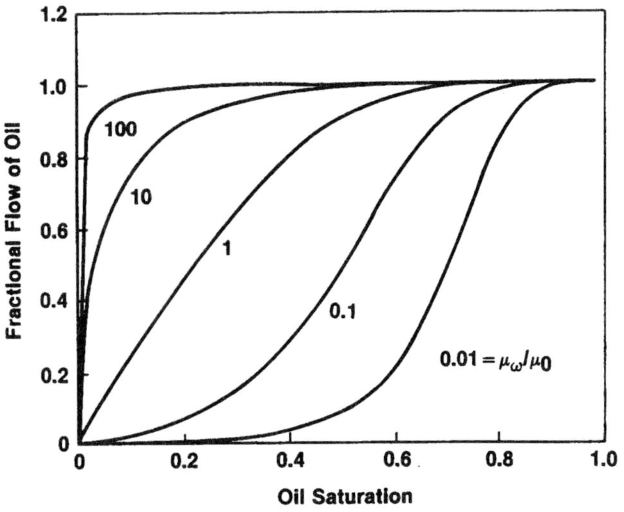

Figure 4. Fractional flow of oil as a function of oil saturation for oils with different water/oil viscosity ratios. To maintain equal fractional flows, more viscous oils require higher saturations.

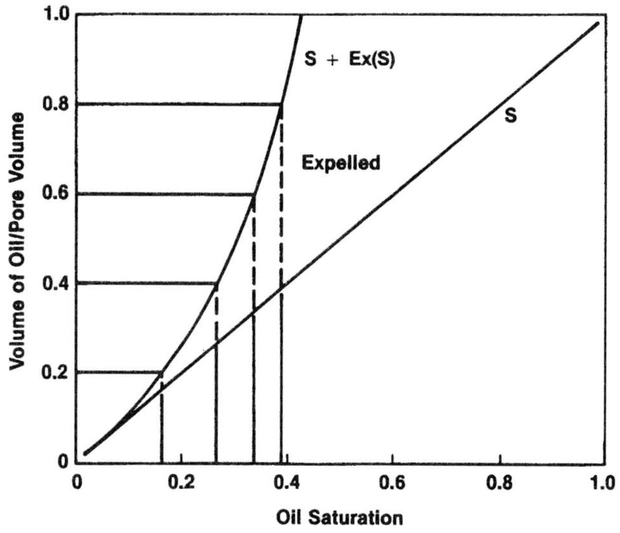

Figure 5. Graphical solution for matrix saturation *S* of a generated oil with a water/oil viscosity ratio of 1. The horizontal lines represent volumetric amounts of conversion, which is equal to the sum of saturation *S* and oil expelled Ex(*S*).

flows at significantly lower saturations. The important conclusion from this graph is that less viscous oils require lower saturations for the same fractional oil flow.

Source Rock Richness and Expulsion Efficiency

To illustrate how source rock richness affects the volume of expelled oil, a graphical solution of equation (10) is presented. The left-hand side of equation (10) is the sum of two functions of saturation, one linear (pore saturation S) and one nonlinear (number of pore volumes expelled Ex(S)). Their sum, S + Ex(S), which is equal to the volume of generated oil ψCI, is plotted in Figure 5, as is the linear function S. A specified volume of generated oil is represented by a horizontal line that intersects the curve, S + Ex(S), determining the value of matrix saturation for this level of conversion. The difference between these two curves, shown by a dashed line in the figure, gives the volume of expelled oil for these values of conversion. The increasing slope of the curve S + Ex(S) shows that expulsion efficiency increases with increasing volume of generated oil. When all other factors are equal, it is easily seen that richer source rocks have higher expulsion efficiencies. Figure 6 reemphasizes this point graphically by expanding the vertical axis to a relative richness of 5.

Viscosity Ratio and Expulsion Efficiency

Because the viscosity ratio plays an important role in the flow of immiscible fluids, the viscosity of generated oil is one of the important variables determining expulsion efficiency. To illustrate this point, S + Ex(S) is plotted for several viscosity ratios in Figure 7. Note that for a given volume of conversion, ψCI = 1 in this example, the volume of expelled oil (length of dashed lines) increases with increasing viscosity ratio, i.e., increasing oil gravity. The viscosity of water at standard pressure and temperature is approximately 1 cp (centipoise) and decreases with increasing temperature.

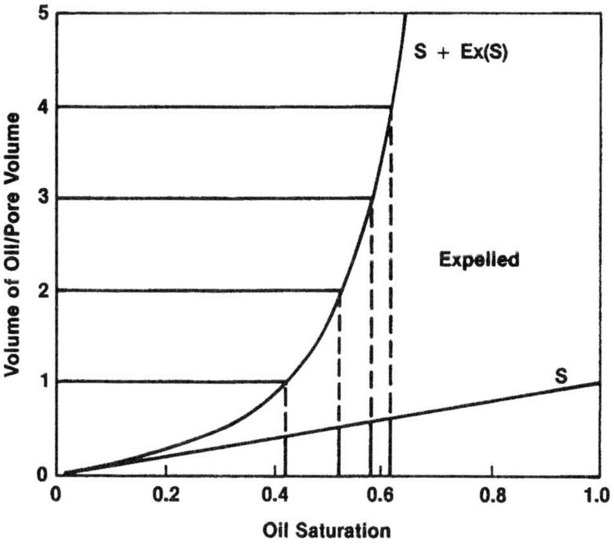

Figure 6. Graphical solution for matrix saturation *S* for a generated oil with a water/oil viscosity ratio of 1. The horizontal lines represent volumetric amounts of conversion, on an expanded vertical scale.

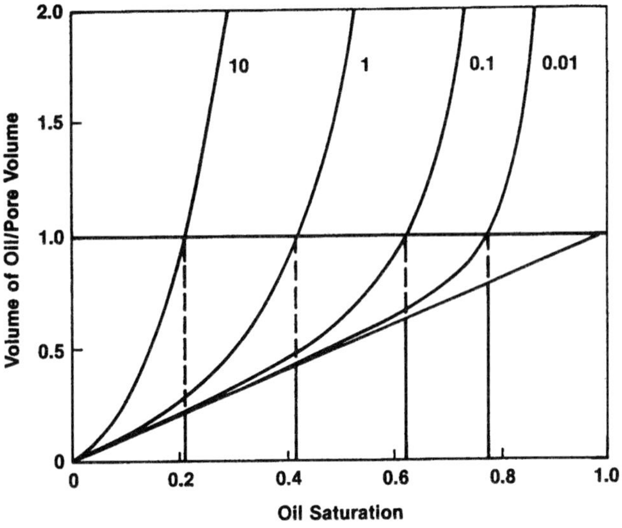

Figure 7. Graphical solution for matrix saturation *S* and volume of expelled oil of generated oils with assumed water/oil viscosity ratios of 0.01, 0.1, 1, and 10.

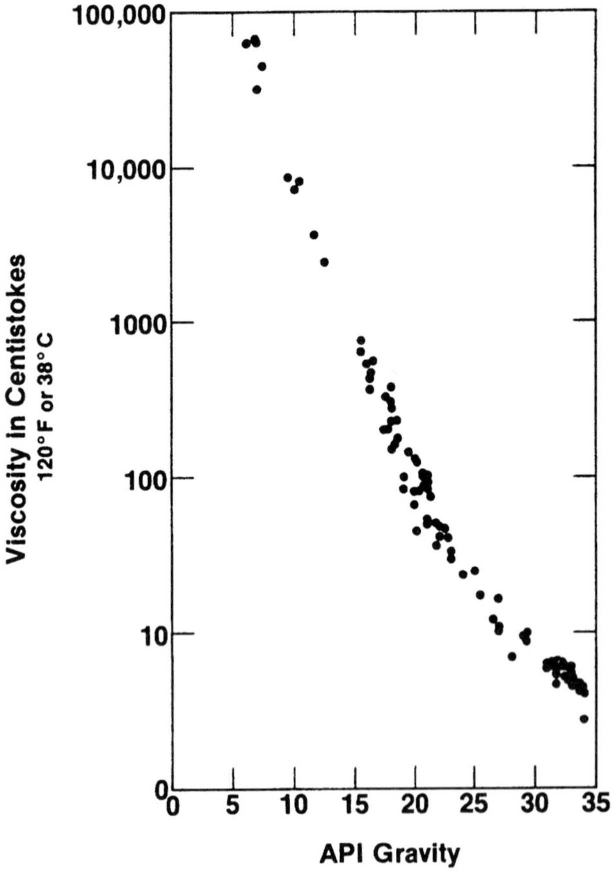

Figure 8. A correlation between viscosity and gravity of offshore, uncontaminated Monterey oils in southern California. The water/oil viscosity ratio would be approximately the inverse of shown viscosities ($m_w \approx 1$ cp at 20°C). (From Baskin, 1991.)

Table 1. Illustrative viscosity ranges of various hydrocarbons and petroleum fractions.

Location or Type	API°	Viscosity μ_o (cp)	μ_w/μ_o[a]
Offshore California A[b]	10°	900	0.0011
Offshore California B[b]	14°	200–400	0.005–0.0025
Kerosene	35–40°	~2.0	0.5
Gasoline	40–45°	~6	1.6
Volatile oil	46°	0.15	7.0
Gas condensate	50°	0.12–0.05	8–20

[a]μ_w = 1 cp at 20°C was used in the computation. At 100°C the viscosity of water is approximately 0.282 cp or about four times smaller.
[b]Reservoir conditions.

For illustrative purposes, ranges of viscosity of several oils under reservoir conditions and other hydrocarbons are presented in Table 1.

Several correlations between viscosity and API gravity have been presented in the literature. One of these, showing how viscosity varies with API gravity for offshore Monterey oils in southern California, is presented in Figure 8. The vertical axis represents measured oil viscosity divided by its specific gravity and is numerically very close to viscosity in centipoise. The vertical scale is logarithmic, illustrating the dramatic decrease in viscosity with increasing API gravity, a feature that is seen in other correlations. Because viscosity also depends on temperature, corrections for temperature differences are important when comparing data.

The important conclusion is that expulsion efficiency depends on the viscosity of generated oil, which in turn depends on oil composition. Heavier oils (low API gravity) will tend to saturate the inorganic matrix before expelling, while lighter oils (high API gravity) will require much lower saturations before expulsion. This effect is clearly important for lean source rocks, i.e., ψCI = 1, where heavier oils would be retained within the matrix while lighter oils would be partly expelled.

Saturation and Expulsion Efficiency versus Conversion Index

Up to this point, the various quantities have been plotted as a function of matrix oil saturation. But as illustrated in Figure 5, the oil saturation itself changes in a nonlinear fashion with conversion index. Figure 9 shows how the saturation S and volume of expelled oil Ex(S) vary with conversion index for four viscosity ratios. If we assume HI = 400 mg HC/g OC, TOC = 10 wt. %, porosity ϕ = 0.10, matrix density ρ_{im} = 2.65 g/cm^3, and oil density ρ_o = 0.9 gm/cm^3, then the relative richness of a source rock is ψ = 0.954. The straight lines in Figure 9 represent the increasing volume of generated hydrocarbons with conversion index, while the curved lines show the increasing

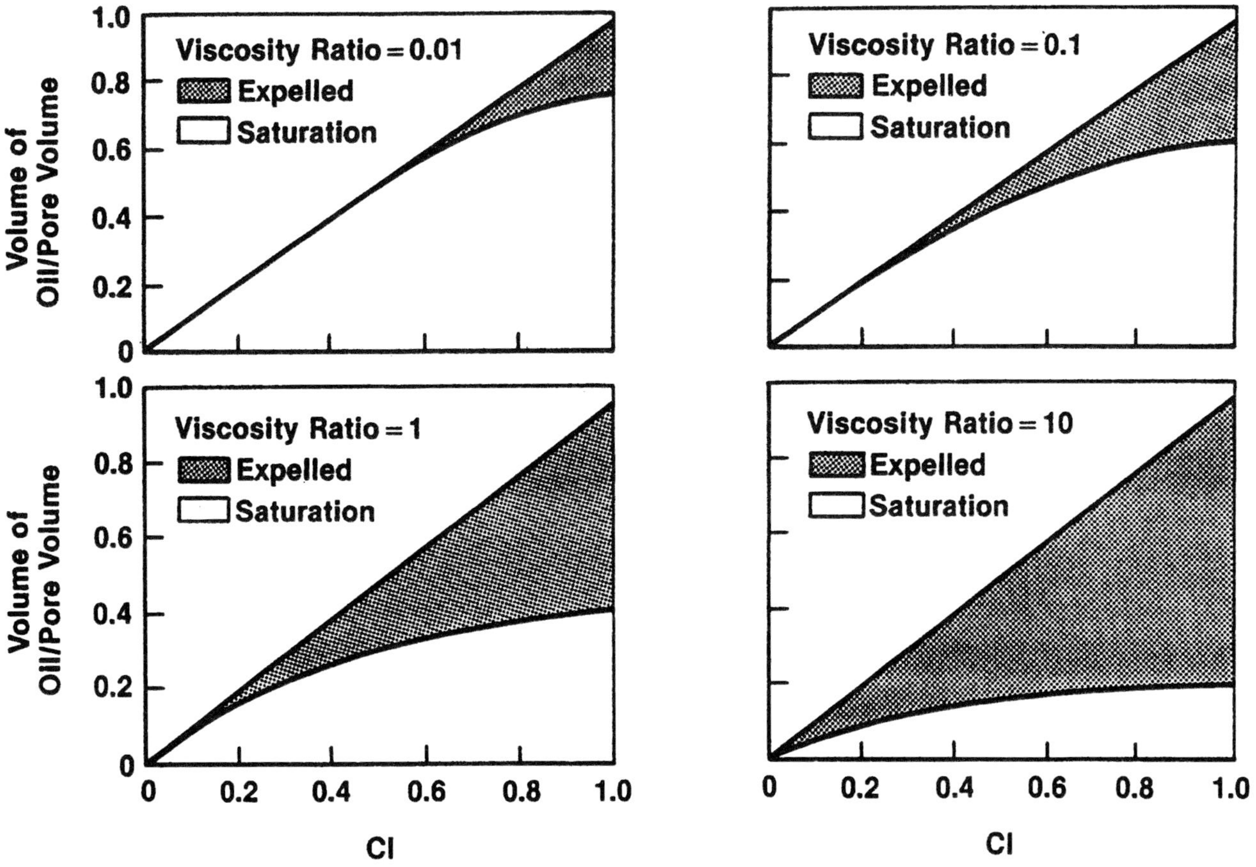

Figure 9. Matrix saturation and expelled oil volume as a function of conversion index for generated oils with four different water/oil viscosity ratios.

saturations of the rock matrix. The difference between these two lines (shaded region) is the number of pore volumes expelled. Note that the saturation continually increases with conversion index and does not reach a limiting value. This is a consequence of the relative permeability model that was used to describe matrix flow properties. It is not necessarily a general result that holds for all source rocks.

Having determined how saturation changes with conversion index, it is possible to compute how the fractional oil flow (expulsion) varies with conversion index. These results are shown in Figure 10 and again illustrate the sensitivity of expulsion to viscosity ratio. Note that if volatile oils or gas condensates (viscosity ratios of 1 to 10) are generated from the organic matter, significant expulsion occurs very early, while if heavy oils are generated, much more conversion is required before significant expulsion begins.

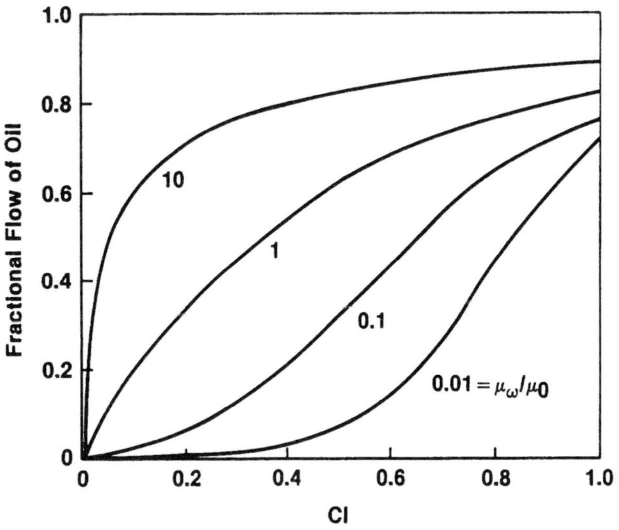

Figure 10. Fractional flow of oil as a function of conversion index for generated oils with four different water/oil viscosity ratios. For equal values of conversion, the fractional flow (expulsion) of lighter oil is greater.

SUMMARY

The three important parameters required for estimating expulsion efficiency by immiscible displacement are (1) relative richness of the source rock (i.e., number of pore volumes of potential oil), (2) relative permeability of the inorganic matrix to oil and water, and (3) viscosity ratio of generated oil to water. The important conclusions concerning expulsion efficiency are as follows:

1. Viscosity of the generated oil greatly affects its rate of expulsion. Specifically, higher viscosity oils require higher matrix saturations before expulsion. Because oil viscosity is closely linked to its API gravity, there is an inseparable connection between expulsion efficiency, oil/gas viscosity, and type of organic matter.

2. The physical properties of the inorganic matrix (porosity and relative permeability) also affect the expulsion efficiency. Porosity microstructure and relative permeability influence the fractional flow of oil and water and thus the expulsion characteristics of source rocks.

Chapter 3

Migration from Source to Trap

William A. England, Alastair L. Mann, and David M. Mann
BP Research
Middlesex, England, U.K.

INTRODUCTION

This chapter discusses the mechanisms by which petroleum migrates from the source rock to the trap. This process is usually described as "secondary migration," which is considered by most authors to start once petroleum reaches a relatively high-permeability carrier bed (Figure 1). We will also consider the processes that involve the movement of petroleum from an organic-rich interval through a lean mudstone to a carrier bed (which may be deeper or shallower than the source). This will involve a consideration of the effects of hydrodynamics and basin dewatering and the development of overpressures.

The influence of phase behavior on chemical composition may be considerable and is caused by the large alterations in pressure and temperature that occur during migration. The effect of petroleum phase behavior on migration and the type of accumulation that results (i.e., gas, gas condensate, or oil) is described in the section on phase behavior. Then we discuss the rate and efficiency of secondary migration. Finally, the section on chemical changes reviews the possible chemical effects that have been postulated to occur during secondary migration.

PROCESS OF MIGRATION

A consensus is emerging that petroleum migration occurs by movement as a separate phase in a water-wet rock matrix (e.g., Durand, 1981; England et al., 1987; Welte, 1988). An alternative theory, that petroleum migrates in aqueous solution, may be rejected on the grounds that (1) the water solubilities of the constituents of petroleum are too low to transport the required masses and that (2) a solution mechanism would *enrich* reservoired petroleums in the more soluble aromatic components, whereas the reverse is observed (McAuliffe, 1980a). However, methane is the most soluble component of petroleum and is thus the most likely candidate for an aqueous migration mechanism. The shallow dry gas deposits of Siberia (du Rochet, 1980) may have migrated as dissolved methane, which exsolved due to the relatively large fractional change in pressure experienced near the surface. Proponents of the "hydraulic theory of migration" (e.g., Toth, 1987) stress the role of water flow as an agent for petroleum migration. We hope to show in this chapter that although regional aquifer flow has a relatively small effect on secondary migration, it is not the major agent for migration (although its role is not insignificant).

Secondary migration is controlled by a competition among *buoyancy* (which always acts vertically), *capillary forces* (which oppose movement of petroleum from coarse- to fine-grained water-wet rocks), and *hydrodynamic forces* (which result from bulk water flow).

In this section, we discuss how to estimate the magnitudes and directions of these three forces in various geological situations. The objective is to predict migration directions and rates from (and through) source beds to the trap. This involves combining the effects of the various forces by means of a "fluid potential" approach.

Fluid Potential

Fluid potential plays a role similar to other potentials in physics and chemistry (such as voltage or potential energy). Material will tend to move from zones of high potential to those of lower potential. Fluid potentials also lend themselves to manipulation by computer on a prospect or basin scale so that they may be contoured to indicate wide-scale fluid movements.

The water fluid potential, Φ_W, is simply equal to the prevailing overpressure, i.e., the amount by which local water pressure exceeds hydrostatic pressure:

$$\Phi_W = P - \rho_W g z \qquad (1)$$

Water tends to move from areas of high overpressure to areas of low overpressure. Φ_W is therefore

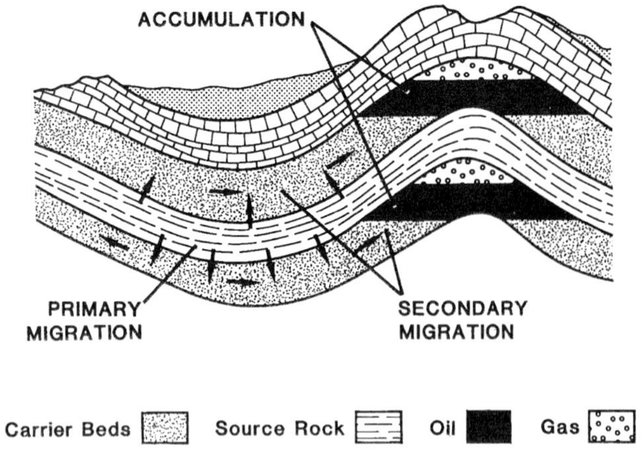

Figure 1. Definitions of primary and secondary migration. (After Tissot and Welte, 1984.)

measured in units of pressure (MPa or psi) and represents energy per unit volume. Some authors (e.g., Hubbert, 1953) define a fluid potential in terms of energy per unit *mass*, giving rise to a factor of ρ_W in the numerical values computed. For petroleum, the petroleum fluid potential (Φ_P) is equal to the water potential plus terms to describe the buoyancy and capillary potentials:

$$\Phi_P = \Phi_W + \underbrace{(\rho_W - \rho_P)gz}_{\text{Buoyancy}} + \underbrace{\frac{2\gamma}{r}}_{\text{Capillary}} \quad (2)$$

where

ρ_W is the water density (typically 1000–1100 kg/m^3)
ρ_P is the petroleum density (typically 400–800 kg/m^3 subsurface)
g is the acceleration due to gravity (9.81 m/s^2)
z is the depth below arbitrary surface (usually surface of sea or water table)
γ is the interfacial tension (typically 0.03 N/m), and
r is the pore radius (m).

Equation (2) neglects the small correction factor of cos β to the capillary term since the contact angle β is close to zero, leading to values of cos β close to unity.

The water and petroleum potentials are not of interest in themselves, however their three-dimensional derivatives, -$\nabla\Phi_P$ or -$\nabla\Phi_W$ define the *direction* and *magnitude* of the vector force driving migration of petroleum or water:

$$\mathbf{F}_W = -\nabla\Phi_W$$
$$\mathbf{F}_P = -\nabla\Phi_P \quad (3)$$
$$\text{where } \nabla = \frac{\partial}{\partial x}, \frac{\partial}{\partial y}, \frac{\partial}{\partial z}$$

The problem of estimating migration directions is therefore one of using equation (3) to define the force acting on petroleum at all points of interest in a basin. The first step is to use data, experience, or a mathematical model of fluid flow to calculate Φ_W and hence Φ_P using equation (2). If these data are contoured, the "equipotential" contours of Φ_P establish the directions in which fluids will migrate. The negative sign in equation (3) shows that flow is from areas of high to low potential. This use of fluid potentials to model the migration and entrapment of petroleum was first described by Hubbert (1953) in his classic paper.

Before considering specific examples, a few useful generalizations can be made on the direction of secondary migration under various conditions of subsurface water flow.

Hydrostatic Environments

In hydrostatic environments, the water pore pressure is given by the equivalent column height of water:

$$P(\text{hydrostatic}) = \rho_W gz \quad (4)$$

From equation (1), it is clear that Φ_W is always equal to zero in a hydrostatic regime, i.e., water is at rest since there is no net force acting on it (and therefore $F_W = 0$). Migrating petroleum will have a potential Φ_P given by:

$$\Phi_P = (\rho_W - \rho_P)gz + \frac{2\gamma}{r} \quad (5)$$

by substituting equations (1) and (4) into equation (2).

Since the force acting on unit volume of petroleum is given by -$\nabla\Phi_P$ (equations 2 and 3), $\Phi_W = 0$.

$$\mathbf{F}_P = (\rho_W - \rho_P)g + 2\gamma\nabla(1/r) \quad (6)$$

The capillary term $2\gamma\nabla(1/r)$ is small except at lithological boundaries where it gives rise to the large forces by which reservoir seals operate and that prevent petroleum from leaving carrier beds.

Neglecting the capillary term (i.e., considering only migration within a homogeneous carrier system), and noting that $\rho_W\, g = -\nabla P$, the vertical pressure gradient in the water, we have

$$\mathbf{F}_p = -\nabla P - \rho_P g \quad (7)$$

This equation defines the buoyancy force, which acts vertically upward in the hydrostatic case. The force is larger for subsurface gases because of their lower density than for petroleum liquids. Figure 2 defines $\mathbf{F}_P$ in terms of the vector addition of forces.

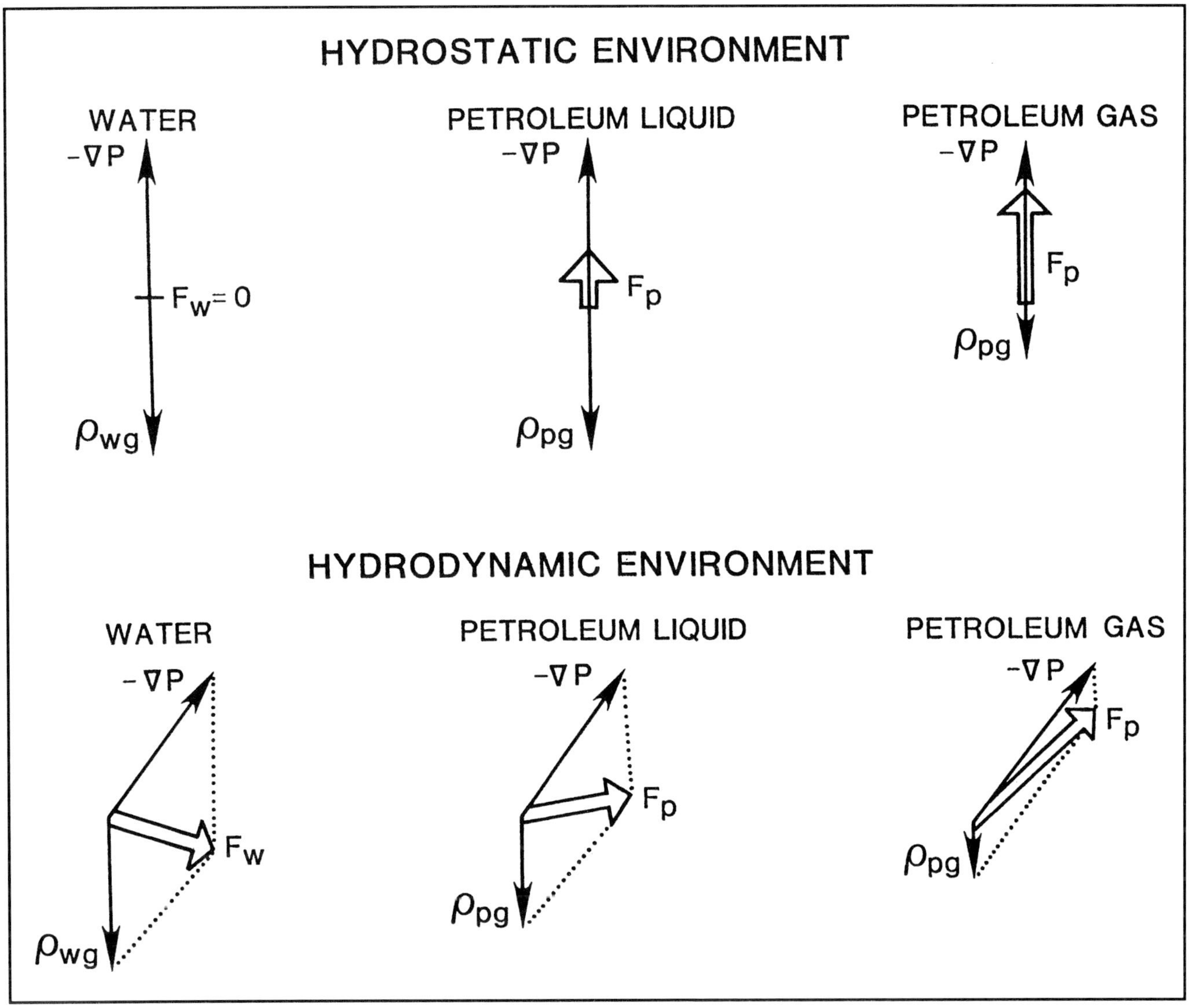

Figure 2. The forces acting on petroleum and water in hydrostatic and hydrodynamic environments.

Hydrodynamic Environments

In hydrodynamic environments, water is in motion, and accordingly, $\nabla\Phi_W \neq 0$. This leads to a net force F_w on water, which may be in any direction. The resulting forces on petroleum liquid or gas (again neglecting capillarity) will also be nonvertical and may also be in any direction depending on the geological circumstances. This is illustrated in Figure 2, which shows that the effect is greater for petroleum liquid than for petroleum gas since the lower density gaseous phase will experience a higher buoyant force. In the following sections, we discuss the circumstances under which petroleum migration directions are significantly different from those expected in hydrostatic regimes.

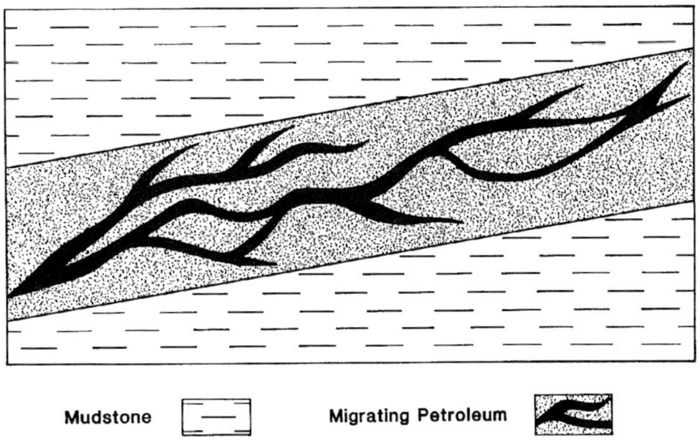

Figure 3. Schematic representation of a tortuous petroleum migration pathway.

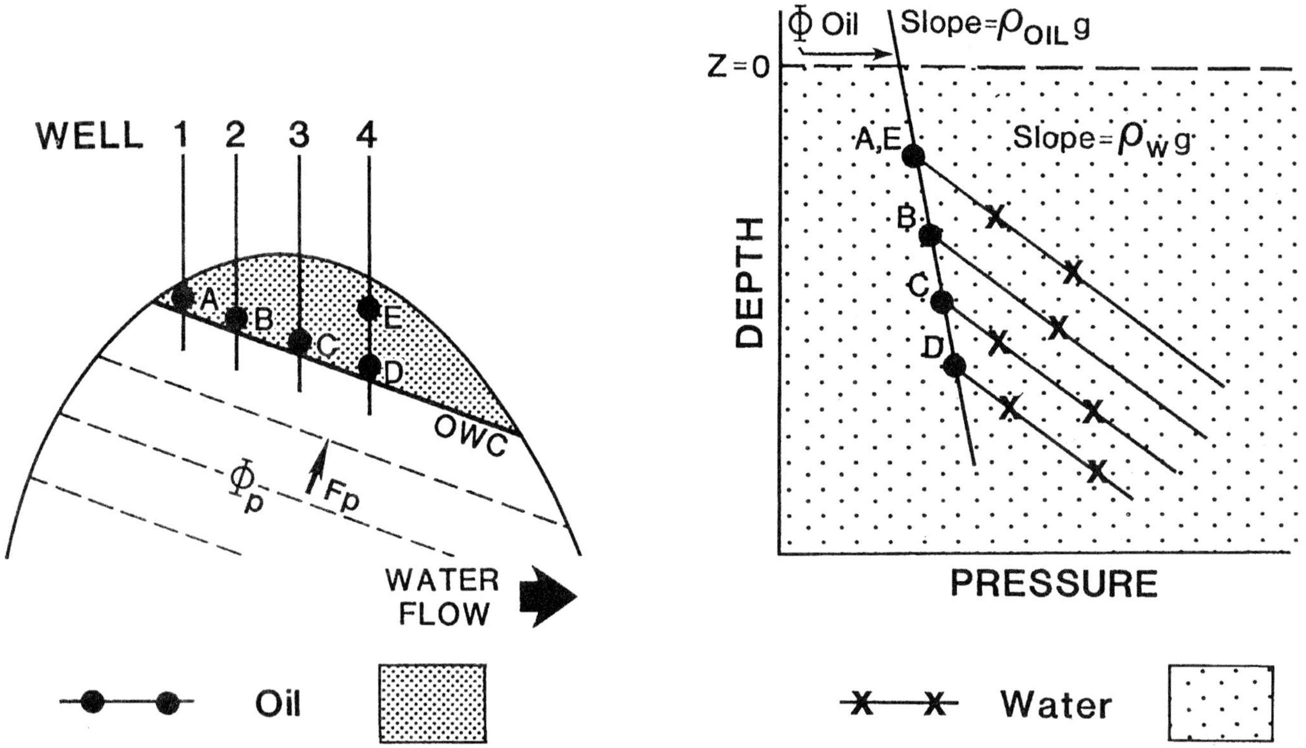

Figure 4. Tilted oil–water contact resulting from water flow in the underlying aquifer and corresponding pressure–depth plots.

PATHWAYS OF MIGRATION (HORIZONTAL AND VERTICAL)

Secondary Migration in a Hydrostatic Regime (Source Interval Juxtaposed with Carrier Bed)

Secondary migration in a hydrostatic regime is the simplest "textbook" situation, in which the organic-rich interval is in direct contact with a regionally extensive carrier bed. It does not matter whether the source rock is above or below the carrier bed since capillary forces (in a water-wet system) will drive petroleum from the finer grained rock.

Once secondary migration proper starts to occur, the only net force (ignoring capillarity) is vertical (Figure 2). Migration will therefore occur updip, focused along the most steeply dipping directions (at right angles to contours of Φ_P). Although capillary pressure differences *within* the carrier bed are relatively small on a 100–1000 m (330–3300 ft) scale, on a local scale (<100 m) they will determine the precise nature of the secondary migration pathway. It has been suggested (England et al., 1987) that migrating petroleum moves as a dendritic network which preferentially picks out a pathway involving the lowest pore entry pressures (i.e., through the coarser beds). The heterogeneities present in most geological situations will thus inevitably lead to a migration pathway that is locally convoluted, but that tends to follow the steepest updip directions *regionally* (Figure 3). This can be described mathematically by a percolation model (Winter, 1988).

Lehner et al. (1988) ignored the disruption caused by local capillary pressure variations in their model of petroleum migration and accumulation, predicting that migration occurred exclusively at the top of the carrier bed. We feel that this approach is too idealized to model secondary migration successfully. It does, however, represent an "end-member" case of a totally homogeneous carrier bed in which all pore size differences are absent.

Secondary Migration in a Hydrodynamic Regime

Figure 2 illustrates the perturbation to F_P, the force experienced by migrating petroleum. Quantitative calculations of the amount by which F_P is deflected from the vertical (e.g., Davis, 1987), indicate that the effect is very small. In other words, the shift in the angle that F_P makes with the vertical is unlikely to be sufficient to initiate downdip migration unless it is greater than the angle of dip of the carrier bed. This is because relatively high-permeability carrier beds cannot usually maintain high enough differences in Φ_W

to cause large shifts in $\mathbf{F_P}$. However, downdip migration should be considered as a possibility in low-dip beds experiencing significant water flows. We will show later that in lower permeability siltstones and mudstones, water movement can indeed cause downward migration.

In a hydrodynamic regime, the style of entrapment will be altered. The main effect is the establishment of sloping hydrocarbon–water contacts (HWCs) (Figure 4). In extreme cases, the hydrodynamic flow may preclude the possibility of entrapment in a structural feature. Dahlberg (1982) has presented a method ("U, V, W mapping") that allows one to combine water pressure and depth maps to predict the degree of perturbation caused to HWCs and to the occurrence of hydrodynamic traps. Hydrodynamic effects appear to be especially common in foreland basins and shallower fields. Note that it is the shift in the *pressure gradient* (∇P) set up by water flow, and *not* the flow itself, that is the cause of hydrodynamic effects.

It is often difficult to determine in practice if one is working with a hydrodynamic system. Since relatively small changes in water potential may cause significantly tilted HWCs, water pressure measurements of great accuracy and precision are required over a large geographical area. For example, a lateral water pressure gradient of only 0.13 MPa/km (30 psi/mi) will cause a HCW tilt of about 40 m/km (200 ft/mi), assuming typical oil and water densities. Thus, with normal measurement errors on depths and pressures, it can be difficult to establish a hydrodynamic regime unless wide well spacings were available. An alternative model for different HWC depths in different wells can always be made in terms of interwell faulting. The question may only be resolved by collecting a large data set of pressure measurements.

Because of the importance of identifying hydrodynamic effects both for exploration and appraisal, it is important to specify a program of water zone pressure tests in addition to normal hydrocarbon zone procedures.

Migration Through Barren Shales and Other Low-Permeability Rocks

The previous two sections have described secondary migration through high-permeability carrier beds, with the usual assumption that the source is juxtaposed with a regional carrier. In practice, the explorer is more often confronted with a play in which the source is vertically separated from the high-permeability carrier by hundreds of meters of barren shale or mudstone. The present section deals with this type of problem and outlines the techniques needed to predict the direction and efficiency of this type of secondary migration. Of particular importance is the identification of situations in which *downward* migration may occur since this has such a great effect on the predicted positions of hydrocarbon accumulations.

In rocks with a permeability of less than 1 millidarcy (md) (predominantly mudstones and tight limestones), significant flow of water or petroleum only occurs if it is driven by large fluid potential gradients ($\nabla \Phi_W$ or $\nabla \Phi_P$). In most geological settings, this only occurs if part of the rock is overpressured, i.e., if $\Phi_W > 0$, as in equation (1).

The predominant cause of overpressuring in sedimentary basins is compaction disequilibrium in which the sediments are loaded (by deposition of more sediments or by thrusting) at a rate faster than fluids can escape and allow the sediments to reach their equilibrium porosity, (Smith, 1971; Keith and Rimstidt, 1985; Palciauskas and Domenico, 1980; England et al., 1987).

Other causes of overpressuring, such as clay dewatering and thermal expansion of water, play a relatively minor role and can be neglected. Our calculations show that in actively generating rich source rocks (with potentials of about 30 kg/ton), the *local* fluid potentials are increased by the volume expansion involved in petrogenesis, but the effect is only of importance in the immediate vicinity of the source rock.

Most methods of calculating water potentials resulting from compaction disequilibrium rely on complex computer models such as those discussed in the section on basin-scale modeling. However, a much simpler method has been devised by Mann and Mackenzie (1990). This model uses the empirically derived expression relating generated water potential to sediment deposition rates (dh/dt).

$$\left|\frac{d\Phi_W}{dz}\right| = \frac{C}{k}\frac{dh}{dt} \tag{8}$$

where $\left|\frac{d\Phi_W}{dz}\right|$ is the modulus of the water potential gradient,
C is a constant,
K is the intrinsic permeability, and
dh/dt is the average depositional rate of the top 2 ± 0.5 km of sediments.

The deposition rate in a sedimentary basin usually only changes slowly in a lateral direction. Thus, in the absence of any perturbation due to the presence of high-permeability rocks, overpressure gradients ($\nabla \Phi_W$) are essentially vertical. Since buoyancy also acts in a vertical direction, the flow of both petroleum and water is predominantly vertical within mudstones or other low-permeability sediments.

Downward flow of petroleum can only occur where the overpressure decreases with increase in depth and the overpressure gradient $d\Phi_W/dz$ exceeds the

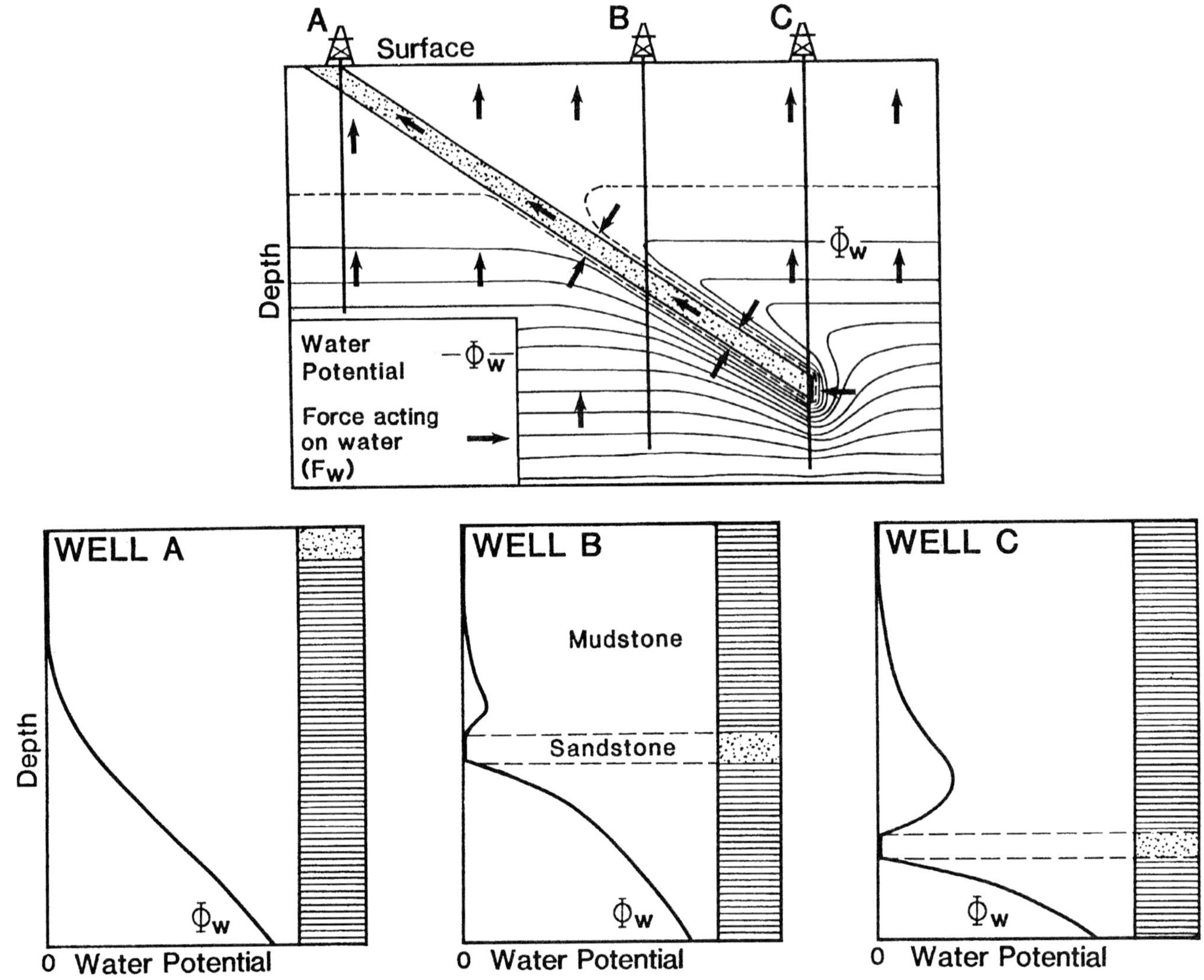

Figure 5. Water potentials, Φ_W, resulting in three wells that penetrate an outcropping, tilted sandstone in an otherwise mudstone sequence.

buoyancy gradient (ρ_W - ρ_P)g (see Figure 2). Vertically downward migration of petroleum is generally restricted to clay-rich mudstones having a porosity of less than about 35% and very low permeabilities of less than about 1 μd (10^{-18} m^2) (Mann and Mackenzie, 1990). Downward migration through mudstones is only possible if they are underlain by permeable rocks that are able to depressurize by lateral fluid flow.

An example of downward fluid flow in mudstones resulting from updip lateral flow in an outcropping, tilted sandstone is shown in Figure 5. The sandstone is normally pressurized at the surface, and because of its high permeability, it only has a very low overpressure even at its greatest depth. The effect of this sandstone is to cause a downturn in the water potentials particularly above the deeper parts of the sandstone. The water potentials through three wells (A, B, and C) penetrating the dipping sandstone are shown in Figure 5. In wells B and C, it can be seen that water flows into the sandstone both upward from the underlying mudstone and downward from the overlying mudstone. The water then flows updip in the sandstone to the surface.

To achieve downward fluid flow, it is not necessary for a permeable rock to be in pressure communication with the surface but only that pressure can be dissipated to a shallower depth. An example of fluid flow in a buried, tilted sandstone is shown in Figure 6. The sandstone has an almost constant overpressure and therefore causes a decrease in the overpressure of the mudstone surrounding the downdip end of the sandstone and an increase in the overpressure of the mudstone surrounding the up-dip end. The overpressure in the mudstone surrounding the mid-point of the sandstone is relatively unperturbed by the presence of the sandstone. The water potentials are therefore as

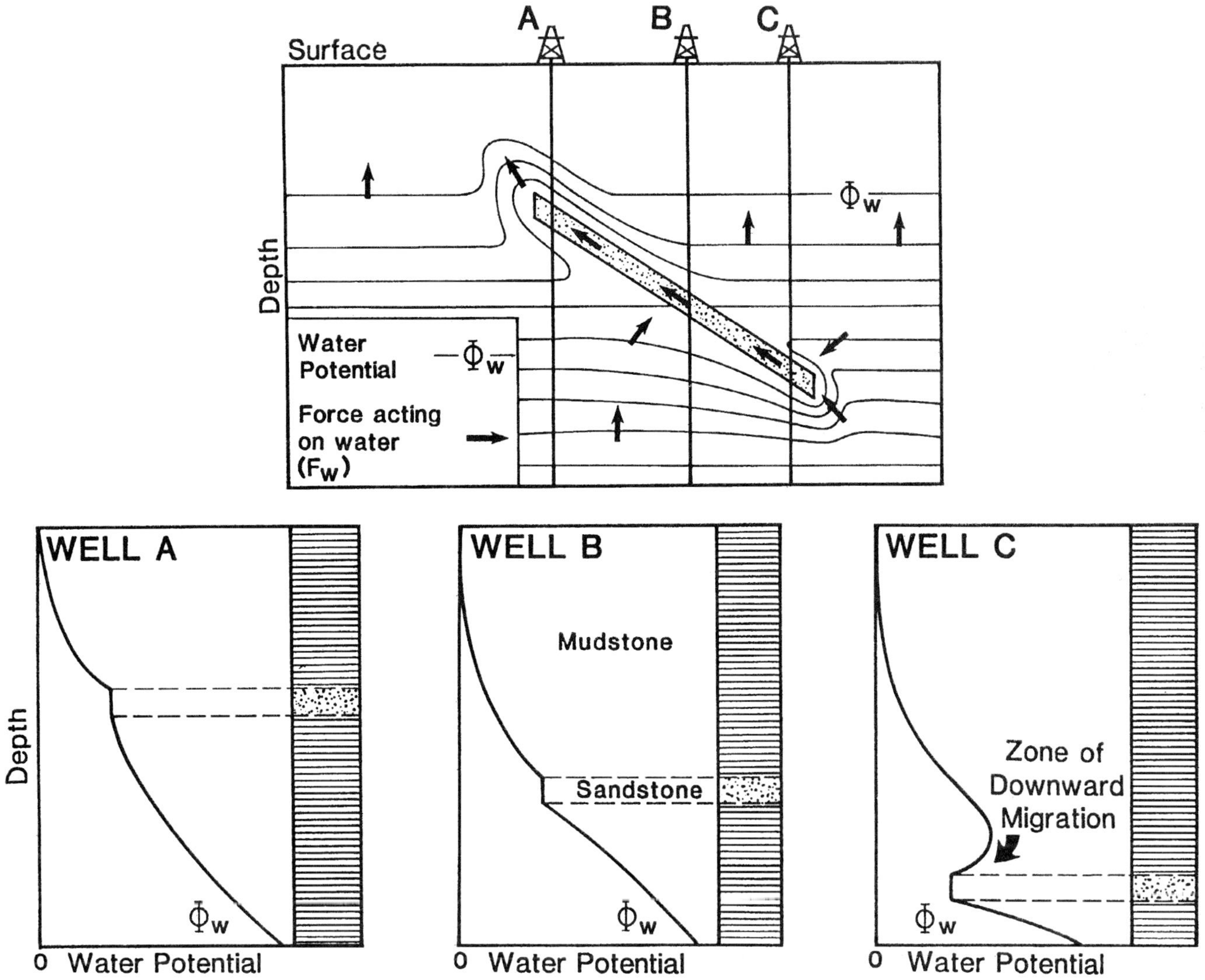

Figure 6. Water potentials, Φ_w, resulting in three wells that penetrate an extensive, buried, tilted sandstone lens in an otherwise mudstone sequence.

shown in Figure 6 from which it can be seen that there is a downward water flow from the mudstone overlying the downdip end of the sandstone (well C).

If the downward overpressure (water potential) gradient exceeds the buoyancy gradient, petroleum migration will also be downward. Detailed modeling has shown that downward migration of petroleum in mudrocks is limited to a distance of about 300 m. Downward migration beyond a distance of 500 m is extremely unlikely to occur.

Migration Resulting from Seal Failure

From the previous discussion it should be clear that petroleum flow is predominantly vertical in fine-grained, low-permeability mudstones and laterally updip in coarse-grained, higher permeability rocks. However, petroleum can only migrate from a coarse-grained rock into a fine-grained rock if the capillary potential is overcome (or if fractures are present). Where the sediment sorting is poor or the mudstones are particularly silty, only a very small capillary barrier may need to be overcome before petroleum can migrate from coarse-grained into fine-grained rocks. A small petroleum accumulation may build up in each layer of coarse-grained rock, and then the seal may be breached, allowing further flow. This is shown schematically in Figure 7. Due to the deflections in the migration paths caused by the presence of the sandstone lenses, some reservoirs may be unable to receive any petroleum charge and may therefore be dry structures even if they have good reservoir quality and good seals.

In some geological situations, the designation of lithotypes into carrier versus seal beds to produce a conventional "layer-cake" picture breaks down. Reality may better be represented by a somewhat random sequence of laterally discontinuous sandstones, silt-

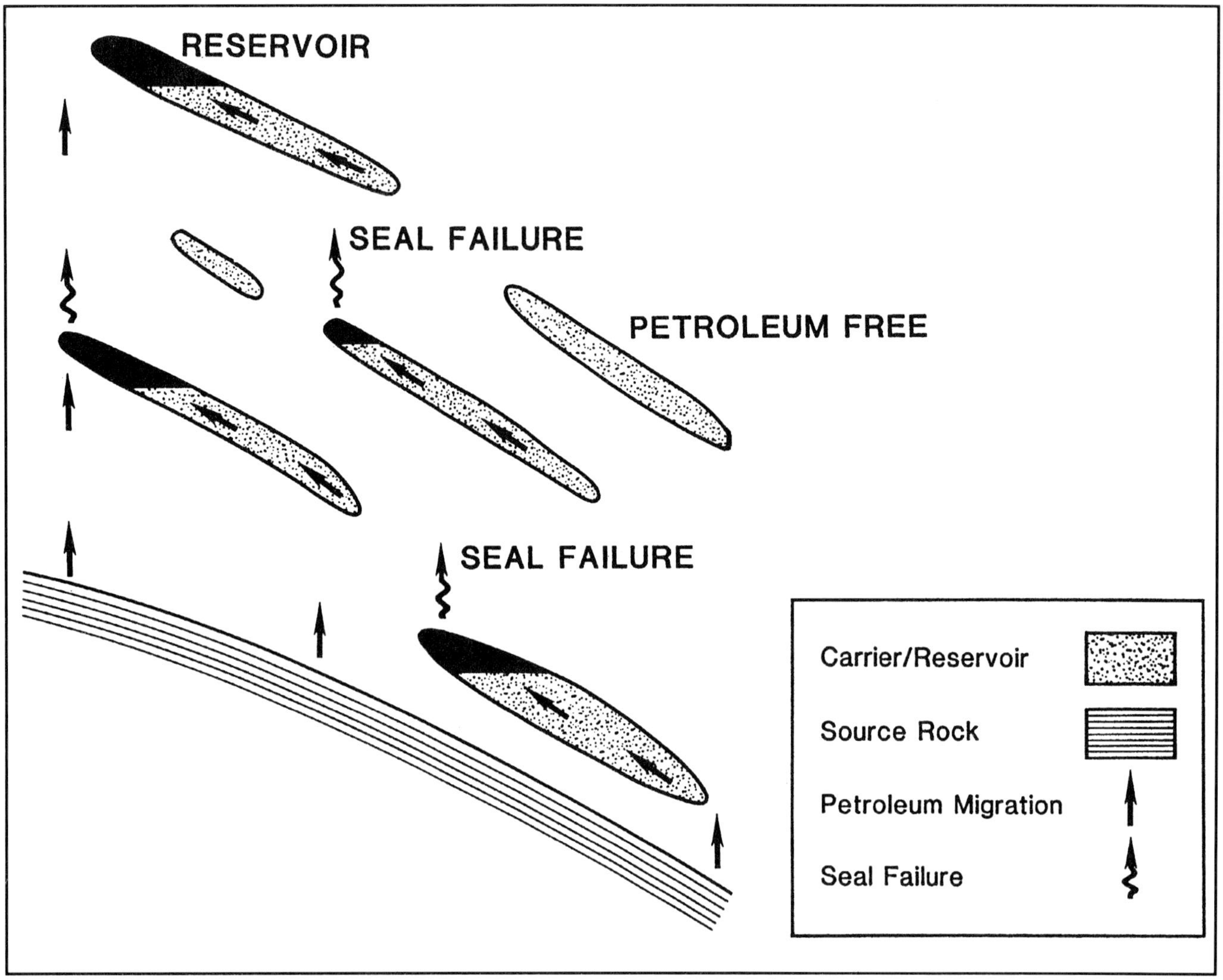

Figure 7. Migrational effects due to a series of poorly sealed sandstone lenses.

stones, and mudstones together with a set of source rock intervals. This type of situation is most frequently encountered in deltaic environments.

Due to the great thickness of mudstones (assuming rapid sedimentation), updip transmission of overpressure through any extensive dipping carriers may be sufficient to cause seal failure by fracturing. Periodic occurrences of this by deeper generated gas migrating upward may lead to chemical changes in the reservoired oil, as proposed by Thompson (1988).

Basin-Scale Migration Modeling

The previous sections were based on a combination of one-dimensional modeling of water and petroleum potentials, Φ_W and Φ_P, extended to two dimensions by geological reasoning (e.g., in the case of dipping carriers).

Several groups are working on two-dimensional computer models of water and petroleum migration in compacting sedimentary basins. Good examples include work in the Gulf of Mexico reported by Bethke (1985, 1986) and by Ungerer et al. (1988c) on the use of the THEMIS program to model migration in the Norwegian sector of the North Sea. A basin-scale approach has also been reported by Yalcin and Welte (1987). The main problems inherent in this approach are the need to assemble large amounts of permeability data and the treatment of faults. In many stages of exploration, the simpler one-dimensional approaches, typified by Mann and Mackenzie (1989) and Dutta (1988), are more than adequate. Later on, in the final stages of exploration, sufficient data to justify a two-dimensional approach may be available.

PHASE BEHAVIOR: CONDENSATION OF LIQUIDS OR EXSOLUTION OF GAS DURING MIGRATION

The phase behavior of petroleum plays an important role during migration because of the different physical properties of subsurface petroleum gases and liquids. These differences usually lead to the preferential migration of the majority phase and thus have a profound influence on the composition of the petroleum reaching the trap. This section will briefly describe petroleum phase behavior, and how it can be predicted from limited compositional information, such as that available during exploration. The resulting effects on the prospect will be described for a number of cases.

Petroleum Phase Behavior

Petroleum in the subsurface can exist in either a gas or liquid phase. The factors controlling this are the pressure, temperature, and composition of the system. Sophisticated methods have been developed by petroleum engineers for predicting petroleum phase behavior from detailed chemical analysis. However, in advance of drilling, only very limited compositional data are available to the explorer. From an appreciation of the maturity and nature of the source rock kitchen, it is possible to predict the gas–oil ratio (GOR), G_F, of the petroleum entering the secondary migration pathway (Mackenzie and Quigley, 1988). G_F is the ratio of components that are gaseous when produced at surface conditions to those that are liquid. As a good approximation, one can assume that C_{1-5} are gases at the surface and C_{6+} (i.e., hydrocarbons with six or more carbon atoms) are liquid. Thus, G_F is defined by their relative quantities:

$$G_F = C_{1-5} / C_{6+} \qquad (9)$$

For geochemical purposes, it is convenient to express G_F as a mass ratio (kg/kg) rather than as the more familiar volume ratio (scf/bbl or m^3/m^3). It is intuitively obvious that petroleums with very high values of G_F will exist as subsurface gases, while very low values will be associated with subsurface liquids (i.e., that produce oil reservoirs if suitable traps are encountered). However, a more detailed approach is necessary in many cases in which phase changes are anticipated during migration. This is described in the following paragraphs.

It is possible to predict the phases expected for a given petroleum composition, G_F, by making use of the results of engineering correlations and experimental results published in the literature (England et al., 1987). These correlations treat the many components of petroleum in terms of just two components: gas (C_{1-5}) and oil (C_{6+}). In petroleum engineering parlance, it is a "black oil" model.

These correlations are illustrated in Figures 8 and 9. Figure 8 shows the variation in the *saturated* GOR, G_L, of petroleum liquid as a function of pressure and temperature. The black oil correlation of Glasø (1980) was used to derive it, assuming the oil and gas properties shown. The correlation is probably accurate to about ±15% up to 30 MPa (4350 psi). Significant errors may arise for high-CO_2 gases or oils whose chemistries depart from those on which the calibration was based. G_L represents the maximum amout of C_{1-5} that a subsurface petroleum liquid is predicted to contain before a separate gas phase is formed. In other words, the G_L correlation shows the maximum amount of C_{1-5} that will dissolve in petroleum liquid as a function of pressure and temperature. Figure 8 shows that G_L increases rapidly with pressure, with the effect of temperature being less marked. (This is a reflection of the familiar increase in gas solution with pressure in, for example, carbonated drinks).

Highly compressed C_{1-5} has different properties than those that are familiar for gases at the surface. The main difference is its ability to dissolve significant quantities of higher molecular weight material at high pressure and temperature. The relevant parameter is the condensate gas ratio, or CGR, defined as the mass ratio of C_{6+} to C_{1-5} present in a subsurface petroleum (or its reciprocal, the corresponding gas–oil ratio, G_V):

$$\mathrm{CGR} = 1 / G_V = C_{6+} / C_{1-5} \qquad (10)$$

where C_{6+} and C_{1-5} are measured in units of mass. The significance of this for petroleum migration has been discussed in detail by Price et al. (1983), who performed a set of experiments on the ability of high pressure and temperature methane to dissolve C_{6+} components in the gas phase. The data of Price et al. are the basis of the correlation shown in Figure 9, which shows how high temperature and pressure are needed to dissolve significant quantities of C_{6+} in the gas phase. Thus, rich gas condensate discoveries are only likely to be made in the deeper parts of a basin where temperatures and pressures are sufficiently high to maintain a reasonably high CGR, provided that the source rock can provide sufficient C_{6+}. It should be emphasized that the CGR correlation is much more sensitive to the detailed compositions of the oil and gas. Such factors as API gravity, C_{1-5} composition, and the presence or absence of CO_2 are all known to cause major perturbations (of about 50–100%).

After setting up correlations to predict the mutual solubilities of C_{1-5} and C_{6+}, they can be used to predict the effects of phase changes induced by the pressure and temperature variations experienced during

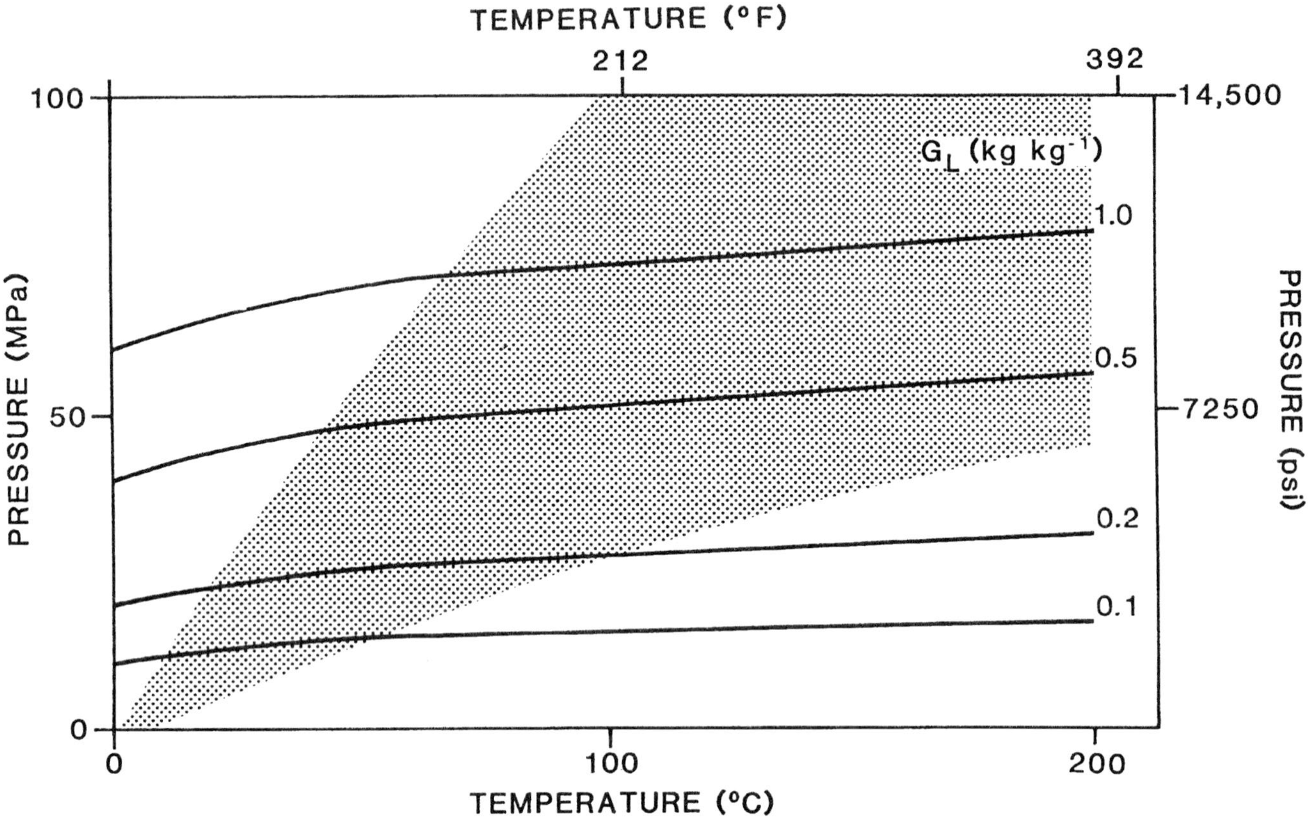

Figure 8. Variation in the *saturated* gas–oil ratio, G_L, of petroleum liquid as a function of pressure and temperature. The diagram was calculated using Glasø's (1980) correlations for oil with a surface density of 800 kg/m^3 and gas with a surface density of 0.8 kg/m^3. The stipled area delimits the typical subsurface range of pressures and temperatures.

secondary migration. The method is based on comparing the GOR generated by the source rock kitchen, G_F, with the GORs defining the mutual solubilities of C_{1-5} and C_{6+} calculated at the pressure and temperature of interest (usually that of the migration pathway or the trap). Three situations are possible:

1. If $G_F < G_L$, only a liquid phase will be present whose GOR is equal to that leaving the source rock kitchen, G_F.
2. If $G_F > G_V$, only a gaseous phase will be present, whose GOR is equal to that leaving the source rock kitchen, G_F.
3. If $G_V > G_F > G_L$, both a liquid and a gaseous phase will be present in thermodynamic equilibrium. The GOR of the liquid phase will be G_L and that of the gas phase will be G_V. In general, only one phase will be mobile and migrate toward the trap.

This simple approach is based on the description of petroleum as a mixture of oil and gas. A more complex approach has been used by Heum et al. (1986) in which they used the following components: C_1, C_{2-5}, C_{6-14} and C_{15+}. The following examples describe the effects on petroleum composition of phase changes brought about by the reduction in temperature and pressure involved in updip petroleum migration for a variety of typical cases.

Case 1

A source rock kitchen with a rich oil-prone source rock composed of labile kerogen will expel petroleum with a GOR (G_F) of about 0.2 kg/kg, or about 1000 scf/bbl, provided that the threshold for oil to gas cracking has not been exceeded. This threshold is at about 150°C for normal heating rates. (More details of this approach have been given by Mackenzie and Quigley, 1988.) Figures 8 and 9 show that as long as the petroleum does not experience pressures below ~30 MPa (4350 psi), only a liquid phase will be present since above 30 MPa $G_F < G_L$. For a normally pressured basin, this depth corresponds to about 3 km (9800 ft). Any further migration updip is predicted to lead to the exsolution of bubbles of gas that will probably be lost along the migration pathway. The resulting reservoir is predicted to contain gas-saturated oil with a GOR given by the G_L correlation (see Figure 8). If the exsolved gas becomes mobile, an oil leg with a gas cap is predicted.

A closely related possibility is for a reservoir to be filled with liquid at a depth of, say, 3 km and subse-

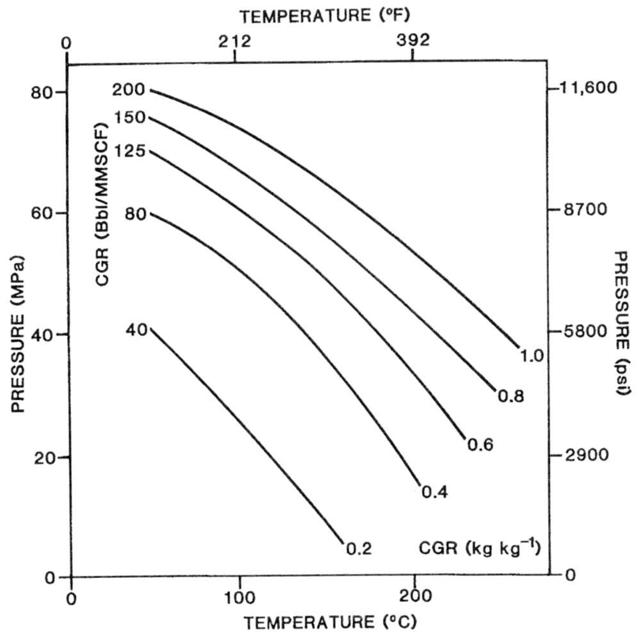

Figure 9. Variation in the predicted condensate gas ratio (CGR) for an oil-saturated gas as a function of pressure and temperature, using experimental data from Price et al. (1983) involving equilibrating mixtures of Spindletop crude and methane.

quently uplifted to half that depth. The pressure and temperature reduction is expected to lead to exsolution and to the formation of a gas cap, with the related risk of the oil being displaced updip from the target. For a case history of how this approach has been applied in the Norwegian sector of the North Sea to explain the widely varying GORs of the petroleum discoveries there (i.e., oil, gas condensate, and gas fields), the reader is referred to England and Mackenzie (1989).

Case 2

A source rock kitchen that contains an organically lean source rock composed of labile kerogen cannot expel a liquid phase during the oil generation window because sufficient petroleum pore space saturation for expulsion is not achieved. If increased temperatures are subsequently experienced, oil to gas cracking will begin and a gaseous petroleum phase will be expelled in preference to a liquid phase.

In these circumstances, the phase behavior of petroleum will dominate the composition of the expelled material rather than the kerogen type. Figure 9 allows one to estimate the *maximum* possible CGR of the expelled petroleum as a function of the prevailing temperature and pressure. In effect, as the more mobile gas phase is expelled from the source rock, it becomes saturated with respect to the immobile C_{6+} that was generated earlier (but was not expelled). As long as sufficient C_{6+} is present, Figure 9 shows that rich gas condensates, with CGRs greater than 100 bbls/MMScf (per million standard cubic feet) are only likely to occur at sufficient depths to give pressures and temperatures exceeding around 100°C (212°F) and 40 MPa (5800 psi). As a gas condensate migrates updip it will lose C_{6+} by condensation of a minority liquid phase in the carrier bed; the remaining gas phase will become drier. Similarly, if a reservoir containing a gas condensate is subjected to a lowering of pressure and temperature due to uplift, the formation of a light oil leg is expected (providing the volume of the trap is sufficient to accommodate the increased gas cap volume).

A more detailed discussion with special reference to gas condensate exploration has been presented by England and Price (1988). Note that although the discussion presented in this chapter is qualitatively correct, quantitative conclusions are only as reliable as the correlations used to derive them.

Compositional Effects of Phase Changes

We have previously concentrated on the gross C_{1-5}/C_{6+} composition, which is important for determining the economics of a hydrocarbon prospect. In this section the effects of phase changes on the detailed geochemical composition of petroleum are discussed. An understanding of this is essential to combining the traditional geochemical influences of source rock maturity with the influence of phase changes (which are often ignored).

Very little work has been published in this area. The most important paper is that of Price et al. (1983) which discussed the geochemical changes observed when a relatively light (45° API) oil (Spindletop, Texas) was equilibrated with pure methane at high temperatures and pressures in the presence of water. Thompson (1988) has extended this work to include light hydrocarbon analyses and has described a set of systematic compositional trends caused by phase changes. Thompson's work concentrates on the Gulf of Mexico, where faulted, poorly sealed, stacked reservoirs encourage the movement of gaseous petroleums through oil-filled structures. The migrating gas phase (which will be a rich gas condensate if the pressure and temperature are high enough) will preferentially remove the lower molecular weight, saturated compounds, leaving behind a highly altered oil of lower API gravity. This leads to the following effects on the remaining oil, named "evaporative fractionation" by Thompson (1988):

1. increased aromaticity (within any molecular weight range),
2. reduction in the concentration of gasoline-range hydrocarbons as a fraction of total oil, and
3. a decrease in the ratio of paraffins to napthenes for any given carbon number.

Summary of Phase Behavior

An appreciation of the influence of phase behavior is important in predicting the distribution of oil, gas condensate, and gas fields from the source rock properties and basin thermal history. Phase effects become especially important where high temperatures and pressures are encountered. Other authors who have discussed the combined influence of sourcing and phase behavior include Larter (1988), Mackenzie and Quigley (1988), and Heum et al. (1986).

RATE, DISTANCE, AND EFFICIENCY OF MIGRATION

Petroleum continues to flow down a potential gradient until it reaches a potential minimum (i.e., it reaches a trap) or the available petroleum is lost within the pores of the carrier. The rate of migration may be estimated from Darcy's law:

$$q = -\frac{k}{\mu}\frac{d\Phi}{dz} \qquad (11)$$

where
q is the volume flow rate ($m^3/m^2/s$),
k is the permeability,
μ is the viscosity (Pa/s), and
$d\Phi/dz$ is the fluid potential gradient.

Typical water permeability values are

Sandstones: 10^{-12}–10^{-15} m^2 (1 d–1 md)
Limestones: 10^{-14}–10^{-17} m^2 (10 md–10 μd)
Mudstones: 10^{-18}–10^{-22} m^2 (1 μd–0.1 nd)

Mudstones appear to be particularly impermeable to water because of polar interactions between the water molecules and the clay surfaces. It therefore seems reasonable to assume that sandstones and limestones will have broadly similar permeabilities to petroleum and water, whereas mudstones will be perhaps 10–100 times more permeable to petroleum than to water.

Rate of Migration

During updip lateral flow in a coarse-grained carrier, such as shown in Figure 3, the petroleum potential under hydrostatic conditions is given by:

$$\frac{d\Phi_P}{dz} = (\rho_P - \rho_W) g \sin\theta \qquad (12)$$

This formula allows one to make order-of-magnitude estimates of migration rates. Typical values are in the range of 1-1000 km per million years in sandstones and 0.01–10 km/m.y. for limestones.

Vertical migration rates in mudstones are more difficult to estimate, as the fluid potential gradient is dominated by the overpressure gradient, $d\Phi_W/dz$. However, migration rates calculated from the positions of known accumulations suggest petroleum flow rates of about 1 km/m.y. These estimates show that in most geological situations, the rate of migration is not a significant factor. Once a petroleum-saturated migration pathway has been established, the actual time taken to reach the trap is not significant compared to the rate at which petroleum is supplied by the source rock. Generation is thus likely to be the rate limiting step and not migration.

Efficiency of Migration

There is a finite limit to the distance over which petroleum can migrate from its source to its trap. This limit is determined by the volumetric losses in the rock necessary to saturate the secondary migration pathway depicted in Figure 3. The loss may be thought of as analogous to the volume of fluid needed to fill an empty pipeline before flow is received at the exit. Because migrating petroleum preferentially utilizes the parts of the migration pathway with the lowest pore entry pressures, it does not necessarily pass through every part of the carrier bed. England et al. (1987) have presented arguments to suggest that less than 10% of the gross rock volume is actually used for migration. Within the oil-saturated pathway, a critical pore saturation to achieve flow is about 50%. These two factors can be multiplied to suggest that an overall loss factor of approximately 5% of the total migration path pore volume is a reasonable estimate.

A more satisfactory approach is to deploy case histories to estimate migration efficiency. If the total volume of petroleum generated by the source rock kitchen can be estimated and the total reserves in the basin are known, the migration efficiency may be estimated by the difference. Naturally, this is a somewhat error-prone procedure because it involves subtracting two numbers that both have large errors in them. Assumptions also have to be made concerning petroleum that is not identified or that leaked to the surface through seal failure, faulting, or other processes.

The results of this type of approach have been reported by Mackenzie and Quigley (1988). The volume of petroleum reaching a prospect, V_C, is equal to the volume expelled from the source rock, V_E, minus the volume lost in secondary migration, V_L:

$$V_C = V_E - V_L \qquad (13)$$

Assuming that a geochemical model has been used to give V_E, we can estimate the volume lost during migration, V_L, on the assumption that it is related to the pore volume of the carrier bed (or drainage area), V_D, as previously outlined:

$$V_L = \phi V_D f \quad (14)$$

where f is the volumetric loss factor and has a range of about 1–3%. Macgregor and Mackenzie (1987) found that a value of 2% was satisfactory in a case study in the Malacca Straits. Parameter f is crucial because under some circumstances, the volume of petroleum predicted to reach the trap by equation (13) will be extremely sensitive to its exact value (i.e., if V_L is nearly the same size as V_E). In other circumstances, where migration distances are small, migration losses will not be important.

The THEMIS program (Ungerer et al., 1988c) models the efficiency of secondary migration by making assumptions about the relative permeability curves for oil and water. This is essentially an approach similiar to equation (14) in which the migration loss is assumed to be related to the overall pore volume of the drainage area. Future work in this area must be directed toward replacing the empirical models with more detailed, physically based ones. The effect of faults and fractures should also be considered.

CHEMICAL CHANGES IN COMPOSITION DURING SECONDARY MIGRATION

The exact composition of a petroleum measured in the laboratory is the result of a complex interplay of many factors that determine and subsequently modify its properties. The processes can be subdivided into those that operate within the source rock and those that take over after the source rock and petroleum have become separated. A source rock can affect petroleum properties in two ways: (1) through the nature of the organic matter present within it (for example, the presence of abundant plant epidermal tissue may give rise to oils with high wax contents) or (2) through the nature of the associated lithology (for example, high-sulfur oils generally result from carbonate-, phosphorite-, or silica-rich sediments rather than from aluminosilicates). Furthermore, the absolute abundance of oil-prone material within the sediment may also be important (since organically lean sediments are unlikely to generate sufficient quantities of petroleum for efficient expulsion in the liquid phase to take place (Mackenzie and Quigley, 1988).

Petroleum composition may then be modified either during expulsion (primary migration) from the source rock or during secondary migration. Finally, prior to any alteration that may occur during production, there is a range of processes that can occur within the reservoir itself. All these alter the make-up of the reservoired fluids. They include biodegradation, water washing, gas deasphaltening, gravity segregation, oil to gas cracking, and gas souring.

This section discusses what is perhaps the least documented link in this chain of events—those changes that occur *during* secondary migration. The main processes that need to be considered are

1. the effects of phase changes during migration to regions of lower temperature and pressure,
2. the effects of biodegradation and associated water washing, and
3. what is termed *geochromatography*, a more subtle effect brought about by chromatographic separation of petroleum as it migrates past a stationary inorganic medium.

The first of these processes has already been discussed in detail (see section on phase behavior). The second is handled in detail in Chapter 4 (Palmer, this volume) and will only be briefly summarized here. Our attention will thus be mainly directed toward the third process—geochromatography.

Biodegradation and Water Washing

Both biodegradation and water washing are mechanisms more commonly associated with alteration to petroleum in the reservoir rather than along the migration pathway. A more complete description of the controls on and effects of these processes are given in Chapter 4 of this volume, which is concerned specifically with in-reservoir processes.

Of the two processes, biodegradation is by far the better documented (e.g.,Winters and Williams, 1969; Bailey et al., 1973a; Hunt, 1979; Connan, 1984; Goodwin et al., 1983; Seifert et al., 1984). Generally, there is good agreement about both the controls and the ensuing effects on petroleum. Attack by aerobic bacteria operating at relatively low subsurface temperatures (typically <70°C) results in the oxidation of petroleum, ultimately to carbon dioxide. In some cases, this attack has taken place on a massive scale, producing, for example, the extensive tar sands of western Canada and Venezuela.

Degradation of petroleum is usually systematic, with successive destruction (among the saturates fraction) of the *n*-alkanes, branched alkanes, isoprenoids, alkylcyclohexanes, and polycyclic alkanes. The relative rate at which aromatic material is altered, with respect to the saturated compounds, may vary depending on precise conditions. The most resistant compounds tend to be the polycyclic biomarkers, particularly the aromatic ones (Seifert et al., 1984), which enables correlation and maturity determination to be carried out on all but the most degraded oils.

Of the bulk parameters, trace minerals and asphaltenes tend to rise, as does viscosity and pour point. Conversely, the API gravity, saturates content, and light hydrocarbon content tend to fall. In general, none of these changes is desirable in terms of production or economic value of the reservoired petroleum. If conditions are right (i.e. coexisting oil and bacteria below 70°C with a supply of fresh, oxygenated, nutrient-rich water), then these processes can alter petroleum during migration as well as in the reservoir. However it may be extremely difficult to distinguish just where that attack has taken place. (It is worth noting, however, that if severe degradation occurs during migration, then viscosity may increase to such an extent that the rate of migration may become negligible.)

If temperatures are too high for bacteria to flourish, yet water is still flowing past an oil (either reservoired or in transit), then water washing may occur. Again, systematic alteration takes place, mainly via selective removal of more water-soluble material. These effects are most pronounced for light ends of petroleum and particularly for aromatic and heteroaromatic compounds (McAuliffe, 1979; Price, 1976; Lafargue and Barker, 1988). For a given carbon number, cyclization and aromatization both increase solubility in water. Simulated laboratory studies of the effects have been reported by several authors (see Lafargue and Barker, 1988, for a review), although field examples are less common. Those that have been reported include a study of oils from northwest Palawan, Philippines (Palmer, 1984), where many of the observations were in good agreement with those made in the laboratory. Similar findings have also been made by Itzan (1985) using a suite of naturally and artificially altered oils from the Misener Formation, Oklahoma. Although it is again difficult to distinguish between effects taking place in the carrier bed versus the reservoir, it should be noted that all oils are to some extent water washed as a result of their passage through water-wet sandstones.

Geochromatography

The idea that petroleum composition might be altered due to movement through a carrier bed is over 30 years old, dating back to the review by Silverman (1965). Central to the argument is that processes akin to laboratory chromatographic separation of material can take place in the subsurface (hence the term *geochromatography*). These result from the juxtaposition of an initially water-wet stationary phase (in the form of a carrier bed consisting of variable amounts of sand, carbonate and clay minerals) and a mobile petroleum phase. Because some components in petroleum are more polar than others, differing degrees of adsorption to the solid phase would be expected, resulting in separation of the moving oil. Lighter, less polar material would be expected to migrate more efficiently than heavier, more polar fractions.

Strictly speaking, *chromatography* is defined as the separation of a sample that is injected as a distinct "pulse" into a "carrier" stream of solvent flowing past a stationary adsorbent. Since these conditions are not met in petroleum migration, geochromatography more closely resembles a continuous adsorption process. A better analogy is the removal of selected compounds from mixtures by passing them through packed beds. Examples include the removal of H_2S and H_2O in the gas processing industry by use of ZnO or zeolites, respectively. However, for the purpose of this review, we will adopt the usual meaning of geochromatography to imply chemical changes due to selective adsorption during secondary migration.

However, as discussed by Carlson and Chamberlain (1986), it is far from generally accepted that this type of process actually succeeds in altering composition. One significant problem is that its effects can easily be confused with other nonmigrational processes such as facies variation within the source, or differing levels of thermal maturation. It is possible that some of these other variables may cause more significant changes that can, except under vary favorable conditions, mask the effects of geochromatography. Despite this controversy, the process has been studied in some detail, both from a more theoretical standpoint (Carlson and Chamberlain, 1986), via laboratory simulation (Bonilla and Engel, 1986) and from a variety of field studies on a range of scales (Leythaeuser et al., 1983b, 1984). The results provide a useful description of what might be expected under such circumstances.

The attraction of simulating subsurface processes in the laboratory is that all except one variable can be kept constant, thus the effects of either source or maturity variation can be eliminated while those due to migration can be selectively studied. Furthermore, field observation of such processes is often hindered by sampling difficulties. Although oils and related source rocks can be sampled and analyzed, systematic sampling of migration pathways is generally less easy. There are one or two notable exceptions to this where short-range migration may be examined by analysis of cored sequences of interbedded sandstones and shales (Leythaeuser et al., 1983b). However, under these circumstances, it is likely that primary migration is also a factor.

The published results of a laboratory simulation (Bonilla, 1986; Bonilla and Engel, 1986) suggest that geochromatography does result in a systematic change in oil properties. In particular, lower molecular weight *n*-alkanes migrate most effectively, thus skewing the original carbon number distribution. However, this effect is found to be less significant at higher temperatures, with coarser grained media, and for lower viscosity oils. The authors point out that these experimental results agree with field observations (Durand, 1983), particularly that less fractionation is apparent when migration has taken place through fractures or clean sands (Mackenzie et al., 1983c). In addition, these experiments revealed some slight changes in carbon isotope ratios during migration. Both saturates and NSO fractions were found to have become isotopically lighter with increasing migration distance, although the effect was both small (at most 1‰) and diminished with increasing particle size of the inorganic migration medium.

The effects on gases were not reported in this study, but they have been well documented elsewhere (see, e.g., Galimov, 1974; Fuex, 1980; James, 1983; Schoell, 1983a, 1984). In general, the consensus seems to be that there is no change in the concentration of stable isotopes of methane with migration, although there is one reported case study suggesting otherwise (May et al., 1968; Runge, 1980). The dominant controls on isotope ratios appear to operate during generation of gas (Schoell, 1983a; Clayton, 1988) and not during its subsequent migration.

It is less clear whether or not geochromatography affects the distribution of biological marker compounds used widely in assessment of thermal maturity and in oil–oil and oil–source correlation (see Mackenzie, 1984, for a review). It was originally proposed by Seifert and Moldowan (1978) that with increasing distance of migration, an increase was seen in the abundance of $5\alpha(H),14\beta(H),17\beta(H)$ steranes ("$\alpha\beta\beta$ steranes") with respect to their $5\alpha(H),14\alpha(H),17\alpha(H)$ isomers ("$\alpha\alpha\alpha$ steranes"). In other words, anomalous maturities might be inferred from such parameters as a result of migration processes. A similar effect has also been reported by other workers (King and Claypool, 1983; Vincent et al., 1985). This observation was attributed to a chromatographic process involving mineral grains as the stationary phase and petroleum as the liquid phase, resulting in more effective migration of the $\alpha\beta\beta$ isomers.

Theoretical support is provided by sterane–clay mineral adsorption free energy (AFE) calculations (Carlson and Chamberlain, 1986). These indicate that $\alpha\beta\beta$ steranes do in fact "stick" less effectively to minerals such as montmorillonite than do $\alpha\alpha\alpha$ isomers. Thus, of the two groups, $\alpha\beta\beta$ steranes would be expected to migrate more efficiently. It is worth pointing out, however, that the same authors suggest that primary rather than secondary migration may provide a better explanation for these observed enrichments. Also, other lithological effects in the source rock may also enhance the relative abundance of the $\alpha\beta\beta$ steranes independently of thermal maturation (ten Haven et al., 1986).

Other migration perturbations to biomarker distributions have been reported, again giving rise to anomalous values of maturity parameters. Philp and Gilbert (1982, 1985a) reported anomalously low values of hopane and sterane isomerization parameters for oils from the Gippsland basin, Australia. They attributed these to oils passing through sandstone interbedded with immature brown coals, such that the migrating oil "extracted" the coals, picking up their abundant and low maturity distribution of biomarkers.

In a study of oils from southeastern Hungary, Sajgo (1984) attributed anomalously low values of sterane aromatization maturity parameters to preferential migration of monoaromatic as opposed to triaromatic, steranes. (Interestingly, he reported that there was no apparent migrational effect on the $\alpha\alpha\alpha$ and $\alpha\beta\beta$ saturated steranes). The magnitude of the effect was related not to the migration distance, but to the migration pathway, i.e., the effects were reduced if movement was through clean sands or faults and were greatest if through clay-rich horizons. Again, it becomes debatable whether these observations are related to primary or secondary migration. Similar results have also been reported for oils from the Mahakam delta (Hoffman et al., 1984) and for oils from the Shengli field, onshore China (Shi et al., 1982).

In a separate study of the Mahakam delta, Vandenbrouke et al. (1983) demonstrated that light hydrocarbons (C_6–C_9) migrate more efficiently than heavier material in the Handil field. They postulated a vertical chromatographic process that results in a distribution of oils with increasing API occurring in increasingly shallow reservoirs, with much methane seeping to the surface and being lost to the atmosphere. This should be compared with the work of Thompson (1988) on gas flushing in deltaic systems, in which the heavier more aromatic compounds are not carried as efficiently as the lighter hydrocarbons because of their phase behavior (see section on phase behavior). The relative ease with which light hydrocarbons can migrate (and whose distribution may be altered during migration) has been reported many times (Schaefer et al., 1983; Whelan et al., 1984; Saenz, 1984), particularly in the study of DSDP sediments, for which porosities are still high and migration a facile process. However, caution must be exercised when trying to extrapolate these results to secondary migration in carrier beds. It is clear that some of these reported changes are associated with expulsion of compaction waters and hence are more appropriately classified under the heading of primary migration.

Discussion

A range of processes has been described here, all of which serve to alter the chemical composition of petroleum during secondary migration. Of these, it is likely that the one having the most dramatic effect in practice is phase separation when significant partitioning of low and high molecular weight material may take place. Of the others, the next most significant are biodegradation and water washing, although these are perhaps more important factors in in-reservoir alter-

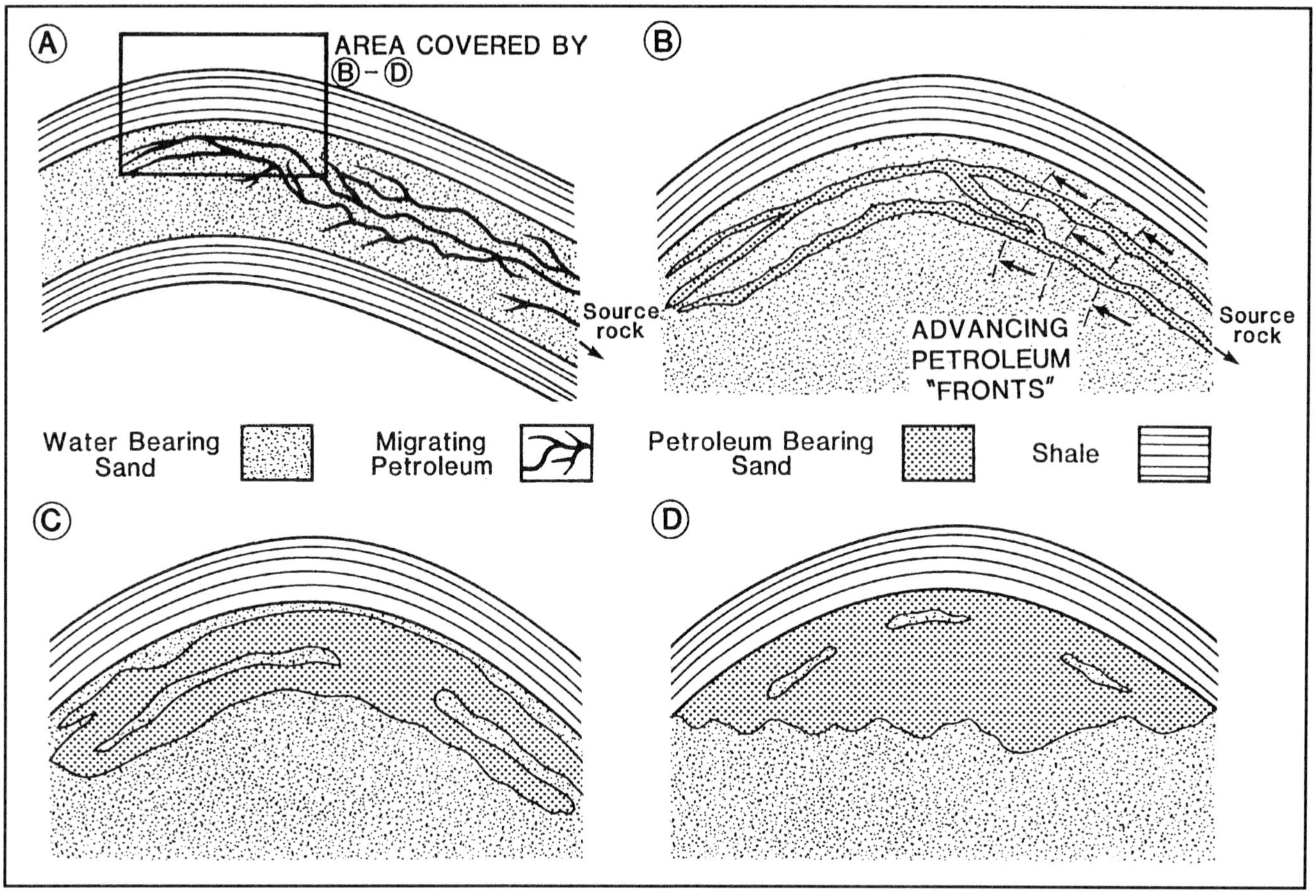

Figure 10. Schematic diagram showing reservoir-filling sequence.

ation. As was commented earlier, the effects are likely to be similar no matter where they actually occur. Furthermore, if severe degradation were to occur during migration, it is possible that movement of petroleum may cease all together.

The last process, that of geochromatography, is still the most enigmatic. Although there is clear evidence (from theoretical and laboratory simulation work) that it can take place, the scale of its effect is harder to assess. From the available evidence, it is probable that in most cases its effect will be small. Variations in the nature of the source rock and differing degrees of thermal maturity are likely to cause much bigger spreads in petroleum properties. Furthermore, it is clear from what has been published that many of the processes that can generally be attributed to migration are in fact related to primary migration (expulsion) and not strictly to secondary migration. Thus, of the "links in the chain" alluded to at the start of this review, secondary migration, although important, is perhaps the one having the least impact on ultimate petroleum composition.

TRAP FILLING AND MIXING

Nature of Reservoir Filling

Figure 10 shows how a reservoir fills with petroleum by displacing the water originally occupying the pore space. As was discussed in the previous section, it is believed that migrating petroleum only makes use of the coarser beds of the carrier system where the capillary entry pressures are lowest. This is illustrated in the top part of Figure 10, in which the migration system entering the trap is shown as a tortuous pathway that avoids much of the bulk rock volume of the carrier. This implies that petroleum initially flows into the trap through only a small percentage of the gross pore volume of the trap, the zones with the lower pore entry pressures (England et al., 1987). Within the migration pathways themselves, petroleum saturation will exceed the threshold saturation necessary for flow; in the rest of the rock, petroleum saturations will be negligible. Petroleum will not necessarily fill from the crest down, except in the unusual case of an exceptionally uniform reservoir rock.

Because a petroleum trap is a geological dead-end for petroleum, the petroleum saturation of the reservoir rock grows as more petroleum arrives; the resulting increased buoyant pressure causes the petroleum to displace water from ever smaller and smaller pores.

Fresh petroleum will thus advance into the structure as a continuous series of fronts, since it cannot bypass the main flow paths. It arrives at the side of the structure closest to the source beds and displaces the petroleum that arrived earlier, moving it away from the source beds. As Figure 10 indicates, during most of the filling process, the lack of reservoir-wide oil-saturated rock will prevent large-scale mixing. Therefore, one expects initially to find the most mature petroleum nearest to the sourcing system. This "inherited" composition difference will persist over a long period unless it is subsequently eliminated by the in-reservoir mixing processes discussed next.

Gravity Segregation

Gravity segregation is the process by which the earth's gravitational field establishes concentration gradients within a static petroleum column. This results in components with lower than average densities (such as C_{1-5}) having higher concentrations toward the top of the column. Conversely, components with higher than average densities (such as C_{10+}) tend to have lower concentrations at the top of the column. Thus reservoir fluid properties will vary with depth in a systematic way. In an oil reservoir, for example, the solution gas–oil ratio (GOR) and bubble point pressure will increase toward the top of the column.

The exact magnitude of the effect is hard to model accurately. The gravitational gradients computed by Sage and Lacey (1939) were smaller than observed in practice because they could not (at that date) easily include the effects of non-ideality. Using more modern "equation of state" techniques, Schulte (1980) showed that a more realistic simulation could be achieved. Although agreement between prediction and observation was not perfect, the importance of considering complex nonideality was clearly demonstrated. Hirschberg (1988) has stressed the role of asphaltenes, and Montel and Gouel (1985) have shown that gravitational effects increase dramatically near critical points, i.e., at high pressure and temperature.

Experience seems to indicate that larger gravitational gradients are observed in gas condensates and volatile oil reservoirs, presumably because of the greater proximity of their critical points to reservoir pressures and temperatures. For example, the gravitational gradient in the Statfjord field was so large that its fluid type changed from gas condensate to volatile oil within one well (Schulte, 1980).

Thermal Segregation

Theoretically, the thermal gradient in a column of petroleum should also cause a compositional gradient to be set up. However, modeling its size is extremely difficult and requires the estimation of many unknown parameters (Holt et al., 1983).

RESERVOIR MIXING OF DIFFERENT SHOTS OF PETROLEUM (RATES AND MECHANISMS)

As previously mentioned, reservoirs will often fill from one side. They will therefore not be completely oil saturated (allowing significant mixing) until rather late in their filling history. If the source rock kitchen is composed of a single source rock type, increasing burial will cause it to expel a petroleum of ever-increasing maturity. This changing maturity will manifest itself by differences in geochemical maturity indicators, as well as bulk properties such as GOR and CGR. A petroleum accumulation will therefore "inherit" a fluid composition that varies across the field. An additional complication may be the presence of two or more source rock types in the drainage area of the trap. The rates and mechanisms by which some or all of these differences are eliminated will now be discussed, since they determine what is observed.

There are two main types of mixing process that can operate on petroleum accumulations:

1. density-driven overturning (including thermal convection), and
2. molecular diffusion.

Thermal Convection

The differences in fluid density necessary to drive convective mixing could in principle be set up by *vertical* temperature gradients. However, this type of thermally driven convection is not usually important in petroleum reservoirs because the temperature gradient and reservoir properties do not permit the critical Raleigh number to be exceeded (England et al., 1987). In addition, gravitational segregation tends to oppose the process.

Thermal convection will always occur when a *horizontal* temperature gradient is present. However, its rate is usually insignificant on a geological time scale. An exception occurs for low viscosity fluids in large structures with large lateral temperature gradients. This case has been discussed by Moser (1986).

Other Types of Density-Driven Overturning

As previously outlined, lateral compositional differences often result from the filling process. A variation in GOR and CGR usually implies a difference in the subsurface density of the reservoir fluid due to the different effective densities of the oil and gas. This will set up a lateral density gradient across the field, which

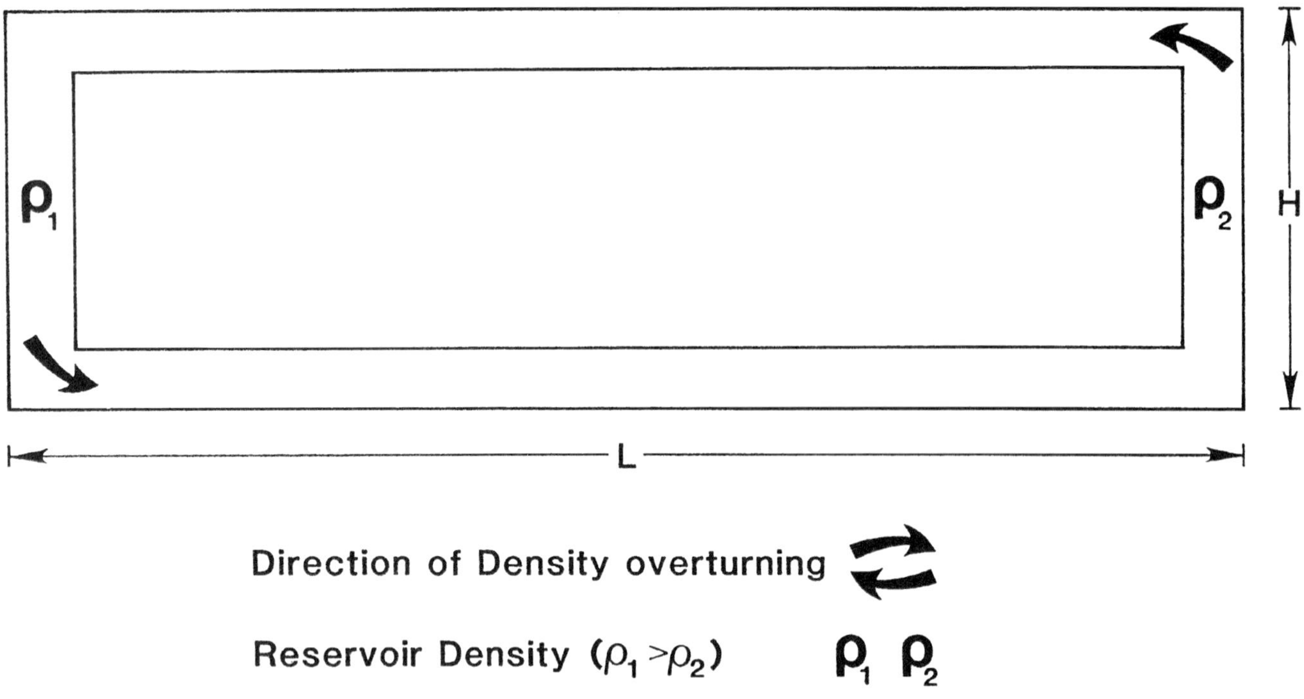

Figure 11. Schematic pathway for density-driven overturning in a reservoir containing fluids of unequal density at the same depth.

is clearly mechanically unstable with respect to overturning, as shown in Figure 11.

The rate at which this occurs can be estimated from the reservoir's solid and fluid characteristics. The basic method has been outlined by England and Mackenzie (1989) and allows an order of magnitude estimate of the time scale for this process. A typical overturning pathway 70 m high and 2000 m wide was assumed for the calculation together with reasonable oil properties. It is made by estimating the *effective* reservoir-wide permeability, $<k>$, for the overturning pathway shown in Figure 11. The reservoir-wide average permeability, $<k>$, is the appropriately weighted average of the permeabilities encountered over a putative density-driven overturning pathway. It should include the effect of shales and reservoir anisotropy and heterogeneity, if present. The parameter $<k>$ is usually at least 100 times smaller than core plug permeabilities. A $<k>$ value of 100 md represents an "excellent," well-connected reservoir; 10 md would characterize a "good" overall reservoir permeability; and 0.1–1.0 md represents more poorly connected, lower quality reservoirs. The pressure difference due to the lateral density gradient is then combined with Darcy's law to calculate the time required. It should be remembered that these calculations represent order of magnitude estimates and should be treated as such. Ideally, the specific rock and fluid parameters for any particular reservoir should be used to recalculate Figure 12 for specific applications.

The results, shown in Figure 12, suggest that except for "good" or "excellent" oil reservoirs, it is possible that the density-driven mixing time scale, τ_{DENS}, will be slow compared to the time available since filling, τ_{FILL}. Naturally the calculated timscales will vary considerably depending on the reservoir structure, fluid type (gas, oil, or heavy oil), and viscosity. The important point is that the order of magnitude calculation suggests that it is indeed possible to observe reservoirs at present that have lateral density gradients (and have therefore not reached mechanical equilibrium). These gradients occur despite pressure communication across the reservoir, as defined by well testing using normal techniques. This implies that any inherited compositional differences should not have been completely eliminated. It should also be noted that density-driven overturning is a "one off" movement of material, unlike thermal convection (if it occurs), which is continuous.

Molecular Diffusion Mixing

Diffusion acts to reduce (and ultimately eliminate) differences in chemical potential. It is caused by random molecular motions. Horizontal differences in chemical potential—assuming constant temperature and pressure—are governed by differences in composition. Given sufficient time, diffusive mixing would

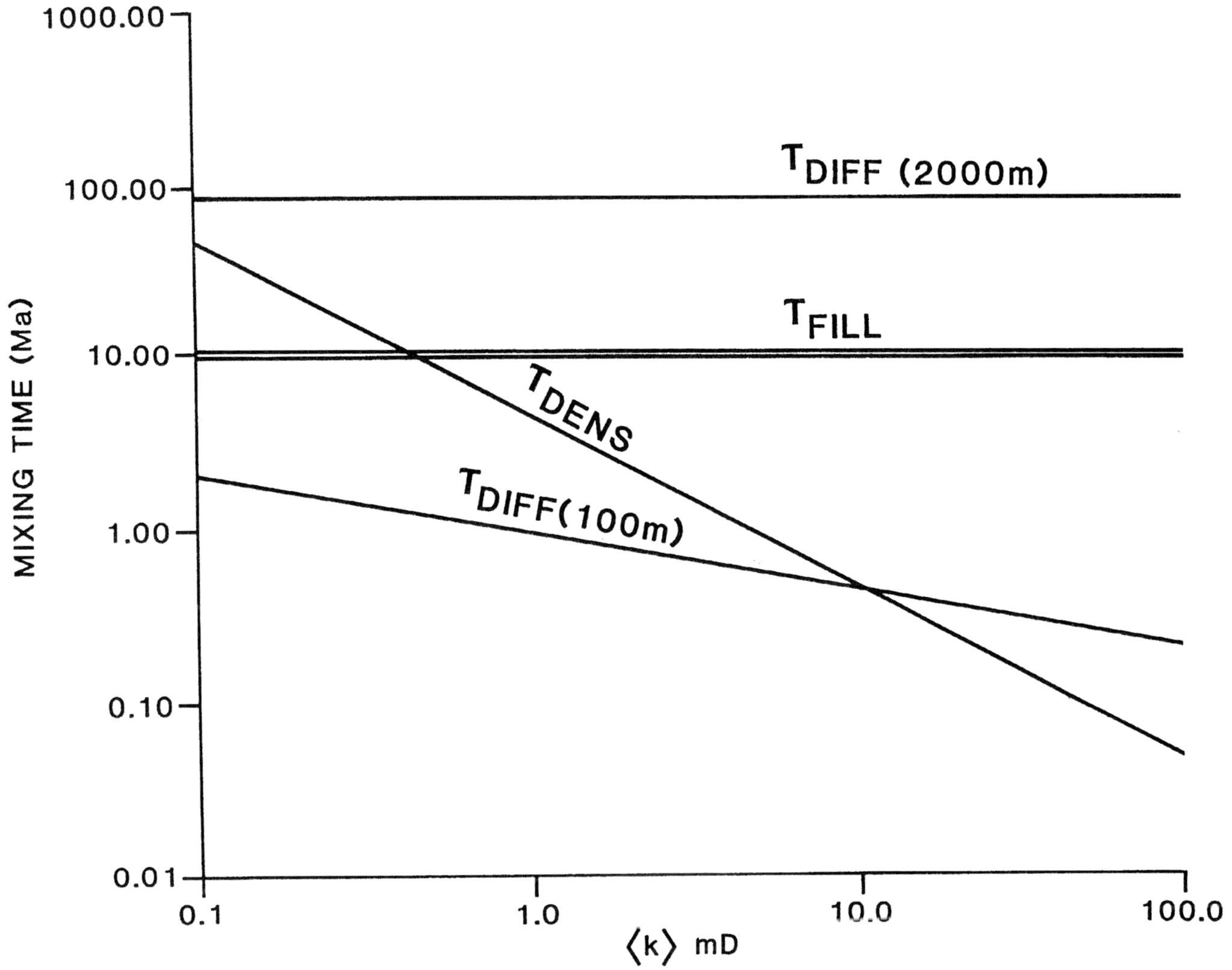

Figure 12. Order-of-magnitude estimates of in-reservoir mixing time scales for an oil reservoir. Details of assumptions are given in England and Mackenzie (1989). τ_{FILL} is the available time since filling, τ_{DENS} is the density overturning time scale, and τ_{DIFF} (100 m) is the time scale for diffusive mixing over 100 m.

eventually cause the composition at any given depth to be identical, defining a state of lateral chemical equilibrium. (We will show later, however, that this is a very slow process.)

Gravitational segregation means that *vertical* chemical equilibrium does not correspond to constant composition with depth, as discussed in the section on gravity segregation. In fact, molecular diffusion is the mechanism by which the gravitational gradients are established.

The time scales of diffusion, τ_{DIFF}, can be estimated to an order of magnitude by the method given in England et al. (1987). The calculation shows that over the distance typical of an oil column height, i.e., 100 m (330 ft), diffusion is fast compared to the elapsed time since filling. In all cases, τ_{DIFF} (100 m) is less than 1 m.y. However, because τ_{DIFF} increases with the *square* of the distance involved, diffusion time scales become very long over distances greater than about 2000 m (6500 ft), the typical interwell distances. The timescales involved are summarized in Figure 12. Even given the uncertainties in the exact diffusion rates applicable to the subsurface, it is clear that diffusive mixing will be an important process within individual wells, but an exceedingly inefficient one on a well-to-well basis.

Summary of Mixing Rates

This section has shown that reservoirs will often inherit a pronounced compositional gradient that may not necessarily be removed by in-reservoir mixing. Order of magnitude estimates for the relevant geological mixing rates can be used to predict the likely effects

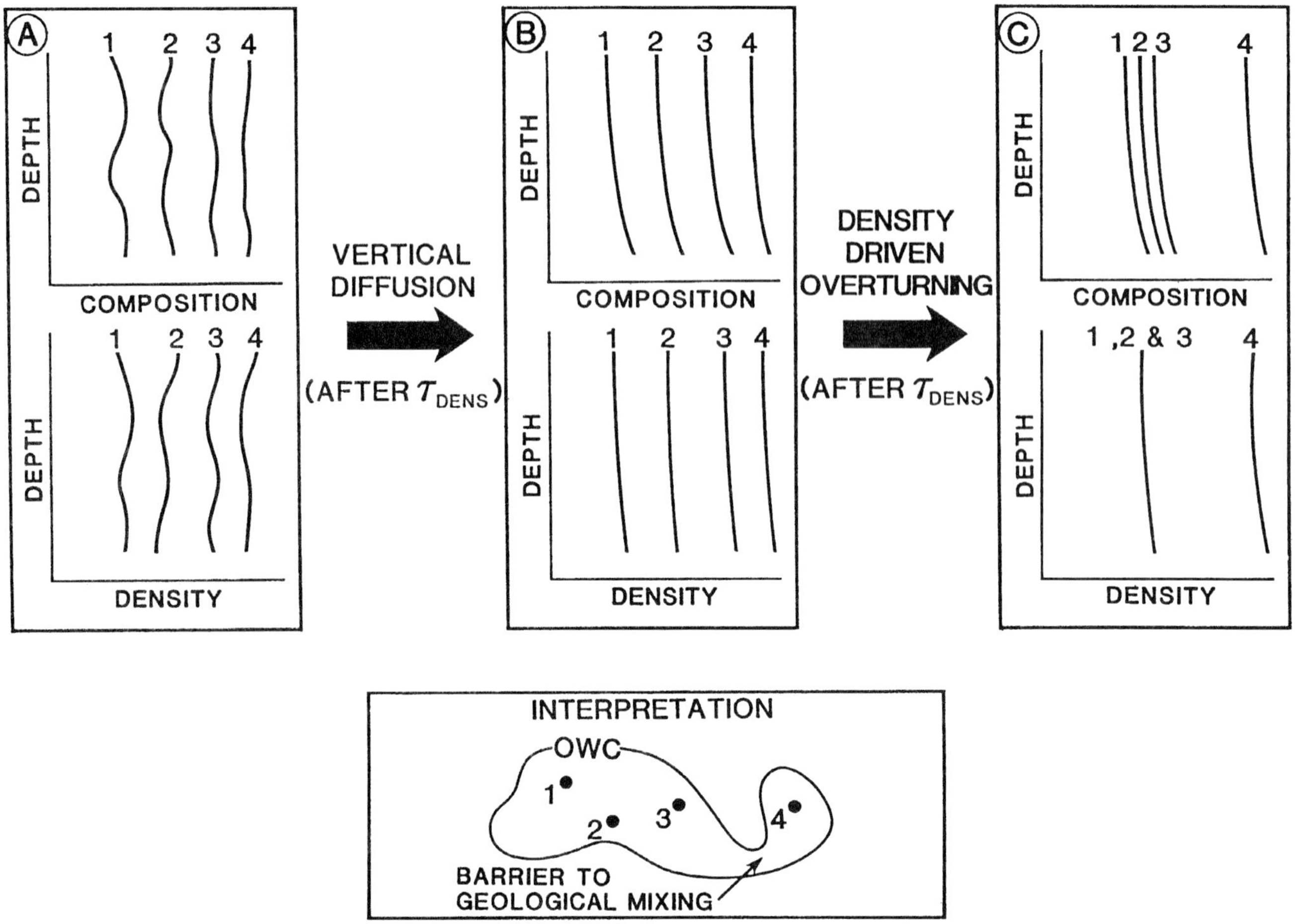

Figure 13. The effect of in-reservoir mixing processes on the composition and density of a hypothetical reservoir. Wells 1, 2, and 3 are assumed to be separated from well 4 by a barrier to mixing on a geological time scale.

in real reservoirs. This is shown graphically in Figure 13, in which the first diagram shows the situation immediately after the reservoir filled (i.e., achieved a reasonable amount of interconnected, saturated rock). Each well has an individual compositional profile reflecting the filling history. After a time period, τ_{DIFF} (100 m), sufficient for diffusion to occur *within* individual wells, gravitational segregation will give rise to a gradual compositional change within each well (Figure. 13b). However, because of slow interwell diffusion, each well has a different average composition.

In reservoirs (or parts of reservoirs) in which density-driven mixing is slow compared to τ_{FILL}, this situation will persist to the present. If density-driven mixing does occur, the lateral difference in density will be eliminated as the reservoir fluids reach mechanical equilibrium. Surprisingly, a remnant of the compositional differences may persist (i.e., in wells 1, 2, and 3) (see England and Mackenzie, 1989). Moser (1986) has considered the effects of simultaneous thermal convection and diffusion on a gas cap with a large lateral thermal gradient and has modeled the process mathematically.

CASE STUDIES

Forties Field

The Forties reservoir is a giant oil field located in the United Kingdom sector of the North Sea, (Blocks 21/10 and 22/6). Figure 14 shows the field outline with all wells marked for which petroleum engineering information is available. It is clear that there is a significant lateral gradient in the bubble point pressure across the field, with the highest pressures located in the southeastern area of the accumulation. Because bubble point pressure increases with gas–oil ratio (GOR), the data indicate that the higher GORs are found in the southeastern part of the field. Although production from the main part of the field started some years earlier than in the southeastern part, no significant effects due to production are expected because the reservoir pressure has always been maintained above its bubble point. In addition, the GOR of the main field oil appears to have remained steady, as measured in new wells.

The geochemical setting of the Forties field is shown in Figure 15. The mature Kimmeridge Clay source rock

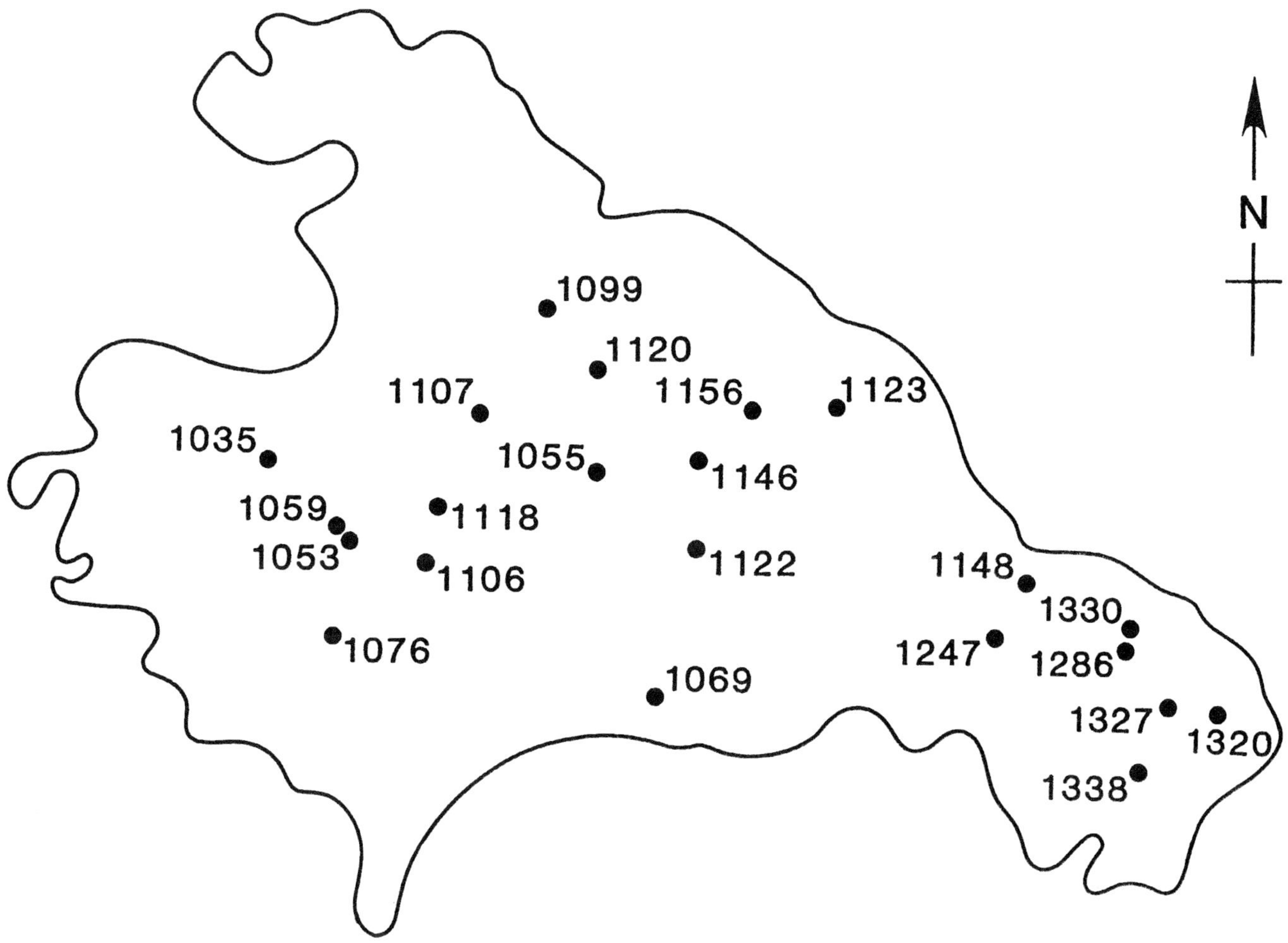

Figure 14. Bubble point pressures (in psi) from the Forties field.

is effectively divided into two source kitchens. The southernmost one initially feeds petroleum into the southeastern end of the field, while the remainder of the field is filled from the north. Thermal modeling and current temperature measurements indicate that the southern source rock kitchen is about 10°C hotter than the northern source rock kitchen. It is therefore expected that the southeastern part of the field would have initially received the majority of its oil from a more mature source rock kitchen. Since there is a general trend for GOR to increase with source rock maturity, it is reasonable to expect that the southeastern corner of the Forties field initially received a petroleum charged with a higher GOR than the rest of the field. This suggests that the present-day observation of a lateral GOR gradient across the field is a reflection of its filling history from two source rock kitchens of slightly different maturity. The "inherited" difference persists to the present day because of slow mixing between the southeastern and main parts of the field.

As previously mentioned, diffusion is ineffective as a process for eliminating compositional differences (such as GOR) between the southeastern and main parts of Forties field because of the large interwell distances. The only other process that needs to be considered is density-driven mixing. This has evidently also been a geologically slow process compared with the time available since filling, which is about 10–20 m.y. for this field. This can be verified by plotting the experimentally measured subsurface oil densities against depth for all available wells (Figure 16). The density data show that at any given depth (allowing for the slight gravitationally induced gradient), there is a distinct difference in measured density between the southeastern part of Forties and the remainder of the field. These data clearly show that field-wide density-driven mixing has *not* occurred to any great extent since a mechanically unstable density distribution is currently observed (see Figure 11). This lack of mixing has preserved the inherited compositional difference between the southeastern and main parts of the field.

The observation of a mechanically stable depth/density trend in the main part of the field (the triangular points in Figure 16) implies that this area has undergone efficient density-driven mixing. The southeastern part of the field also has its separate mechani-

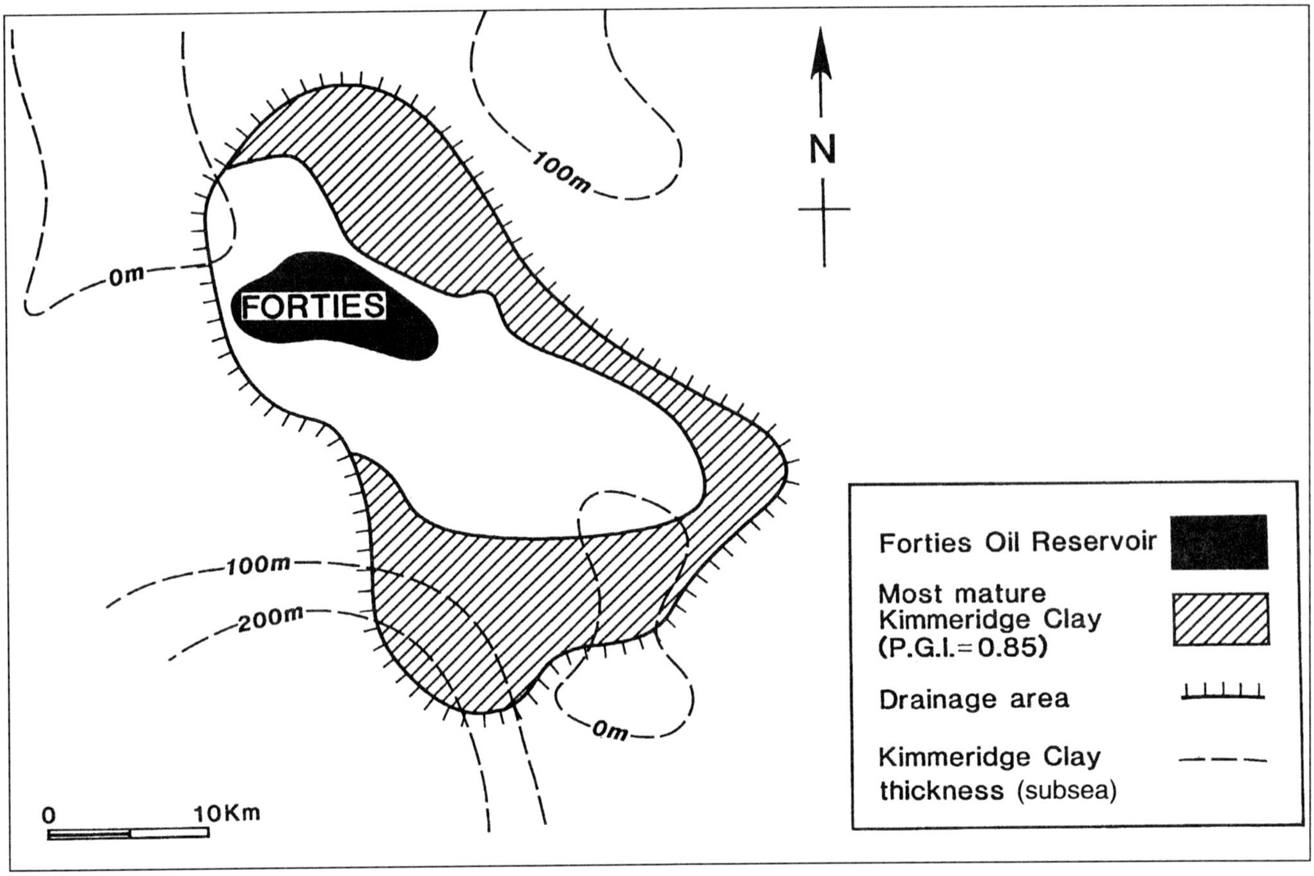

Figure 15. Relative positions of Forties reservoir and source rock kitchens.

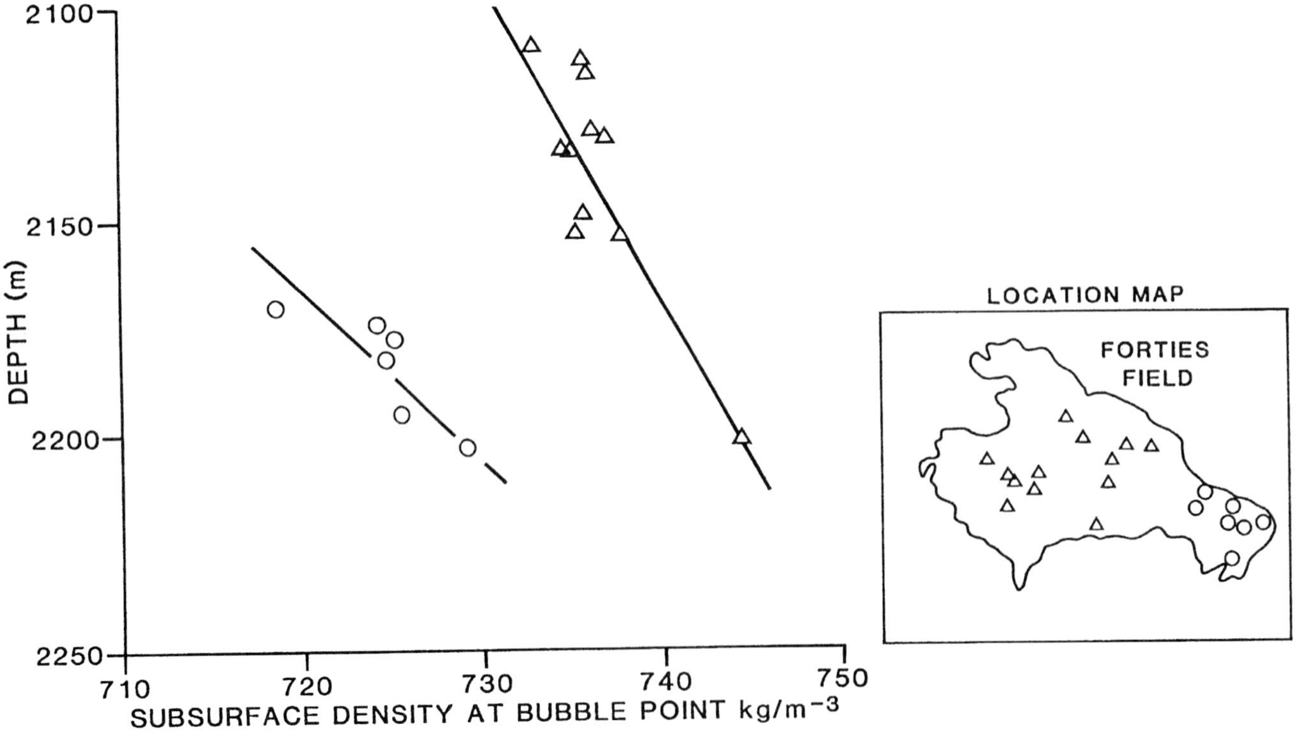

Figure 16. Subsurface densities from the Forties field; lines are an aid to the eye only.

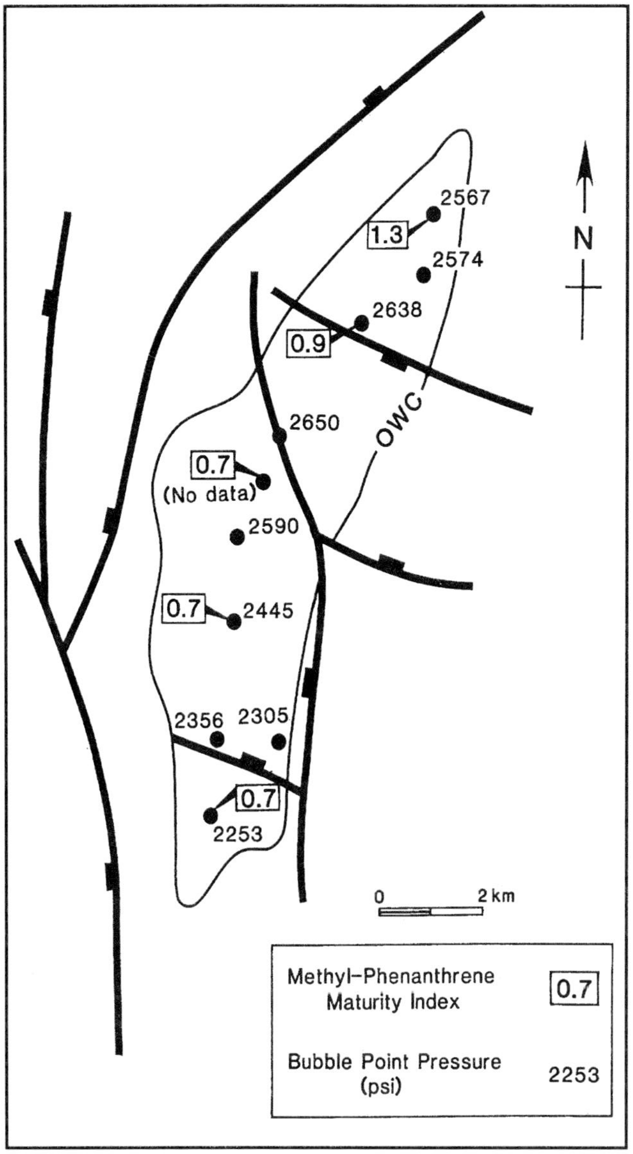

Figure 17. Bubble point pressure and methylphenanthrene (MP) maturity indicator [(3-MP + 2-MP) ÷ (9-MP + 1-MP)] from the Magnus field.

cally stable depth/density trend. As indicated in Figure 13, it is nevertheless possible to have different GORs and still achieve the same density. Complete chemical homogenization has evidently not occurred even in the main part of the field since a lateral GOR gradient is discernible in the main part of the field (see Figure 14). However, the GOR gradient is much more pronounced between the southeastern part of Forties and the remainder of the field than within the main part of the field itself.

The lack of efficient field-wide mixing between the southeastern and main parts of Forties is a reflection of its internal reservoir structure. Although the core plug permeabilities in Forties reach values in excess of 1000 md, the larger scale reservoir quality deteriorates toward the southeast due to increasing heterogeneity. This is attributed to a change from a predominantly channeled lobe depositional environment to an increasingly unchanneled lobe environment. The averaged permeability for a pathway such as Figure 11 is therefore expected to be small and density-driven mixing times correspondingly long.

A detailed geochemical study of oils from the field revealed a small but statistically significant increase in maturity of the oil in the southeastern part of Forties, consistent with the model presented here. Further details can be found in England (1990).

Magnus Field

The Magnus field is a large black oil accumulation in the United Kingdom sector of the North Sea (Blocks 211/12 and 211/7) (Wood, 1988). As in the Forties reservoir, a statistically significant trend in GOR was observed during appraisal. These data are plotted in Figure 17, which shows that the bubble point pressure is about 2600 psi in the northern part of the field, but only 2300 psi in the southern part. Since GOR correlates with bubble point pressure, this data indicates that the northern part of the field contains more dissolved C_{1-5} than the southern part.

As in the case of Forties field, the lateral change in GOR may be related to the geological setting of the field. Filling occurred from a source rock kitchen buried to the east of the Magnus field. Most of this source rock kitchen is in the middle of the oil generation window. However to the northeast, a portion of the source rock interval plunges downward, reaching maturities corresponding to the onset of oil to gas cracking. The higher GOR observed in the northern part of the present-day Magnus accumulation is therefore consistent with the northern part *initially* receiving an additional contribution of gassier, more mature petroleum. This inherited compositional gradient has persisted to the present day because of geologically slow in-reservoir mixing.

As mentioned earlier, diffusion is an inefficient interwell mixing mechanism. The lack of mixing between the northern and southern parts of Magnus field is due to the geological structure of the reservoir, which is divided into depositional units that are separated by nearly continuous shales and extensive faulting (Wood, 1988). This type of structure would allow only very slow density-driven mixing, since the low-permeability shales would disrupt any density overturning pathway such as that shown in Figure 11. The measured subsurface densities show a small difference between wells from the north and the south, although not as great as in the Forties field.

Figure 17 also plots the values of a methyl-phenanthrene molecular maturity parameter (see Radke et al., 1982) measured from oil samples taken from each well.

The higher values in the north confirm an increased oil maturity in that part of the field. The observation of more mature oil in the northern part of the Magnus confirms the filling and mixing model previously described. In other words, it appears that the Magnus field originally inherited a maturity and GOR difference due to the presence of a more mature source rock kitchen to the north of the field. Lack of subsequent in-reservoir mixing processes and the geological structure of the reservoir with its relatively isolated lobes has preserved the compositional differences to the present day.

CONCLUSIONS

The application of geochemistry to reservoir phenomena is a new field of research recognizing that significant lateral compositional gradients are frequently observed in petroleum accumulations (England et al., 1987). It is hoped that increasing cooperation between geochemists, production engineers, and geoscientists will increases the efficiency of oil and gas appraisal and production in the future.

The main conclusions of this chapter can be summarized as follows:

1. Significant lateral and vertical differences in petroleum composition are often observed in oil and gas accumulations. When a reservoir fills from one side, the most mature petroleum is found nearest the sourcing direction.
2. The lateral composition variations are "inherited" from the reservoir filling process and often persist over geological lengths of time due to the slow rate of field-wide mixing.
3. Vertical compositional differences are caused by gravitational segregation and are set up by the relatively rapid vertical diffusion.
4. In well-connected reservoirs containing gases or low viscosity oils, mechanical equilibrium is rapidly attained via density-driven overturning. It is notable that field evidence shows that this does not necessarily lead to complete chemical homogenization. Thus, it may be dangerous to presume the presence of sealing faults or barriers to flow in general from lateral compositional gradients alone.

Acknowledgments *We are grateful to the numerous geoscientists and engineers who have contributed their knowledge and experience to develop and improve the material presented here. We thank BP for permission to publish.*

Chapter 4

Effect of Biodegradation and Water Washing on Crude Oil Composition

Susan E. Palmer
Amoco Production Company
Tulsa, Oklahoma, U.S.A.

INTRODUCTION

The study of crude oil geochemistry becomes difficult when crude oils are altered by microbial action (biodegradation) and/or water washing. These processes can alter parameters used to compare oils when determining genetic relationships (oil–oil correlation), depositional environments, and time of oil generation (i.e., thermal maturity of the source rock at the time of oil generation and expulsion). Much of this knowledge comes from petroleum geochemists who have been documenting case histories of such occurrences through the years (e.g., Winters and Williams, 1969; Bailey et al., 1973a, b; Rubinstein et al., 1977; Connan et al., 1975, 1980; Seifert and Moldowan, 1979; Rullkotter and Wendish, 1982; Volkman et al., 1983; Momper and Williams, 1984; Palmer, 1984; Williams et al., 1986)

In addition, microbiologists and petroleum geochemists have studied the action of bacteria on petroleum in the laboratory (e.g., McKenna, 1972; Horowitz et al., 1975; Jobson et al., 1979; Connan, 1981; Goodwin et al., 1983). Some workers have isolated products of bacterial metabolism of crude oils or classes of hydrocarbons (e.g., Gibson, 1976; Higgens and Gilbert, 1978; Cripps and Watkinson, 1978; Cain, 1980; Mackenzie et al., 1983). Connan (1984) illustrates the metabolic products recognized by some of these workers and others. Products of aerobic degradation are most often organic acids and CO_2. Anaerobic bacteria can live on the metabolites of the aerobes but do not grow on hydrocarbons. Thus, the impact of anaerobic bacteria on oil biodegradation is only slight.

Alteration of crude oil by water washing has been indirectly studied in the laboratory through determination of water solubilities of individual hydrocarbons and mixtures of several hydrocarbons and by studying compositional changes of whole crude oils (e.g., McAuliffe, 1966, 1980; Bailey et al., 1973b, Price, 1976; Eganhouse and Calder, 1976; May et al., 1978a, b; Lafargue and Barker, 1988). Extensive review articles covering work done prior to 1985 on biodegradation and water washing have been prepared by Milner et al. (1977) and Connan (1984). These two reviews give an overview of the effects of biodegradation and water washing gleaned from the many exemplary papers found in the literature. Also, Lafargue and Barker (1988) have reviewed and discussed data obtained from laboratory water washing experiments and present their own observations and conclusions.

The purpose of this chapter is to compile and present the results of these studies and review articles in an abbreviated form to aid the explorationist in understanding the effects of biodegradation and water washing. In this regard, the reader is directed to the literature for details of individual studies. Results of more recent studies (1985–1988) are also included here. Examples of parameters and hydrocarbon distributions demonstrating the effects of biodegradation and water washing are given to aid the reader in recognizing altered oils. Because many geochemical laboratory groups have developed their own ways to portray data, some of these techniques will be referenced in this discussion. Organic geochemical parameters and types of hydrocarbon classes referred to here are defined in the Glossary at the back of this volume.

GEOLOGICAL CONSTRAINTS AND PHYSICOCHEMICAL CONDITIONS FOR BIODEGRADATION AND WATER WASHING

The processes of microbial degradation and water washing of crude oils occur when certain conditions are met. Milner et al. (1977) and Connan (1984) have outlined the requirements for both processes in their review articles. Their findings are summarized here.

Biodegradation occurs in surface seeps and in relatively shallow reservoirs, e.g., 4000–6000 ft (1220–1830 m) or less. Although anaerobic bacteria can survive on partially degraded oils, both case histories and microbiological studies show that aerobic bacteria are the major agents of crude oil degradation. Aerobic bacteria can grow in relatively cool reservoirs (i.e., below 80°C, or 176°F) that are invaded by oxygen-charged waters. In addition to dissolved oxygen, nutrients such as nitrate and phosphate must be present and the salinity of the water must be less than 100–150‰. Also, unless special cases such as oxygenated microenvironments exist, the amount of H_2S in the oil must be very low, as it is toxic to aerobic bacteria. Thus, cool, shallow reservoirs that are flushed by oxygenated, nutrient-rich fresh water can be expected to contain oil that is being actively biodegraded.

However, biodegraded oils are also present in deeper reservoirs. In areas where tectonic activity causes subsidence of reservoirs, biodegraded oils are found preserved far below the arbitrary 6000-ft cutoff for bacterial activity. Thus, cases where biodegraded oils occur in reservoirs of, e.g., 8000–10,000 feet (2440–3050 m) have been reported. Also, with regard to maximum depth of reservoirs and ongoing degradation, lower thermal gradients can permit deeper occurrences. Connan (1984) points out that aerobic bacteria degrade oil at the oil–water interface. In such cases, the lower part of an oil accumulation will be degraded rather than the upper portions, unless more than one oil–water contact is present. Some reservoirs have multiple oil–water contacts and different hydrological regimes that could lead to a complex and perhaps confusing array of degraded and undegraded oils. Lafargue and Barker (1988) mention that hydrodynamically tilted oil–water contacts are indicators of actively flowing waters and delineate areas where degradation is occurring.

Physical and chemical processes other than biodegradation and water washing (e.g., fractionation of light and heavy ends during migration and in-reservoir maturation) can add to the difficulty of understanding the transformation of oil from its initial state to the time it is recovered in a discovery well or surface seep. Some of these other processes could be mistaken as biodegradation or water washing. Thus, a good understanding of the postgeneration history of an oil is of major importance for establishing cause and effect relationships.

Because water is a necessary ingredient for biodegradation, the process of water washing generally accompanies biodegradation. In spite of many studies that have attempted to isolate the effects of water washing from microbial alteration, questions concerning the actual cause of alteration of a crude oil's geochemistry, especially specific parameters, are still open for discussion. Conditions favorable for water washing exist during oil migration if oil is passing through a water-wet carrier bed and reservoir system. However, Lafargue and Barker (1988) have suggested that water washing during migration must be minimal because highly water soluble molecules, namely, benzene and toluene, are present in most oils. They suggest that most water washing occurs after accumulation where, given the proper conditions, biodegradation is also occurring. Water washing can, however, take place outside the temperature, oxygen, and salinity constraints of biodegradation. Price (1976) and Lafargue and Barker (1988) have noted that the solubility of crude oil components increases markedly at higher temperatures. These results demonstrate that water washing can occur in zones where microbial activity is precluded by high temperature. The work of Price (1976) shows that high salinities (over 270‰) cause exsolution of hydrocarbons; thus, salinity may control the occurrence of water washing. Price's work is supported by the results of Lafargue and Barker (1988)

EFFECTS OF BIODEGRADATION AND WATER WASHING ON CRUDE OIL COMPOSITION

Biodegradation

Biodegradation produces heavy, low API gravity oils depleted in hydrocarbons and enriched in the nonhydrocarbon nitrogen-, sulfur-, oxygen-bearing (NSO) compounds and asphaltenes (see Glossary for definition of terms). Water washing usually accompanies biodegradation, removing the more water-soluble hydrocarbons, especially the lower molecular weight aromatics such as benzene and toluene. It also aids in concentration of the heavier molecules in the residual oil. Studies demonstrating the selective loss of the gasoline-range hydrocarbons (i.e., the less-than-C_{15} fraction) by water washing and mild biodegradation have been reviewed by Milner et al. (1977). It is not always clear whether the composition (i.e., relative amounts of saturates, aromatics, and naphthenes) of light hydrocarbons is altered by biodegradation, water washing, or evaporative loss. For example, in microbiological experiments using these volatile compounds, unless one collects and identifies metabolites, the loss of these compounds could have causes other than biodegradation (i.e., evaporation and water washing). It is clear, however, that biodegraded and water-washed oils generally are depleted in less-than-C_{15} hydrocarbons; are enriched in sulfur, NSOs, and asphaltenes; and have low API gravities. Enrichment in NSOs, asphaltenes, and cyclic hydrocarbons relative to *n*-paraffins causes an increase in the optical rotation of an oil. Winters and Williams (1969) and Momper and Williams (1984) demonstrate that optical activity is a useful parameter for indicating degrees of biodegradation.

In addition to these gross compositional changes, biodegraded oils are most readily recognized by low concentrations of *n*-paraffins relative to branched (e.g., the

C_{19} and C_{20} isoprenoids, pristane and phytane) and cyclic hydrocarbons (naphthenes and aromatic hydrocarbons). For example, the weight percent of *n*-paraffins relative to naphthenes and aromatics is approximately 2–15 wt. % in the C_{15}+ fraction of biodegraded oils (Figure 1). Gas chromatograms of whole crude oils show that low molecular weight components, e.g., C_{10} to C_{14} *n*-paraffins, are depleted first; the C_{15+} *n*-paraffins are then attacked (e.g., Williams et al., 1986). Gas chromatographic patterns of C_{15+} saturated hydrocarbon fractions (Figure 2) of biodegraded oils contain low amounts of *n*-paraffins relative to pristane, phytane, and naphthenes. Thus, the loss of *n*-paraffins relative to branched and cyclic hydrocarbons (in conjunction with low API gravity and enrichment in percent sulfur, NSOs, and asphaltenes) is the most common parameter alluded to as an indicator of biodegradation.

Perhaps the focus on the use of the saturated hydrocarbon fraction (e.g., *n*-paraffins, branched paraffins, and naphthenes such as steranes and terpanes) in the application of oil geochemistry has led to a better understanding of the effects of bacterial action on this fraction. However, aromatic hydrocarbons can also be degraded by bacteria.

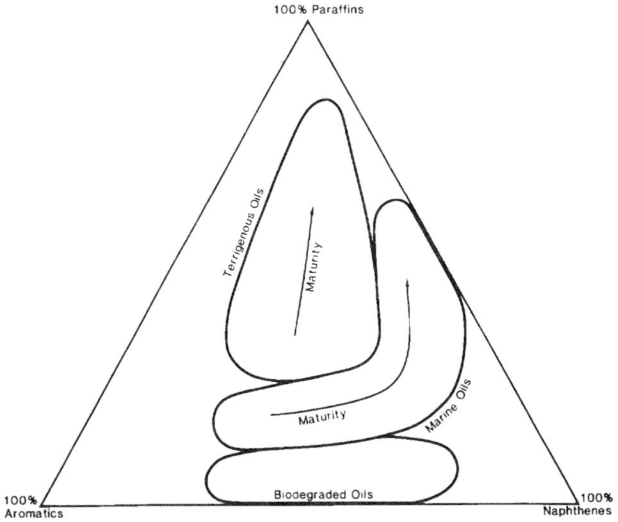

Figure 1. Gross C_{15+} hydrocarbon composition of crude oils in terms of percent abundance of paraffins, naphthenes, and aromatic hydrocarbons. Biodegradation removes paraffins leaving an oil enriched in aromatic and naphthenic hydrocarbons.

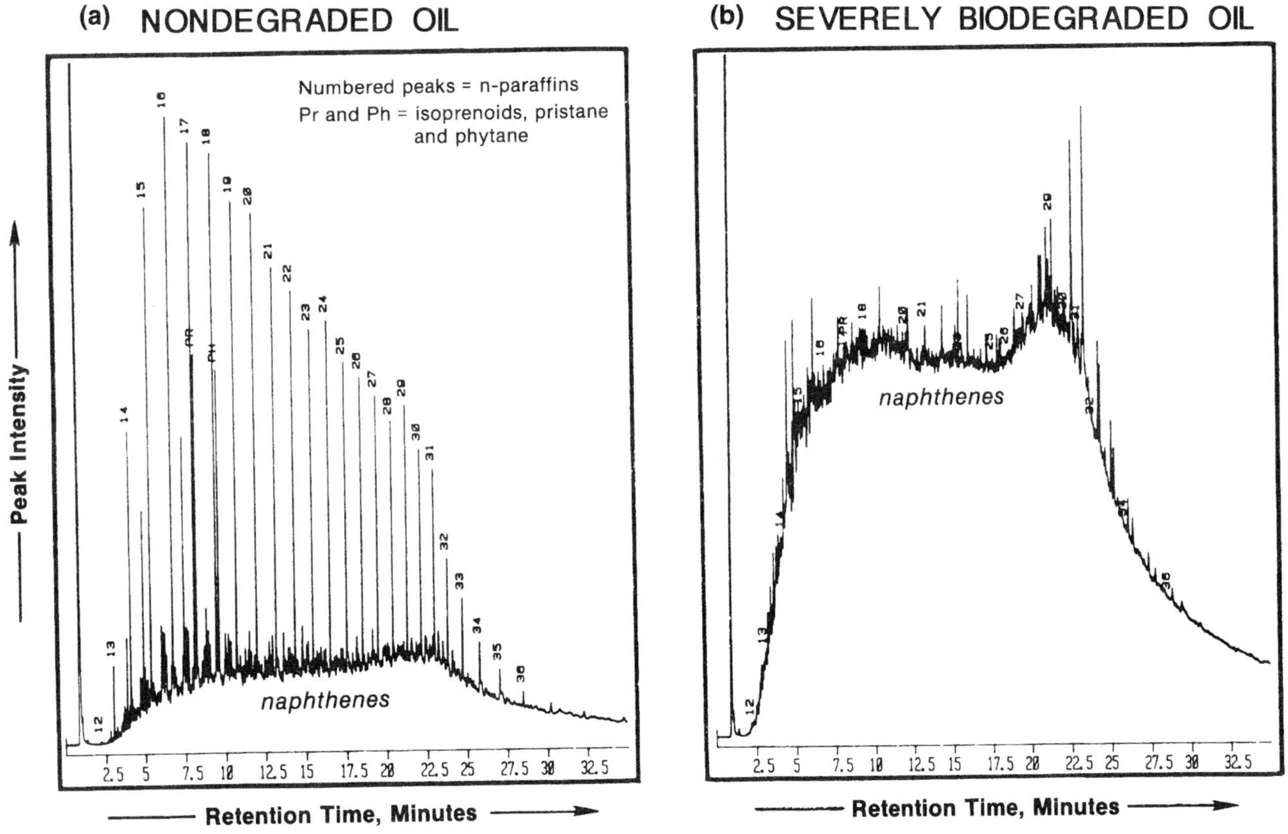

Figure 2. Effect of biodegradation on the saturated hydrocarbon fraction of crude oils. (a) Gas chromatogram of C_{15+} saturated fraction of a nondegraded oil contains prominent *n*-paraffin and branched paraffins. (b) Chromatogram of C_{15+} saturated fraction of a severely biodegraded oil contains primarily naphthenes; the paraffins have been removed.

Connan (1984) lists examples where aromatic fractions are altered and concludes that more in-depth studies, such as laboratory cultures of aromatic hydrocarbon-metabolizing bacteria, are needed. More documented field examples are needed of biodegradation of the aromatic fractions of reservoired crude oils and the types of bacteria that attacked these oils.

Aromatic Hydrocarbons

The major classes of aromatic hydrocarbons that are altered by bacteria are those with paraffin side chains, such as the alkylbenzenes (single-ring aromatics). Two- and three-ring aromatics are more resistant to bacterial attack than are the single-ring aromatics (Connan, 1984). Thus, alkylbenzenes are depleted in moderately biodegraded oils. In a study of a sequence of biodegraded oils from south Texas, Williams et al. (1986) showed that some dimethylnaphthalenes (two-ring aromatics not to be confused with the class of saturated ring compounds, the naphthenes) are removed prior to others. In line with earlier findings of Volkman et al. (1984), this study showed the selective removal of specific dimethylnaphthalenes (2,6-, 2,7-, 1,3-, 1,7-, and 1,6-) relative to other homologs. In contrast, removal of ethylnaphthalenes prior to dimethylnaphthalenes is indicative of water washing (Eganhouse and Calder, 1976).

Wardroper et al. (1984) showed a loss of the C_{20} and C_{21} triaromatic steranes (i.e., four-ring compounds with three aromatized rings) relative to C_{26} to C_{28} homologs during degradation. These authors suggested that the C_{20} and C_{21} triaromatic steranes are depleted because of water washing rather than biodegradation. However, the lower solubility of C_{20+} hydrocarbons may preclude water washing. In the same study, C_{20} and C_{21} monoaromatic steranes were not depleted possibly because they are less water soluble than triaromatics (McAuliffe, 1966). Recognition of the loss of the C_{20} and C_{21} triaromatics relative to C_{26} through C_{28} is important because the ratio of C_{20} and C_{21} versus C_{26} through C_{28} triaromatic steranes is used to assess relative oil maturity (i.e., timing of oil generation). Connan (1981) showed that even the sulfur-containing aromatics can be removed from severely biodegraded oils (e.g., asphalts from the Aquitaine basin). It was suggested that anaerobic sulfate-reducing bacteria (rather than aerobic bacteria) attacked these usually resistant compounds.

This brief discussion of the effects of biodegradation of aromatic hydrocarbons shows that much remains to be learned about alteration of oils in the subsurface. Indeed, the causes of alteration and the distribution of the affected oils in the subsurface are not always straightforward.

Saturated Hydrocarbons

As previously mentioned, the biodegradation of the saturated hydrocarbon fraction has been more extensively studied than that of the aromatics. In more detailed discussions of biodegradation, changes within classes of saturated hydrocarbon compounds are considered. The effects of degradation on distributions of compounds used in determining genetic relationships among oils (oil–oil correlation) and thermal maturities must be understood. Other parameters used in correlation, such as stable carbon isotopic composition, can also be influenced by degradative processes.

Removal of saturated hydrocarbon compound classes in order of their increasing resistance to biodegradation and a scale of degrees of biodegradation are presented in Volkman et al. (1984). Oil biodegradation in general follows the path outlined as follows, but deviations are frequently observed. Thus, other workers have provided slightly different scales based on their own suite of samples. These deviations remind us that oil transformation is the result of a complex process and that some factors might not be known for a given case. Mild to moderate effects of biodegradation can be readily detected in gas chromatograms of the saturated hydrocarbons, but more extensive degradation (i.e., of the naphthenes) requires gas chromatographic–mass spectrometric analysis (GCMS); these data are usually displayed as single ion mass chromatograms.

Volkman et al. (1984) indicate *initial* or *mild* biodegradation as the removal of low molecular weight *n*-paraffins (e.g., gasoline-range *n*-paraffins), which is most readily observed on whole-oil gas chromatograms. Moderate biodegradation is marked by a nearly total loss of *n*-paraffins. At slightly higher levels of biodegradation (*moderate* to *extensive*), branched paraffins (pristane and phytane) and single-ring naphthenes are removed (Figure 2). In the aromatic fraction, alkylbenzenes are depleted and selective removal of dimethylnaphthalenes occurs during moderate biodegradation. *Extensive* biodegradation is indicated by removal of two-ring naphthenes (C_{14} to C_{16} bicyclics), detected by changes in mass chromatograms of the m/z 123 ion. *Very extensive* biodegradation is denoted as loss of a group of four-ring naphthenes, the C_{27} to C_{29} "normal" steranes (Figure 3). Of particular importance is the selective removal of the 20(*R*)-5α(H)-steranes, which are ratioed against the 20(*S*)-5α(H)-steranes to assess the maturity level of an oil (i.e., timing of oil generation and expulsion of an oil from its source rock).

Severe biodegradation is indicated by demethylation of the five-ring naphthenes, the C_{27} to C_{35} hopanes (Figure 4). A methyl group is removed from the "A" ring (ring number 1 out of 5), producing a new series of compounds: the C-10 demethylated hopanes detected by the m/z 177 ion (Seifert and Moldowan, 1979; Rullkotter and Wendish, 1982). The C_{30} to C_{35} hopanes appear to be altered before the C_{27} to C_{29} hopanes. Demethylated hopanes predominate in cases of *extreme* biodegradation, and the C_{27} to C_{29} "normal" steranes are completely absent. Philp (1985a) added an additional biodegradation step, the alteration of the "rearranged" steranes, which is considered to be *very extreme* degradation.

An alternative degradation path for hopanes also appears to exist. A series of C_{26} to C_{30} (and possibly C_{31})

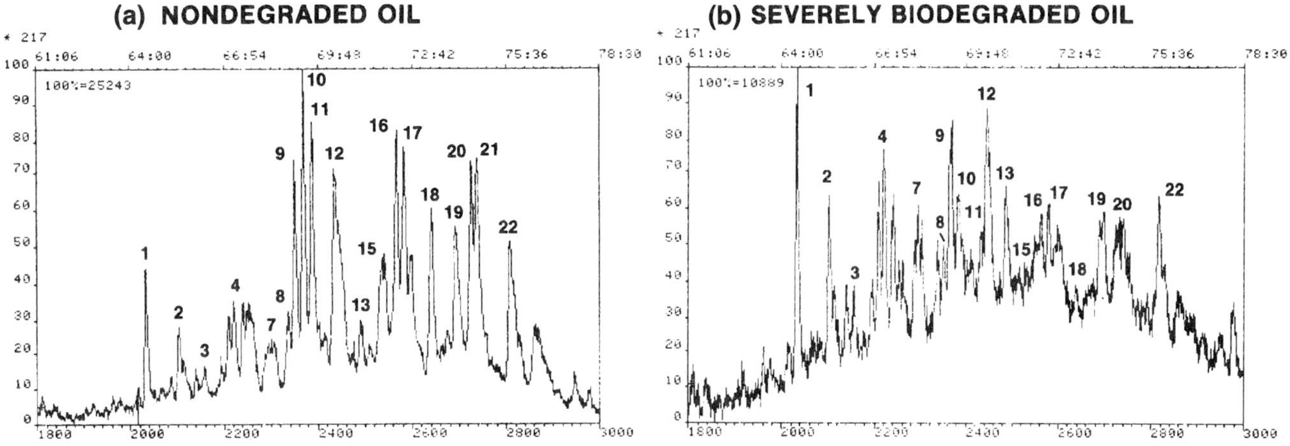

Figure 3. C_{27} to C_{29} distributions (m/z = 217) of (a) a nondegraded oil and (b) a severely biodegraded oil. Normal steranes (peaks 8–11 and 15–22) are consumed by bacteria in (b), leaving an abundance of rearranged steranes (peaks 1–7 and 12–13). See Figure 5 for names of individual steranes.

tetracyclic compounds, the 8,14-*seco*-hopanes, are formed by opening the C ring (ring number 3 out of 5) (Rullkotter and Wendish, 1982). In such cases, demethylated hopanes can also be present and the steranes may be only slightly altered. These examples suggest that various degradative processes can operate to produce severely biodegraded oils. Perhaps certain environmental conditions are required to allow specific bacteria to grow on oils.

The C_{19} to C_{26} three-ring naphthenes (tricyclic terpanes) survive extreme biodegradation, although demethylated tricyclic terpanes have been tentatively identified (e.g., Howell et al., 1984; Philp, 1985a). Because of their resistance to biodegradation, tricyclic terpanes have been used for oil–oil correlation in severely biodegraded oils. Their distributions also supply information concerning depositional environments (e.g., Zumberge, 1987).

As previously mentioned, the stable carbon isotopic composition of crude oils can also be altered by biodegradation, although not in a consistent manner. For example, in a 42-day simulated oil biodegradation study, Stahl (1980) observed that the saturated hydrocarbon fraction was enriched in ^{13}C (i.e., more positive $\delta^{13}C$ values), but the isotopic composition of the aromatic hydrocarbon fraction remained unchanged. Sofer (1984) and Momper and Williams (1984) showed that the saturated fraction of naturally biodegraded oils is also enriched in ^{13}C. However, field examples showing no or little change in isotopic composition or changes in both the saturated and aromatic fractions have also been reported (Sofer, 1984). Connan (1984) has reviewed other studies in which the isotopic composition of crude oil fractions other than the saturated hydrocarbons also become enriched in ^{13}C.

Water Washing

Water washing is most readily recognized by changes in the composition of the gasoline-range hydrocarbons because these compounds are more water soluble than the C_{15+} hydrocarbons (McAuliffe, 1966; Price, 1976). For a given carbon number, ring formation, unsaturation, and branching cause an increase in water solubility. Thus, one would expect that when water washing occurs, aromatic hydrocarbons of a given carbon number would decrease first, followed by naphthenes, branched paraffins, and *n*-paraffins. Generally, the loss of benzene and toluene is a good indicator that water washing has occurred. These low molecular weight aromatics are also biodegradable; however, their high water solubilities make them useful indicators of water washing. Other indicators of water washing are the loss of ethylnaphthalenes relative to dimethylnaphthalenes (Eganhouse and Calder, 1976) and possibly (as discussed in the previous section) the loss of C_{20} and C_{21} triaromatic steranes (Wardroper et al., 1984). Experimental water washing studies by Lafargue and Barker (1988) do support the loss of gasoline-range (less-than-C_{15}) aromatic hydrocarbons relative to naphthenes and paraffins in line with the solubility studies previously noted.

An example of the effect of water washing on the C_{15+} hydrocarbon composition involved a field study of Philippine oils having abundant sulfur-containing aromatic hydrocarbons (dibenzothiophenes). Based on the idea that heteroatomic compounds are more water soluble than aromatic, cyclic, branched, and straight-chain hydrocarbons (e.g., Price, 1976), water washing was thought by Palmer (1984) to cause the loss of dibenzothiophene ($C_{12}H_8S$) and methyldibenzothiophene ($C_{13}H_{10}S$) relative

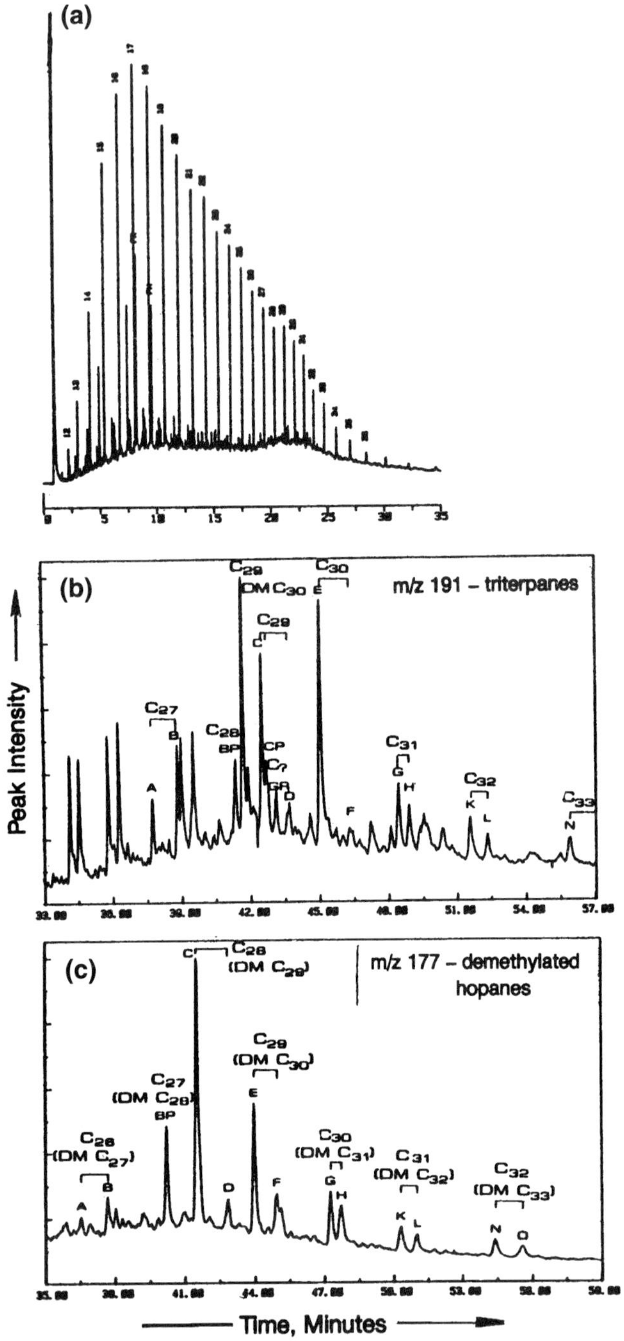

Figure 4. Example of a crude oil containing demethylated hopanes (m/z = 177). The presence of demethylated hopanes is indicative of severe biodegradation. In the case shown here, an early oil became severely biodegraded; migration of a nondegraded oil later filled the same reservoir, mixing with the severely degraded oil. (a) The gas chromatogram of the C_{15+} saturated fraction thus represents hydrocarbons from the nondegraded oil. (b) The demethylated hopanes are initially detected in the pentacyclic triterpane fraction by the large DMC_{30} peak, the demethylated C_{30} hopane. (c) For verification, the complete series of demethylated hopanes are identified by scanning the m/z 177 ion. (From Sofer et al., 1986.) See Figure 5 for names of individual triterpanes.

to phenanthrene ($C_{14}H_{10}$) and methylphenanthrene ($C_{15}H_{12}$) from the C_{15+} aromatic hydrocarbon fraction. Also, because the C_{15+} aromatic hydrocarbon fraction was being altered more extensively than the C_{15+} saturated hydrocarbon fraction, water washing, rather than biodegradation, was proposed as the major agent of oil alteration.

In support of this idea, sulfur-bearing aromatics are not as readily degraded by bacteria as are aromatic hydrocarbons (e.g., Connan, 1981), especially in comparison with *n*-paraffins. Loss of dibenzothiophene relative to phenanthrene in oils not depleted in *n*-paraffins may be a useful indicator of water washing. Unfortunately, not all oils contain sufficient quantities of dibenzothiophenes to allow monitoring of such changes; therefore, this parameter cannot be used universally. Also, Connan (1981) postulated that anaerobic sulfate-reducing bacteria caused the preferential removal of aromatic hydrocarbons (and even the sulfur-bearing aromatics) in asphalts of the south Aquitaine basin in France. Thus, the loss of aromatic hydrocarbons and dibenzothiophenes could be due to special cases of biodegradation rather than to water washing. Therefore, changes in the gasoline-range hydrocarbons remain best suited for studies of water washing. Loss of benzene and toluene from the gasoline-range hydrocarbon fraction appears to be the most useful parameter.

The isotopic composition of crude oils can be altered by water washing (Sofer, 1984). In the case of the Philippine oils, the saturated fraction (especially the naphthenes) became depleted in ^{13}C, resulting in an isotopically light oil (with more negative values). The aromatic fraction, although chemically altered, showed little change in isotopic composition (Palmer, 1984)

SUMMARY

Processes that alter the composition of a crude oil can have a direct bearing on the commercial value of an oil field. Therefore, the petroleum geochemistry community has made a major effort to understand the occurrences and causes of biodegradation and water washing. Biodegradation generally occurs in relatively shallow, cool reservoirs that are charged with oxygenated water. Such conditions allow the growth of aerobic bacteria. Both case history studies of biodegraded crude oils and laboratory-induced changes in oil composition using bacteria have shown that aerobic bacteria are the major agents of oil degradation. Special cases exist in which anaerobic bacteria are the suspected agents of degradation. However, the present understanding is that anaerobic

Figure 5. *(Facing page)* (Top) List of identified triterpanes. (Bottom) List of indentified steranes. Assumes $8\beta(H),9\alpha(H),14\alpha(H),17\alpha(H)$ unless otherwise stated; dia = rearranged.

LIST OF IDENTIFIED TRITERPANES

Peak Designation	Molecular Formula	Molecular Weight	Triterpane Identification
A	$C_{27}H_{46}$	370	18α(H)-22, 29, 30-trisnorhopane
B	$C_{27}H_{46}$	370	17α(H)-22, 29, 30-trisnorhopane
BP	$C_{28}H_{48}$	384	17α(H), 18α(H), 21β(H)-28, 30-bisnorhopane
DM	$C_{29}H_{50}$	398	demethylated hopane at A/B ring
C	$C_{29}H_{50}$	398	17α(H), 21β(H)-30-norhopane
D	$C_{29}H_{50}$	398	17β(H), 21α(H)-30-normoretane
OL	$C_{30}H_{52}$	412	Oleanane
E	$C_{30}H_{52}$	412	17α(H), 21β(H)-hopane
F	$C_{30}H_{52}$	412	17β(H), 21α(H)-moretane
G	$C_{31}H_{54}$	426	17α(H), 21β(H)-30-homohopane (22S)
H	$C_{31}H_{54}$	426	17α(H), 21β(H)-30-homohopane (22R)
I	$C_{30}H_{52}$	412	gammacerane
J	$C_{31}H_{54}$	426	17β(H), 21α(H)-30-homomoretane
K	$C_{32}H_{56}$	440	17α(H), 21β(H)-30, 31-bishomohopane (22S)
L	$C_{32}H_{56}$	440	17α(H), 21β(H)-30, 31-bishomohopane (22R)
M	$C_{32}H_{56}$	440	17β(H), 21α(H)-30, 31-bishomomoretane
N	$C_{33}H_{58}$	454	17α(H), 21β(H)-30, 31, 32-trishomohopane (22S)
O	$C_{33}H_{58}$	454	17α(H), 21β(H)-30, 31, 32-trishomohopane (22R)
Q	$C_{34}H_{60}$	468	17α(H), 21β(H)-30, 31, 32, 33-tetrakishomohopane (22S)
R	$C_{34}H_{60}$	468	17α(H), 21β(H)-30, 31, 32, 33-tetrakishomohopane (22R)
S	$C_{35}H_{62}$	482	17α(H), 21β(H)-30, 31, 32, 33, 34-pentakishomohopane (22S)
T	$C_{35}H_{62}$	482	17α(H), 21β(H)-30, 31, 32, 33, 34-pentakishomohopane (22R)

LIST OF IDENTIFIED STERANES

Peak	Molecular Formula	Molecular Weight	Sterane Identification
1	$C_{27}H_{48}$	372	13β(H), 17α(H) diacholestane (20S)
2	$C_{27}H_{48}$	372	13β(H), 17α(H) diacholestane (20R)
3	$C_{27}H_{48}$	372	13α(H), 17β(H) diacholestane (20S)
4	$C_{27}H_{48}$	372	13α(H), 17β(H) diacholestane (20R) +
	$C_{28}H_{50}$	386	13β(H) 17α(H)diaergostane (20S)
5	$C_{28}H_{50}$	386	Rearranged C_{28} sterane
6	$C_{28}H_{50}$	386	Rearranged C_{28} sterane
7	$C_{28}H_{50}$	386	13β(H), 17α(H) diaergostane (20R)
8	$C_{27}H_{48}$	372	5α(H) cholestane (20S) + 5β(H) cholestane (20R)
9	$C_{27}H_{48}$	372	5α(H), 14β(H), 17β(H) cholestane (20R) +
	$C_{29}H_{52}$	400	13β(H), 17α(H) diastigmastane (20S)
10	$C_{27}H_{48}$	372	5α(H), 14β(H), 17β(H) cholestane (20S)
11	$C_{27}H_{48}$	372	5α(H) cholestane (20R)
12	$C_{29}H_{52}$	400	Rearranged C_{29} sterane
13	$C_{29}H_{52}$	400	Rearranged C_{29} sterane
14	$C_{29}H_{52}$	400	Rearranged C_{29} sterane
15	$C_{28}H_{50}$	386	5α(H) ergostane (20S)
16	$C_{28}H_{50}$	386	5α(H), 14β(H), 17β(H) ergostane (20R) + 5β(H) ergostane (20R)
17	$C_{28}H_{50}$	386	5α(H), 14β(H), 17β(H) ergostane (20S)
18	$C_{28}H_{50}$	386	5α(H) ergostane (20R)
19	$C_{29}H_{52}$	400	5α(H) stigmastane (20S)
20	$C_{29}H_{52}$	400	5α(H), 14β(H), 17β(H) stigmastane (20R) + 5β(H) stigmastane (20R)
21	$C_{29}H_{52}$	400	5α(H), 14β(H), 17β(H) stigmastane (20S)
22	$C_{29}H_{52}$	400	5α(H) stigmastane (20R)

bacteria can live on the metabolites of aerobic bacteria and cannot degrade crude oil directly.

Biodegradation produces a heavy, low API gravity oil depleted in hydrocarbons and enriched in nonhydrocarbons. The most commonly used indicator of biodegradation is the loss of the *n*-paraffins relative to the branched paraffins, naphthenes, and aromatic hydrocarbons. This is because bacteria consume *n*-paraffins prior to branched paraffins, naphthenes and aromatic hydrocarbons. The loss of *n*-paraffins is readily observed in whole-oil gas chromatograms and in chromatograms of the C_{15+} saturated hydrocarbon fraction.

Petroleum geochemists are also concerned with recognizing the effects of biodegradation on correlation parameters. Bacteria consume hydrocarbons used in correlation (e.g., *n*-paraffins, pristanes, phytanes, steranes, and pentacyclic terpanes) and alter the isotopic composition of an oil. These changes must be recognized in oil geochemical studies when one is determining genetic relationships among oils, proposing source rock types and depositional environments based on oil composition, and determining the thermal maturity of the source rock at the time of oil generation.

Most correlation parameters are based on the saturated fraction. Therefore, a greater emphasis has been placed on understanding how biodegradation affects the saturated fraction than the aromatic fraction. Microbial studies, however, show that the aromatic hydrocarbons can also be biodegraded. More case studies need to be performed to gain an appreciation for bacterial attack on these compounds.

Because water is present during biodegradation, water washing and biodegradation can occur simultaneously. In such cases it is difficult to determine whether biodegradation and/or water washing has altered the oil. Water washing can also occur in deeper reservoirs where lack of oxygen and high temperatures prohibit the growth of aerobic bacteria. Water washing is enhanced at higher temperatures but high salinities cause exsolution of hydrocarbons that would reduce their removal by water.

Although other agents need to be considered, water washing is the suspected agent that degrades the aromatic fraction. This idea is based primarily on hydrocarbon solubility studies; aromatic hydrocarbons are more water soluble than saturated hydrocarbons (*n*-paraffins and naphthenes). Because smaller molecules are more water soluble, water washing is most readily recognized by changes in the composition of the gasoline-range hydrocarbons (less-than-C_{15} fraction). The loss of benzene and toluene relative to less-than-C_{15} paraffins and naphthenes is the most useful parameter for this fraction.

Loss of C_{15+} aromatic hydrocarbons, especially the more water-soluble sulfur-containing components, due to water washing has also been proposed. In cases where evidence for biodegradation of the C_{15+} saturated fraction is lacking, depletion of the C_{15+} aromatic fraction is attributed to water washing. However, in cases of extensive biodegradation, as demonstrated by the destruction of the saturated hydrocarbons including naphthenic steranes and pentacyclic terpanes (hopanes), degradation of the aromatic fraction may also be due to biodegradation.

Chapter 5

Modeling the Maturation and Migration of Petroleum

Alan K. Burnham and Jerry J. Sweeney
Lawrence Livermore National Laboratory
Livermore, California, U.S.A.

INTRODUCTION

The objective of this chapter is to outline the modeling of petroleum maturation and migration to support petroleum exploration efficiency. Chapters by Jacobson, Palciauskas, England, and Palmer have outlined the important phenomena involved in oil generation, migration, and accumulation. These chapters have included conceptual models and, in some cases, quantitative expressions for these processes. While each chapter is useful in its own right, geochemical data become even more valuable when considered together. Ultimately, one can imagine deterministic mathematical models that can make fairly detailed predictions about the occurrence of oil and gas in a basin using seismic data and relatively few exploration wells. Here we give various mathematical descriptions of maturation and migration processes that can be used in deterministic models. Our discussion is weighted toward those approaches that can result in tractable models rather than theoretical rigor. Part of the chapter is devoted to unresolved issues in this evolving field.

KINETIC MODELS OF OIL AND GAS GENERATION

A schematic diagram of oil and gas generation, adapted from Tissot and Welte (1984, p. 94) is shown in Figure 1. To calculate the amount of oil and gas that can potentially migrate to a reservoir, we must have (1) a thermal history of the hydrocarbon-bearing sediments; (2) a knowledge of the quantity and quality of organic matter in the sediments; and (3) kinetic equations to calculate conversion of kerogen to oil and gas, generation of additional gas from residual (refractory) kerogen, and cracking of oil to more gas and refractory kerogen. There are several crucial economic issues to consider. First, is the oil expelled from the source rock before it becomes cracked to gas? Second, is there a good geological structure to which expelled petroleum can migrate and accumulate? Because there appear to be significant differences in the kinetic parameters for different source rocks, which laboratory experiments and data analysis procedures, if any, easily give reliable kinetic parameters?

A prerequisite for all kinetic modeling is a good understanding of the thermal history. Thermal history modeling of a basin requires the complete integration of stratigraphic, structural, and geophysical data. Lithologic and stratigraphic data are used to determine sedimentation rates and the character of the physical properties, such as porosity, permeability, and thermal conductivity, throughout the stratigraphic column. An estimate of the temperature distribution throughout the sediment column through time can be made by decompaction thermal modeling (e.g., Sclater and Christie, 1980). This involves mathematically decompacting the sediments, determining a value of heat flow from the base of the sediment column (which may be time dependent), and calculating changes in the sediment physical properties with time. At least six commercial personal computer programs are available to do this at the current time (BasinMod, Bury, Genex, Matoil, PDI/PC, and YüklerPC). The thermal history model is constrained by the present-day temperature profile in the sediments, by the heat flow at the surface, by the history of the surface temperature, and by the present-day state of thermal indicators within the sediment column. Vitrinite reflectance, the state of apatite fission tract annealing, clay mineral composition, and ratios of specific molecular types all depend on the sediment thermal history and can be used as geothermometers. Factors that complicate the modeling are periods of erosion or nondeposition which may be hard to detect stratigraphically and difficult to quantify. Thermal indicators can be used to constrain thermal histories, but it must be realized that the end state of a single

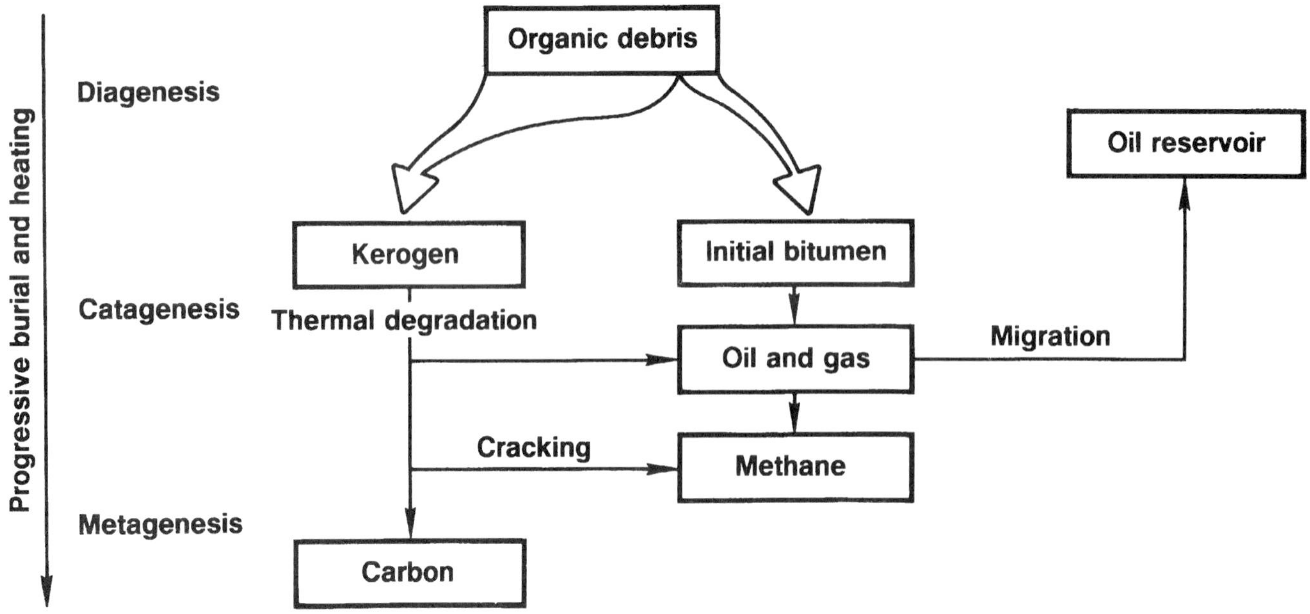

Figure 1. Schematic diagram of the mechanism of petroleum generation and destruction. (Adapted from Tissot and Welte, 1984, p. 94.)

thermal marker can be achieved by an infinite number of different thermal histories and that the thermal dependence of some biomarker indicators is currently under dispute (Burnham, 1989).

Numerous authors have attempted to relate the amount of kerogen conversion quantitatively to time and temperature, which is precisely the aim of the discipline of chemical kinetics. Although directed principally toward oil generation, the following discussion on chemical kinetics also applies to gas generation. The simplest form of chemical kinetics is a first order reaction, in which the instantaneous rate of conversion of component X to component Y is proportional to the amount of reactant:

$$-dX / dt = dY / dt = k(T)X \tag{1}$$

For the special case of constant temperature (T), equation (1) predicts an exponential decrease in X and a corresponding increase in Y. Integration of equation (1) for other than constant T requires a description of the temperature dependence of the proportionality (rate) "constant" $k(T)$. For more than 100 years, workers have used the Arrhenius equation,

$$k(T) = Ae^{-E/RT} \tag{2}$$

where A is the frequency (or preexponential) factor and E is the activation energy. The parameters A and E are independent, so both must be known in order to calculate the amount of reaction for any thermal history. However, for a given amount of reaction at a certain temperature and time, A constrains E and vice versa, so frequently only E is quoted with the implicit constraint that the reaction must occur in a given time frame at either laboratory or geological temperatures.

There has been a long-standing argument over the appropriate value for the activation energy of petroleum generation. Historically, workers who analyzed geological data have obtained "low" activation energies, typically 15 kcal/mol, while laboratory experiments have yielded values typically between 40 and 50 kcal/mol. This discrepancy, along with the difference in composition between laboratory-generated oils and petroleum, caused many to claim incorrectly that petroleum was generated by a distinctly different process than in the laboratory. A value of about 50 kcal/mol is consistent with the relative time scales in lab experiments and nature. It is now clear that the early discrepancies were largely caused by neglecting reactivity distributions, which can cause the apparent activation energy to be substantially lower than the average true value.

Workers at several locations using diverse pyrolysis techniques have shown that it is possible to derive Arrhenius constants from laboratory experiments that extrapolate qualitatively to geological conditions (e.g., Lewan, 1985; Tissot et al., 1987; Sweeney et al., 1987). However, no theoretical argument exists that can either prove or disprove that these kinetic parameters are valid at geological temperatures. The validity of any method must be demonstrated empirically, as

Table 1. Arrhenius rate parameters for the three generic kerogen types for a simple model of oil generation and cracking.

Primary Generation	Type I	Type II	Type III
Potential (mg/g TOC)			
Oil	650	350	50
Gas	70	65	110
Value of *A*			
per second	1.6×10^{13}	3.0×10^{13}	1.6×10^{13}
per m.y	5.1×10^{26}	9.5×10^{26}	5.1×10^{26}
Fraction of Total Potential at Given Value of *E*			
Value of *E* (kcal/mol)			
48	—	—	0.04
49	0.07	0.05	—
50	—	0.20	0.14
51	—	0.50	—
52	—	0.20	0.32
53	0.90	0.05	—
54	0.03	—	0.17
56	—	—	0.13
60	—	—	0.10
64	—	—	0.07
68	—	—	0.03
$T_{50\%}$ at 3°C/m.y. (°C)	145	132	158

Oil Cracking

$A = 1.0 \times 10^{12}$/s $= 3.2 \times 10^{25}$/m.y.
$E = 54$ kcal/mol
Gas fraction = 50%

discussed in the next paragraph. Two questions are still under investigation. What is the easiest, most reliable method for measuring generation kinetics? Also, how do the kinetic parameters relate to kerogen composition? A major goal is to understand the relationship between kinetics and kerogen type so that only a few simple measurements need to be made to estimate reliable kinetic parameters.

Although a variety of techniques are used, there are currently two principal schools of thought on how to measure oil generation kinetics: the open system (e.g., Rock-Eval) and hydrous pyrolysis. The open system typically uses three or so heating rates ranging from a few tenths of a degree Celcius per minute to a few tens of degrees. The data are usually analyzed using an activation energy distribution model, either Gaussian or discrete (Burnham et al., 1987; Ungerer and Pelet, 1987). This assumes that the diversity of reactions involved in oil generation can be represented by a set of independent, parallel reactions with different *E* values but sharing a common frequency factor. A problem with this technique is that the liquid product has many more polar and unsaturated compounds than crude oil, so what process are the kinetic parameters describing? Although the pyrolysate is in some ways more similar to source rock bitumen than to crude oil, open system pyrolysis does involve breaking carbon–carbon bonds as in petroleum formation, and the empirical evidence is that kinetics from this experiment give a good indication of the timing of petroleum generation (Tissot et al., 1987). Hydrous pyrolysis of source rock chips generates an expelled oil distinct from bitumen just as in nature, and the expelled oil is quite similar to the natural product. Although published kinetic parameters derived from the quantity of expelled oil are also reasonable (Lewan, 1985), the method is time intensive and there are questions about the influence of reactivity distributions and mass transport effects on the rate constants (Burnham et al., 1989).

Another issue is whether the kinetic measurements should be made on whole source rock or on isolated kerogens. Adsorption of pyrolysates on minerals and catalytic reactions might influence kinetic parameters. To the extent that these effects occur in nature, they may be desirable to include in the kinetics. However, in open system pyrolysis, it is not obvious that the effects would be the same as in nature. This issue has not been studied enough to be understood, but preliminary evidence indicates that open system pyrolysis sometimes gives significantly different kinetic results for kerogens and whole rocks, especially for lean source rocks.

Several trends are beginning to emerge concerning the relationship between kinetic parameters and kerogen type. Most of the reactivity of well-preserved lacustrine kerogens can be described by a single activation energy between 50 and 55 kcal/mol. Part of this range is attributable to differences in temperature calibration in different laboratories and part is due to variability in kerogens. Marine kerogens have a similar mean activation energy, but require a distribution of ±1–2 kcal/mol about that mean. High-sulfur content in kerogen has been implicated for causing oil formation at lower temperatures (M. Lewan, unpublished work, 1989; Jarvie, 1990), but the relationship to activation energy is in dispute. Terrestrial kerogens are even more complex because of their high-oxygen content (Burnham et al., 1989). Only high-pressure laboratory experiments simulate the early stages of coalification (Monthioux et al., 1985).

There are many kinetic models of varying complexity in the literature. Table 1 gives kinetic parameters for generic kerogen types for the simple model:

$$\text{Kerogen} \rightarrow \text{Oil} + \text{gas} + \text{residue}$$
$$\text{Oil} \rightarrow \text{Gas} + \text{residue}$$

These parameters are derived from inspection of results from three groups (Tissot et al., 1987; Mackenzie and

Quigley, 1988; Burnham et al., 1987; Burnham and Braun, 1989). In reality, gas generated from residual kerogen continues after oil potential has been exhausted. This higher temperature gas comes from elimination of methyl groups from the predominantly aromatic structure of the residual kerogen. Although a separate rate expression is preferred for this source, it has been included in the kinetic expressions for primary generation in Table 1 for simplicity. For type III kerogen, the hydrogen index contributions at 48–54 kcal/mol represent a mixture of oil and gas, but the contributions for 56–68 kcal/mol represent gas generated exclusively from residual (refractory) kerogen. This limitation is not present in the current IFP approach (Espitalié, 1988), which uses separate kinetic expressions for methane, wet gas, light oil, and heavy oil. More complex models are becoming available in commercial basin analysis programs, but are beyond the scope of this chapter.

The primary oil potential and its relationship to the hydrogen index is complex and controversial. Rock-Eval pyrolysis produces much heavy, polar material that is more similar to early mature bitumen than mature petroleum. Hydrous pyrolysis has shown that the maximum quantity of expelled oil is substantially lower than either the peak bitumen yield or the hydrogen index (Lewan, 1985). The potentials in Table 1 include only the hydrogen-rich components that can be expelled as petroleum.

Gas generation can also occur by cracking of oil that has not migrated to a cooler location. Elemental balance requires that roughly equal masses of gas and carbon residue are formed ($2\ CH_2 \rightarrow CH_4 + C$). More generally, oil cracking also has a distribution of reactivity. Some workers have used an activation energy distribution for this (Mackenzie and Quigley, 1988), but using two or three oil components with reactivity increasing with molecular weight spreads the reaction over a temperature range equally well and provides the possibility of predicting the gravity of the oil (Braun and Burnham, unpublished work, 1988). The oil cracking parameters predict that oil is converted to coke and gas between 160° and 210°C (320° and 410°F) in the subsurface, depending on time. Paraffinic oils appear to be more stable than aromatic oil by 10–20°C (18–36°F) (Henderson and Weber, 1985; Ungerer et al., 1988a; Bjoröy et al., 1988). The parameters in Table 1 are qualitatively consistent with the consensus view of oil stability, although some evidence exists (Dominé, 1989) that inhibition of cracking by high pressures may stabilize oil at higher temperatures (greater depths).

Once the kinetic parameters and a thermal history have been determined, the extent of reaction is determined by integrating equation (1) over the thermal history. For an activation energy distribution model, a parallel calculation is done for each component and the results added. There are analytical solutions for equation (1) for the cases of constant temperature and constant heating rate. The latter is closer to most sedimentary histories, but there are usually changes in burial rate and heat flow as well as uplift and erosion that necessitate a more flexible approach. Integration of equation (1) for the more general case is commonly done by breaking the thermal history into a series of isothermal segments. The extent of reaction is then given by the product of the fraction reacted during each segment:

$$X / X_0 = \prod_j \exp\left[-\left(\frac{k_j + k_{j-1}}{2}\right)(t_j - t_{j-1})\right]$$
$$= \exp\left[-\sum_j \left(\frac{k_j + k_{j-1}}{2}\right)(t_j - t_{j-1})\right] \qquad (3)$$

where j is the time segment index. As an aside, the exponent of the second equality is related to the Arrhenius time–temperature index (TTI) approach given by Wood (1988). If the time step size used in equation (3) corresponds to temperature steps greater than about 3°C (5.4°F), or about 100 m (330 ft) of sediment, the integration error becomes unacceptably large. It is even more important to keep the temperature step small if both generation and cracking are considered. As an example of this type of calculation, measured hydrocarbon concentrations in the Uinta basin are compared in Figure 2 to an Arrhenius model that is more detailed than that described in the preceding paragraphs (modified from Sweeney et al., 1987), enabling it to calculate product composition as well as occurrence.

It is instructive to compare the Arrhenius approach to the Lopatin method, otherwise known as TTI, popularized by Waples (1980). TTI is based on the rule of thumb that the rate of a chemical reaction approximately doubles for every 10°C (18°F) rise in temperature. Waples defines the TTI as

$$\mathrm{TTI} = \sum \Delta t_n 2^n \qquad (4)$$

where Δt_n is the time required for the temperature to increase by 10°C (18°F) in the nth interval. The value of n is defined as zero for the interval 100–110°C (212–248°F). Royden et al. (1983) define a continuous function for the TTI as

$$\mathrm{TTI} = \int 2^{[T(t) - 105^\circ\mathrm{C}]/10}\, dt \qquad (5)$$

which gives equivalent results. The TTI calculation is

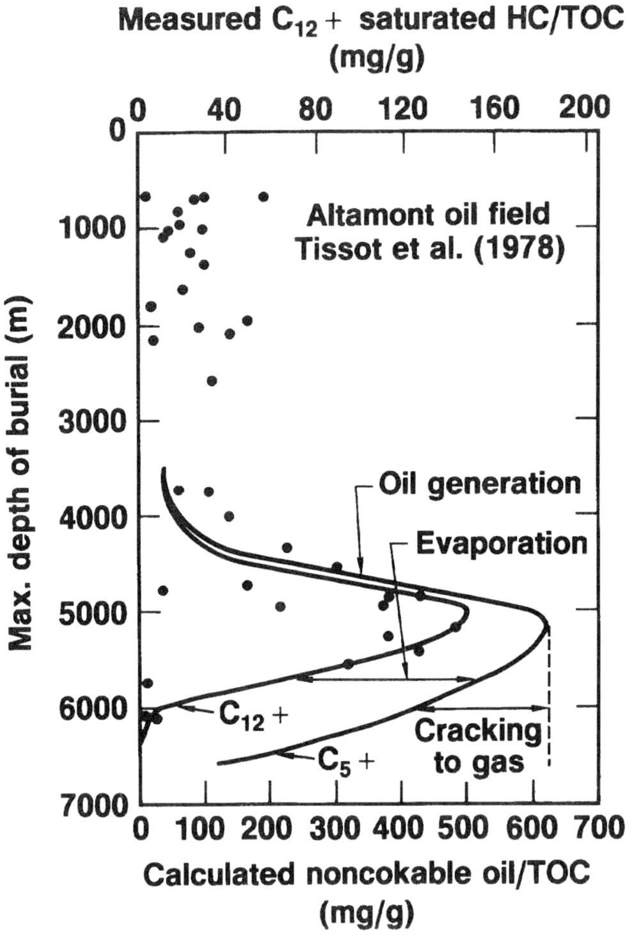

Figure 2. Comparison of measured concentrations of C_{12+} saturated hydrocarbons in the Uinta basin, Utah, with concentrations of uncokable oil calculated by the computer code PYROL. (Modified from Sweeney et al., 1987.)

equivalent to using the Arrhenius equation with an activation energy that increases with temperature. Examples of required *E* values are 20 kcal/mol at 110°C (248°F) and 62 kcal/mol at 400°C (752°F). Although it is founded on a flawed premise, the Lopatin method does work to some extent. Because it is calibrated on geological data, it must by design give a qualitative estimate of geological oil formation, especially for average burial histories and thermal gradients. Waples' TTI value guidelines are as follows: onset of oil generation (15), peak oil generation (75), end of oil generation (160), limit of 40° API gravity oil (500), limit of 50° API gravity oil (1000), and limit of wet gas (1500).

Table 2 shows a comparison of the TTI and Arrhenius methods for generic type II kerogen heated at rates spanning those of geological interest. The calculations correspond to steady burial to 7200 m with a geothermal gradient of 25°C/km (1.4°F/100 ft) over times of 1800, 180, and 18 m.y. The net fraction of oil generated was calculated using the Arrhenius parallel reaction model parameters for marine shale and oil-cracking parameters given in Table 1. The decrease in net oil generated at high temperatures is caused by the cracking of oil to gas. Waples' (1980) guidelines for oil generation agree reasonably well with the Arrhenius calculation at 1°C/m.y. but not at 10°C/m.y. or 0.1°C/m.y. Unmodified, the TTI method overestimates maturity for slowly subsiding basins and underestimates maturity for very rapidly subsiding basins. Some of the limitations of TTI could be overcome by making the relationship between TTI and oil generation depend on heating rate or by decreasing the doubling temperature to 4°C or so. However, this does not overcome the limitation that the correlation between TTI and oil and gas generation is empirical, needs to be established from field measurements, and is different for each distinct kerogen type.

Vitrinite reflectance, either directly measured or calculated by correlation to TTI, is the most common maturity indicator. It is important to realize that because the correlation between vitrinite maturity and TTI is empirical, the same errors will be introduced when using TTI calculations to determine vitrinite–depth profiles. Some workers have claimed that while activation energies of 50 kcal/mol are appropriate for oil generation, TTI is still valid for modeling vitrinite reflectance. This is not true. Vitrinite reflectance is related to its chemical composition, and the change in its composition is governed by chemical kinetics similar to those used in Arrhenius models of oil generation. A good compromise between simplicity and accuracy is provided by a recent Arrhenius model of vitrinite reflectance that uses a very broad distribution of activation energies (Sweeney and Burnham, 1990). Examples of vitrinite–depth profiles calculated using the TTI and Arrhenius approaches for slowly and rapidly subsiding basins are shown in Figure 3.

MIGRATION MODELING

Migration is the series of processes by which oil and gas move from where they are formed to the reservoir. Migration is usually separated into two steps: expulsion from the tight source rock into a permeable carrier bed and migration of the expelled oil to a reservoir. Some use the terms *primary* and *secondary migration* for these two processes, while others restrict primary migration to mean movement of the oil within the source rock matrix and expulsion to mean the dividing line between primary and secondary migration.

The first important issue is to determine what phases are involved. Low solubilities of oil in water and the limited supply of compaction water available at typical depths of generation suggest that oil migration

Table 2. Comparison of TTI and Arrhenius calculations of oil generation and destruction from generic marine sediment undergoing steady burial.[a]

	0.1°C/m.y.			1.0°C/m.y.			10°C/m.y.		
Temperature (°C)	Age (Ma)	TTI	Net Oil (%)	Age (Ma)	TTI	Net Oil (%)	Age (Ma)	TTI	Net Oil (%)
20	1800	0.0	0.0	180	0.0	0.0	18	0.0	0.0
30	1700	0.4	0.0	170	0.0	0.0	17	0.0	0.0
40	1600	1.2	0.0	160	0.1	0.0	16	0.0	0.0
50	1500	2.8	0.0	150	0.3	0.0	15	0.0	0.0
60	1400	6.0	0.0	140	0.6	0.0	14	0.1	0.0
70	1300[b]	12.4	0.0	130	1.2	0.0	13	0.1	0.0
80	1200	25.1	0.3	120	2.5	0.0	12	0.3	0.0
90	1100	50.6	2.0	110	5.1	0.2	11	0.5	0.0
100	1000	102	11.7	100	10.2	1.4	10	1.0	0.1
110	900	204	41.5	90	20.4	7.7	9	2.0	0.9
120	800	408	82.7	80	40.8	29.5	8	4.1	4.5
130	700	816	97.2	80	82	69.1	7	8.2	18.7
140	600	1632	95.0	60	163	94.0	6	16.3	51.1
150	500	3264	77.0	50	326	97.3	5	32.6	86.1
160	400	6529	28.5	40	653	88.8	4	65.3	97.3
170	300	13060	0.1	30	1306	59.3	3	131	95.1
180	200	26120	0.0	20	2612	10.6	2	261	81.2
190	100	52200	0.0	10	5220	0.0	1	522	44.3
200	Present	104500	0.0	Present	10450	0.0	Present	1045	4.2

[a]The Arrhenius calculation uses the primary generation parameters for type II kerogen and oil cracking rate parameters in Table 1. The net oil listed is the fraction of total possible that has been generated minus that cracked to gas.
[b]The three lines trace the onset of oil generation, end of oil generation, and limit for wet gas, respectively, according to guidelines of Waples (1980).

probably occurs as a separate phase from water (Jones, 1981; McAuliffe, 1979). In contrast, CO_2 and hydrocarbon gases are soluble in both oil and water (Price, 1979, 1983; McAuliffe, 1979), so they are partitioned between the two phases at generation depths and temperatures. The pressures in most reservoirs are several times lower than in the source rock, thus along the way, some of the gas and possibly some large polar molecules come out of solution into separate phases.

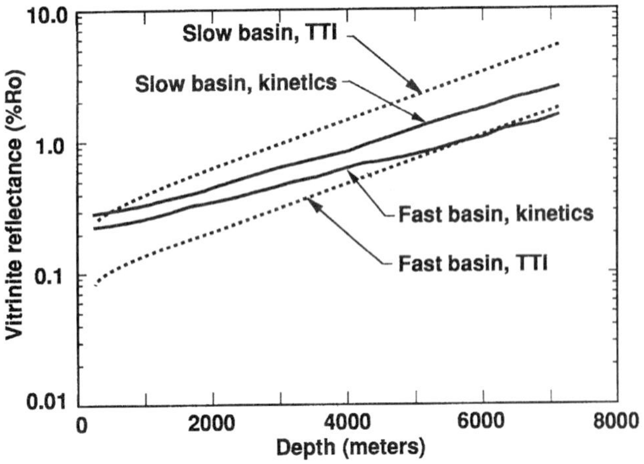

Figure 3. Comparison of vitrinite reflectance calculated by the TTI method (Royden et al., 1983) and Arrhenius method (Sweeney and Burnham, 1990). The slow and fast basins correspond to 0.1° and 10°C/m.y.

For both expulsion and secondary migration, we are faced with problems of multiphase hydrodynamic flow.

Primary Migration and Expulsion

Reexamination of Figure 2 indicates that although the depth of hydrocarbon generation is predicted by the Arrhenius model, the peak concentration of hydrocarbons is about three times smaller than predicted. This difference is due to migration. Other recent measurements (of the amounts and kinds of organic carbon found in source rocks across the oil window) indicate that expulsion of hydrocarbons from rich source rocks is very efficient, especially when they are thin (Cooles et al., 1986; Leythaeuser et al., 1987). Unfortunately, the quantitative estimates have several weaknesses. First, some depend on the assumption that the amount of carbon residue formed geologically is about the same as that formed in rapid, open pyrolysis. Second, the material balance calculations cannot determine whether the material was expelled as heavy oil, light oil, or gas. Finally, direct measurement of residual hydrocarbon content is extremely difficult because most gas and light ends are lost during handling. Therefore, modeling is very important to help estimate the timing of expulsion and composition of expelled fluids. The fundamental issue is not whether hydrocarbons are ultimately expelled from the rock, but the timing of expulsion relative to both trap formation and cracking to gas.

Diffusion and capillary forces contribute to expulsion from thin source beds and on the edges of thick source beds, but overpressure caused by compaction, chemical reactions, and thermal expansion is generally considered to be the principal driving force for expulsion. If the source rock is permeable, the pressure gradient is very small and the volume of water and petroleum expelled is equal to the decrease in pore volume during compaction. If the rock is impermeable, pressure can build to exceed the tensile strength of the rock and fluids can escape through the resultant fractures.

Modeling this process requires knowing, at a minimum, what phases are present and their pressure, volume, and temperature (PVT) behavior. The PVT behavior of each phase depends on its composition. The total composition of all phases can be provided, in principle, by the chemical kinetic modeling described in the preceding section. The amount and composition of each phase and their PVT relations can be calculated from the total composition by either correlations or equations of state (England et al., 1987). The process can be simplified for marine and lacustrine kerogens because the pressures and temperatures are high enough in the region of oil expulsion that only a single hydrocarbon-rich phase and a water-rich phase exist (Hunt, 1979). If the resistance of the source rock to expulsion is considered minor, the amount of expulsion can be estimated merely by calculating, at hydrostatic pressure, the volume in excess of porosity expected for normal compaction. A simple mathematical model using this assumption is given by Ungerer et al. (1988b).

The only remaining problem is to relate, at any particular instant, the relative volumes of petroleum and water expelled to the relative amounts of petroleum and water present. Although this relationship depends on the nature of multiphase flow in porous media, it is generally accepted that petroleum does not migrate until it becomes a continuous phase. Up to that point, only water is expelled. Likewise, water is no longer expelled when it becomes a discontinuous phase. The problem is usually attacked by making the relative permeabilities a function of percent water saturation (Ungerer et al., 1990). Since source rocks are not homogeneous, it may be possible to expel nearly all the oil from organic-rich laminae without displacing any water from barren laminae. It is likely that oil expulsion begins at a lower oil saturation for source rocks than production ceases for reservoir rocks. Momper (1978) gives a widely quoted, though inadequately justified, threshold value for oil expulsion of 0.8 mg/g sediment (15 bbl/acre-ft). Talukdar et al. (1987) gives a much higher value of 200 mg bitumen/g TOC, which corresponds to 10 mg/g rock for an average TOC of 5%. Mackenzie and Quigley (1988) state that 5 mg/g rock is required for "major" oil expulsion to begin.

The ability to create excess volume depends on both organic type and richness (Burnham and Braun, 1990). Lacustrine kerogens generate more excess volume during primary pyrolysis than do marine kerogens, so for the same kerogen content, oil from marine source rocks would tend to undergo more secondary cracking. Likewise, for lean source rocks, primary oil and gas generation does not generate enough additional volume to cause expulsion, so additional volume must be generated by oil cracking. Buoyancy and diffusion may also play a more important role for lean source rocks. Terrestrial kerogens are qualitatively different in that there is far more gas than oil generated from kerogen. In this case, migration of oil by gaseous solution of light ends can be important (Leythaeuser et al., 1989).

If source rock permeability is small, the calculations become more complicated. The total rate of expulsion is proportional to the excess pore pressure times the permeability (Darcy's law). Although calculating the expulsion rate formally requires at least a one-dimensional spatial model, it can be treated by a pseudo one-dimensional model that incorporates permeability and migration distance into a global resistance factor (Braun and Burnham, 1990). The pore pressure builds to achieve a balance between the generation of "excess volume" and the rate of expulsion. The permeability is itself a function of generation and expulsion. The Kozeny–Carman relation is commonly used to relate permeability to porosity. A minority view (McAuliffe, 1979a; Stainforth and Reinders, 1990) is that petroleum does not migrate as a separate phase through conventional porosity. Instead, the petroleum is thought to diffuse through a continuous organic matrix that is formed during the early stages of kerogen breakdown.

If the generation rate is fast enough or the permeability is low enough, the pore pressure will reach the strength of the rock, causing the rock to fracture. Source rock fracturing can be included in pseudo one-dimensional models simply as a pressure relief valve, as shown in Figure 4. As previously discussed, it is necessary to prescribe how the composition of the expelled fluid relates to the composition of the pore fluids. Again, it is normally assumed that the relative permeabilities used in Darcy's law can be expressed as a function of the percent water saturation. The qualitative effect of resistance to expulsion is that petroleum is maintained in the source rock for a longer period of time. This allows more oil to crack to gas. Therefore, oil expulsion efficiency decreases with a decrease in rock permeability and an increase in source rock thickness.

Secondary Migration

Once expelled from its source rock, oil can migrate to a trap. The primary driving force is buoyancy, and a potential hindrance is capillary pressure (Schowalter, 1979). Experiments have shown that migration of a

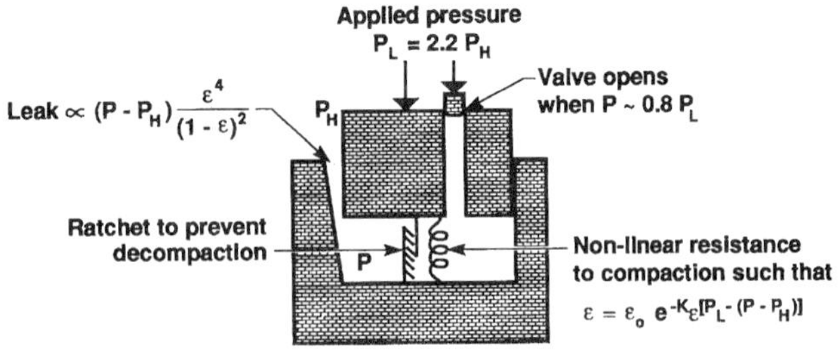

Figure 4. A schematic representation of a simple mechanical model for source rock compaction, fracturing, and oil expulsion. *P* is the total pore pressure, P_H is hydrostatic pressure, P_L is lithostatic pressure, and e is porosity.

continuous oil phase occurs very rapidly compared to generation and occurs through limited conduits (Dembicki and Anderson, 1989). Therefore, movement of oil away from the source rock is potentially a much faster process than generation and probably occurs through relatively narrow fingers through the carrier bed, so that residual saturation in the carrier bed as a whole is relatively small.

The velocity of each phase in multiphase flow is normally considered to be described by a Darcy flow law of the form

$$u_i = -K_i / \mu_i(\text{grad } P_i - \rho_i g) \qquad (6)$$

where u_i is the filtration velocity of phase i, P_i is its pressure, μ_i its viscosity, ρ_i its density, and g the acceleration of gravity. K_i is the effective permeability of the media to phase i, which is normally assumed to be a function of percent saturation, as previously described. *K*, *P*, and *u* are functions of position. Viscosity is a function of temperature and composition. The changes in *K* with depth and lithology are generally assumed to be related to porosity and specific surface area of the rock matrix (e.g., the Kozeny–Carman relation). The difference in the two velocities is proportional to the buoyancy caused by the difference in the two fluid densities. More involved hydrodynamic models include equations for mass balance and conservation of energy.

The difficulty arises if and when oil is not a continuous phase, because capillary forces then become important. If the oil globule is comparable in size to a pore diameter, the buoyancy pressure must exceed the net capillary pressure for the globule to move. The difference in capillary pressure (ΔP) between the oil and water phases is given by

$$\Delta P = 2\gamma \cos\theta / r \qquad (7)$$

where γ is the interfacial tension, θ is the contact angle, and *r* is the pore radius. If the oil globule were large compared to the pore radius, the pore radius were constant, and the rock remained wetted with water, the pore pressures at the top and bottom of the globule would cancel and buoyancy forces would cause the globule to move. However, the pore radius fluctuates with height in a real sediment. Therefore, the oil globule, or slug, must be tall enough that the buoyancy pressure exceeds the difference in pore pressure between the minimum and maximum pore radii in the path of least resistance. Moreover, changes in the surface wetting at the top and bottom of the column could change the contact angle, thereby changing the difference in pore pressure between the top and bottom of the slug. In a sediment where the minimum pore radius decreases with height, the globule would stop migrating when the difference in pore pressure between the top and bottom met or exceeded the buoyancy force. This can produce oil traps that are difficult to detect with seismic techniques. Most migration models greatly oversimplify the capillary effects.

Success in handling secondary migration with the hydrodynamic flow approach depends greatly on the three-dimensional scale of the model, the level of lithologic and structural complexity, and a complete understanding of the physical properties. The models are best applied where the distance from source to trap is short and the properties of the fluids and media are well known.

Attempts have been made to model hydrodynamic flow at the scale of an entire basin, or at least on the scale of several tens of kilometers (e.g., Welte and Yükler, 1984; Ungerer et al., 1988c, 1990; Lehner et al., 1987). Most of these are essentially two-dimensional models that attempt to explain cross-sectional observations of hydrocarbon occurrences by coupling thermal history models and hydrocarbon generation models with the hydrodynamic compaction–fluid flow models. Because of large uncertainties in knowledge about the detailed physical properties of the lithologic media in present-day configurations of basins, extrapolations into the geological past are particularly difficult. Some aspects of modeling are hindered by a lack of knowledge about fundamental physical properties, such as how the permeability of shale changes with compaction or with overpressuring. Also missing is an understanding of how temperature and acidic components might alter rock permeability and porosity, possibly on a basin scale leading to "pressure compartments" (Hunt, 1990). With the models, however, different parameters can be tried and varied to see how they affect the ultimate fate of hydrocarbon accumulation and survival. With modeling one can test the sensitivity of various results to find out which parameters are most important. For example, will the timing of local normal faulting be crucial to whether traps are filled or not? Can the migration be successfully modeled by buoyancy alone or is there a large effect due to compaction-derived fluid flow or a regional flow system? Insights gained from successful application of the model in a well-characterized area can be used to apply the model in less known areas having similar lithologic and structural characteristics.

DISCUSSION

Thermal history modeling and chemical kinetic modeling allow us to estimate the timing of oil generation and expulsion and the amounts and composition of the expelled hydrocarbons. The present state of understanding of secondary migration is such that we can make general estimates of how hydrocarbons are most likely to move in a given structural and stratigraphic framework. At present, however, not enough is known about the interaction of hydrocarbons with the rock matrix within a migration pathway to quantify how much of the hydrocarbons eventually reach a reservoir and the extent to which their composition is modified. In reality, the complexity of the real world may be such that the goal of making complete and accurate mass balances of hydrocarbons within a basin is unachievable. A significant unknown is how much petroleum has seeped to the surface and been lost. However, we must deal with uncertainty in all aspects of exploration, and application of these fundamental principles and techniques can help in assigning probabilities to the variables involved so that we can more effectively rank prospects.

Acknowledgments *This work was performed under the auspices of the U.S Department of Energy by the Lawrence Livermore National Laboratory under contract number W-7405-ENG-48.*

Chapter 6

The Role of Petroleum Geochemistry in Basin Modeling of the Oseberg Area, North Sea

B. Dahl
Norsk Hydro Research Center
Bergen, Norway

A. Yükler
Basin Analysis, Inc.
Carrollton, Texas, U.S.A.

INTRODUCTION

Integrated deterministic basin modeling combining geological, geophysical, geochemical, hydrodynamic, and thermodynamic data is probably the best existing tool for quantitative hydrocarbon predictions. In such an approach, petroleum geochemical data play an important role and provide parameters for model optimization and verification of modeling predictions. In this chapter, the petroleum geochemical aspects of an integrated basin study related to a North Sea case are emphasized.

The modeling software used in this study was an improved deterministic dynamic basin model based on an earlier concept by A. Yükler (Yükler and Welte, 1980). The model can be operated in either a one-, two-, or three-dimensional mode and determines the geological history, paleopressure, paleotemperature and hydrocarbon generation in a sedimentary basin. In a three-dimensional mode, migration and hydrocarbon accumulation locations can be determined. Migration is a volumetric and, thus, a three-dimensional concept. Therefore, modeling of migration in one and two dimensions is only an approximation.

The model describes the physicochemical processes occurring at and after sediment deposition by mathematical terms. The geological processes taking place as a function of time are divided into events, characterized by positive (deposition), zero (nondeposition), or negative (erosion) thicknesses of homogeneous lithologies that occurred during a specific time interval. The sediments deposited are chronostratigraphic units described by lithological composition, present-day porosity, paleobathymetry, and paleotemperatures at the sediment–water interface. The model uses heatflow histories rather than constant geothermal gradients. Heatflow history is determined by iteration until the calculated vitrinite reflectance trend matches the measured trend from a well. This optimization is then checked against 20(*S*)-sterane and 22(*S*)-triterpane isomerization data. Temperature histories are used in a kinetic subroutine modified from Tissot and Espitalié (1975), which calculates the transformation of kerogen to oil and gas.

An organizational scheme for a typical study is outlined in Figure 1. The first step in an integrated basin modeling study is to develop input from a number of different disciplines. Next comes model simulation and optimization. A third stage involves synthesis of model output and regional geochemical maps to yield hydrocarbon volumes generated and expelled as a function of time. Together with structural information, the study at this stage provides predictions of timing and hydrocarbon accumulation locations. Finally, these predictions are verified using actual geochemical reservoir data, if these are available. After this step, iteration back to the simulation and optimization steps is done until a model is developed that adequately explains the data.

OSEBERG AREA: LOCATION AND PETROLEUM GEOLOGY

The Oseberg area described by many authors (Larsen et al., 1981; Badley et al., 1984; Dahl and Speers, 1985; Dahl et al., 1987) is located on the east flank of the

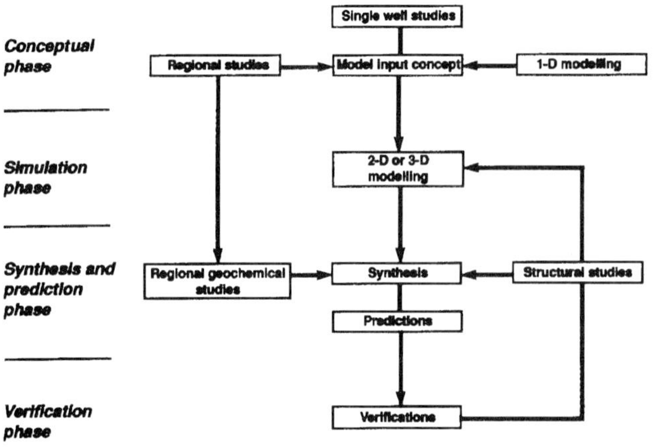

Figure 1. Organization of a typical basin modeling study.

Viking Graben west of Bergen, Norway (Figures 2a and b). The Viking Graben is a north–south trending linear graben that formed during Jurassic time, and its formation is associated with rifting and subsequent sea floor spreading in the central Atlantic (Glennie, 1984). The North Sea basins, including the Viking Graben, started to subside in the Late Cretaceous and continued throughout the Cenozoic. The axis of subsidence closely follows that of the graben systems.

The Oseberg fields consist of several oil- and gas-filled structures located on a large eastward-dipping fault block. The Brage field comprising two oil-filled structures is located east of the Oseberg field on the edge of the Horda platform. The Huldra and Veslefrikk fields are located north of the Oseberg structures. The Huldra field, a gas- and condensate-filled structure, is deeply buried and is located close to the central part of Viking Graben. The Veslefrikk field, which contains mainly oil, is located away from Viking Graben close to the Oseberg Beta South and Beta Sadel structures.

The Middle Jurassic Brent Group sandstones form the main reservoir systems in these fields. However, exceptions exist, and the Upper Triassic Statfjord Formation and the Upper Jurassic Sognefjord Formation are also reservoir rocks in some of the structures, e.g., the Brage field. The truncated Brent sandstones on crestal positions are overlain by Upper Cretaceous shales that are the principal cap rocks over the basinward structures, while organic-rich shales onlap downflank. The structures away from the basin flank, e.g., Brage and Oseberg Beta South, are capped

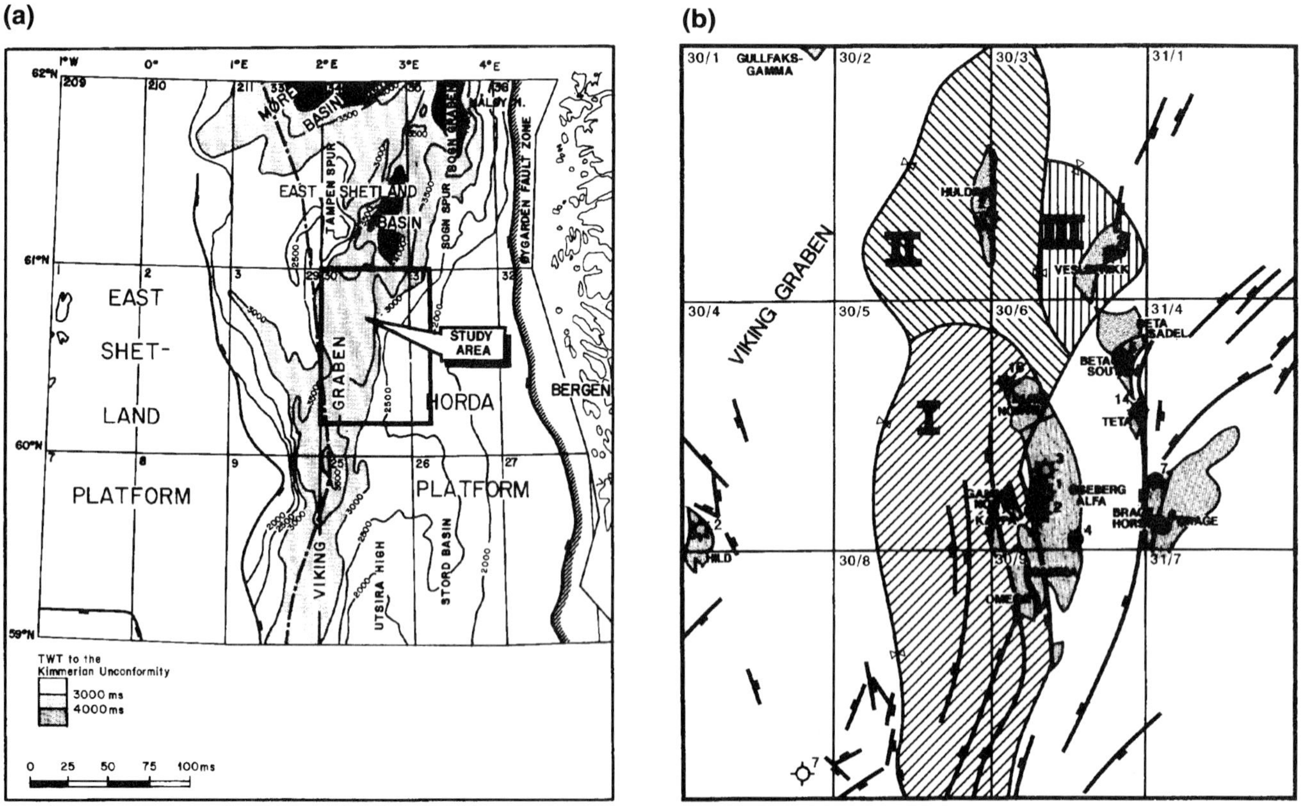

Figure 2. (a) Location of the Oseberg study area. (b) The Oseberg area showing hydrocarbon accumulations and drainage areas.

by the organic-rich Upper Jurassic Viking Group shales. These features seem to have been a controlling factor with respect to migration and accumulation in the area.

MODEL DESCRIPTION

The model input concepts for this study were established from a series of one-dimensional modeling studies and regional stratigraphic assessments. These studies defined 27 isochronous geological events that were identified on the three seismic cross sections that were modeled (Figures 3a and b). The input data are taken from wells and grid points along the profiles. The gridding is selected according to the structural complexity of the cross section. (More faults result in denser gridding.) Two of these east–west cross sections (OSB121 and OSB137) extend from the Viking Graben eastward over the Oseberg Alpha, Alpha North, and Gamma structures. The third cross section (OSB325) extends north–south from the Huldra–Veslefrikk Graben and over the main Oseberg structures.

Figures 4 and 5 summarize the general environmental conditions and lithologies as they relate to geological events over time. The temperatures at the sediment–water interface and the paleowater levels shown in Figure 4 are for the basinal areas. Over the structural areas, the geological framework is different and other values were determined.

Geochemical maturity parameters such as vitrinite reflectance and sterane isomerization played an important role as optimization and checking parameters, respectively, during the determination of heat flow histories in the one-dimensional simulation of the various wells. The model calculates the vitrinite reflectance (% R_o) history for each layer based on a calibrated empirical Lopatin-type kinetic relationship, a subroutine developed by Yükler (1987). The 20(*S*)-sterane isomerization histories are calculated from a first-order kinetic relationship.

$$20(R) \underset{k_2}{\overset{k_1}{\rightleftharpoons}} 20(S) \qquad (1)$$

where $k_n = Ae^{-E/RT}$

and k_n is a rate constant, A is the frequency factor, E is the activation energy, R is the gas constant, and T is the temperature (Kelvin).

An example of results from heat flow optimization against % R_o values (trends) and subsequent checking with sterane isomerization data are shown in Figures 6 and 7. Due to the high sensitivity of the isomerization reactions to small heat flow variations, we were able to determine a small heating pulse of 0.05 HFU for the Eocene. This heating pulse was associated with the Eocene fault reactivation that displaced Paleocene strata in the eastern margin of the Viking Graben. This conclusion is supported by the study of the Tertiary sequences in the same area by Butenko and Østmo (in press). During the simulation studies of single wells, some regional variations in the heat flow pattern were found. Areas closer to the graben axis have experienced a slightly higher heat flux than the platform area. The modeled cross sections were given slightly different heat flow histories along the profile according to the regional variations, as indicated in Figure 4.

(a)

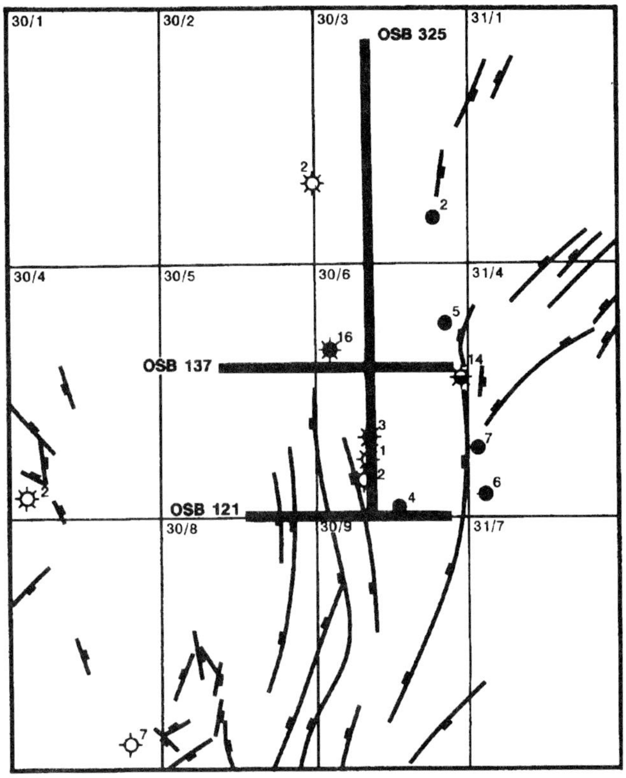

Figure 3. (a) Location of the modeled cross sections in the Oseberg area. *(Figure 3b is on next page.)*

Calibration of Sterane Isomerization Parameters

During the one-dimensional simulation runs and heat flow optimization, the activation energies and frequency factors of the sterane isomerization reactions were calibrated. The calibration was carried out on well data sets of vitrinite reflectance and steranes. After the well was optimized on vitrinite reflectance data, the activation energies and frequency factors were adjusted to match the measured values. Several wells with good quality sterane and vitrinite reflectance data in other parts of the basin were modeled to verify the calibrated sterane isomerization kinetics. For the purpose of comparison, the calibrated kinetic parameters are listed

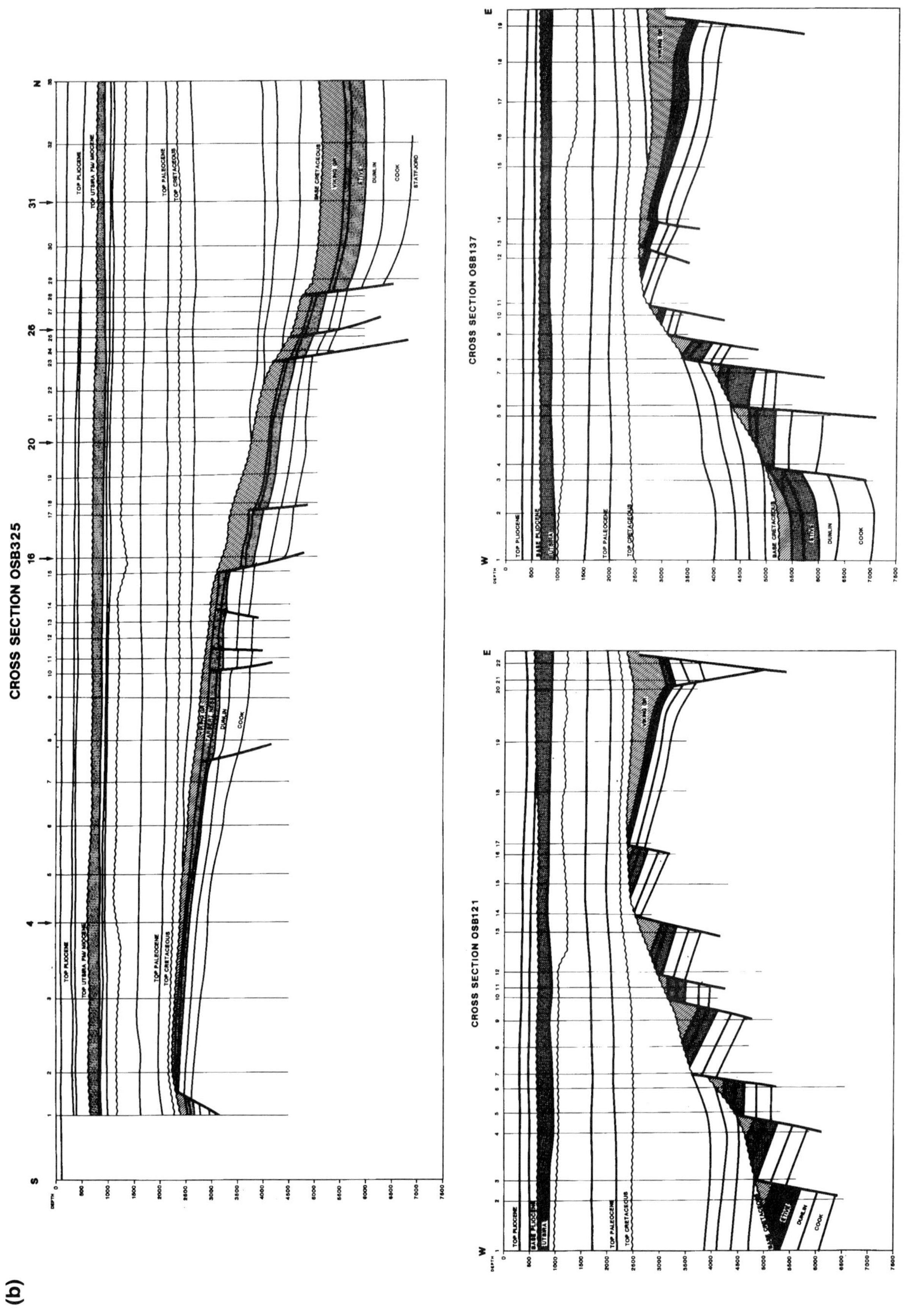

Figure 3. (b) The modeled cross sections with selected pseudowells or grid points.

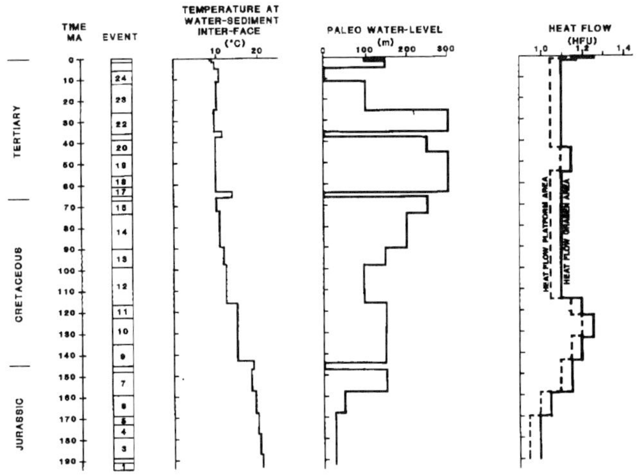

Figure 4. Model input concepts including paleotemperature at the sediment–water interface, paleobathymetry, and heat flow versus time for the basinal area.

in Table 1 together with previously published data.

Calculated sterane values using the same set of input parameters but different kinetic parameters are plotted versus depth after simulation of well 30/6-5 in Figure 8. Because of the higher frequency factor, the parameters of Zumberge et al. (1984) suggest equilibrium at about 1000 m (3300 ft), which is unrealistic and was not included in the plot. Figure 8 shows that even small differences in the sterane isomerization kinetic parameters give large shifts in the computed sterane isomerization trends. It also indicates that laboratory experiments to achieve kinetic parameters can be hazardous when likely experimental errors are considered. In fact, this suggests that laboratory experiments should be carried out only to obtain the approximate parameter values and that field case calibration studies have to be conducted to achieve the "correct" values.

Two-Dimensional Model Simulation and Optimization

Optimization is carried out through iterations until the calculated thickness of the various layers matches that of the depth converted seismic layers. The calculated % R_o, sterane, and temperature data along the profile are checked against the data from the wells at or close to the grid points along the profile. In this study, calculated temperatures at the grid points intersecting the Oseberg reservoirs were always within ~5°C of the measured reservoir temperatures.

Synthesis and Prediction

After basin simulation, the model results must be integrated with other geological and geochemical data to optimize the predictive power of this approach and to suggest locations of trapped petroleum. Petroleum

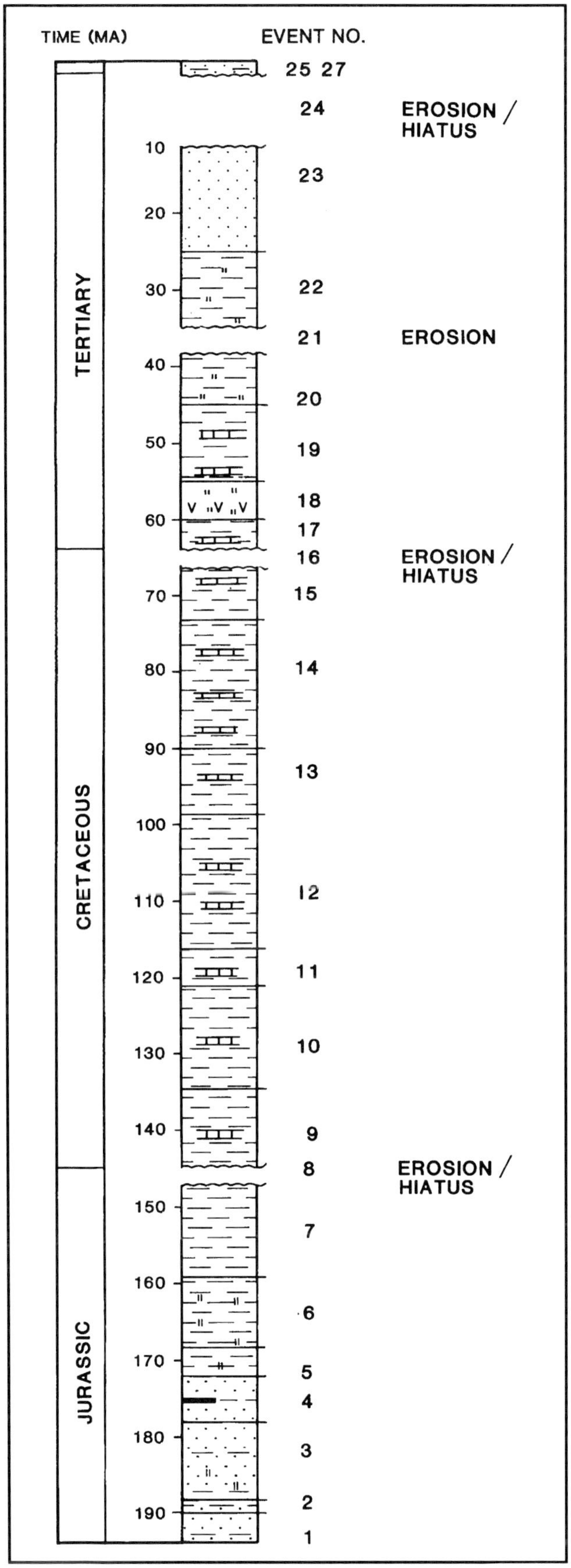

Figure 5. General input concepts for lithologies and events.

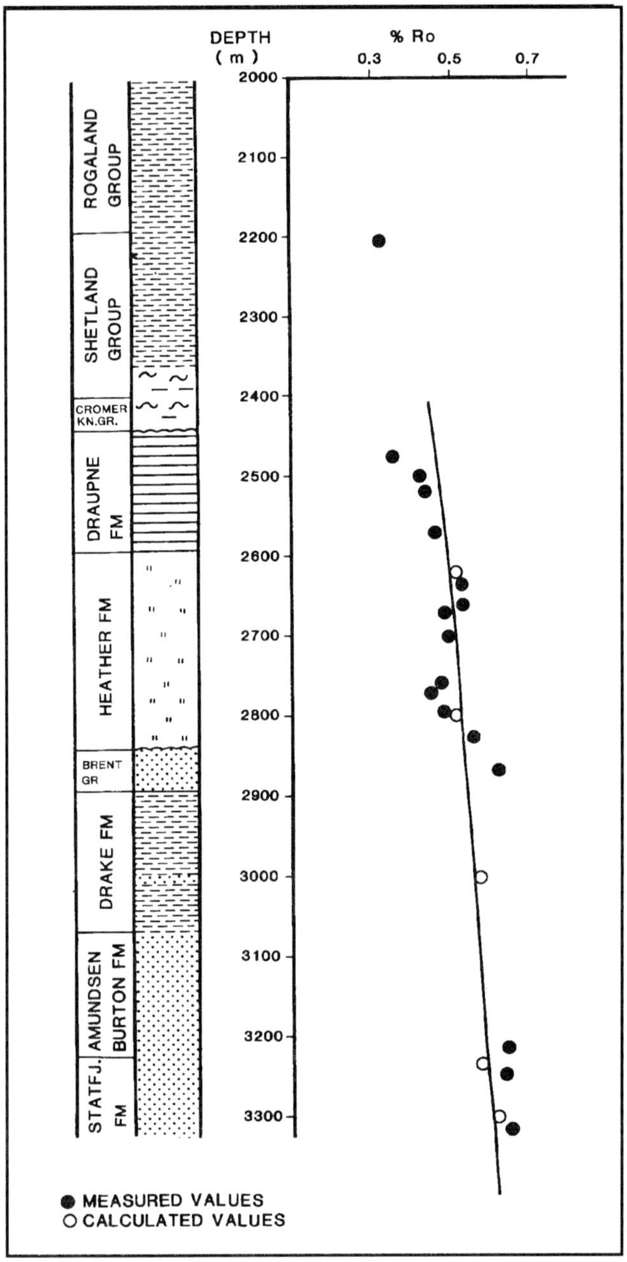

Figure 6. Calculated and measured vitrinite reflectance (% R_o) in the 30/6-5 well.

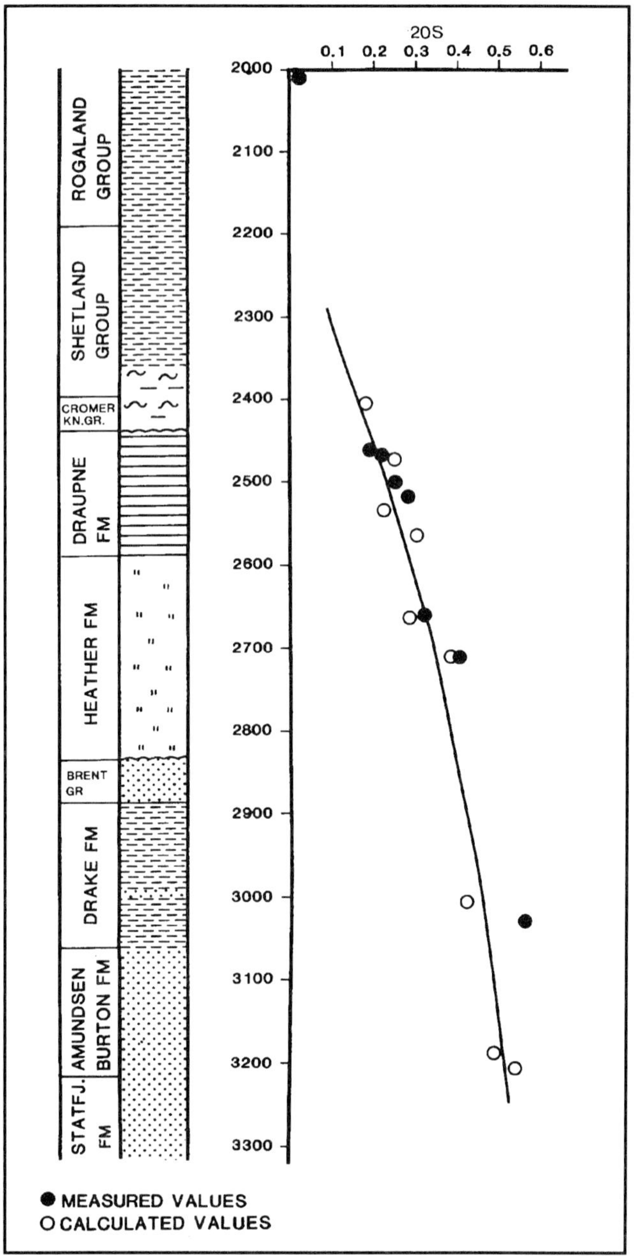

Figure 7. Calculated and measured sterane isomerization [% 20(*S*)] ratios in the 30/6-5 well.

geochemistry plays a significant role at this stage. One of the important aspects is the calculation of generated oil and gas volumes available for (secondary) migration in the drainage area of a given prospect. These calculations are based on regional source rock evaluation of average hydrocarbon potential, average distribution of oil- and gas-generating components, and source rock thickness. In addition, this stage provides the timing of oil and gas volumes generated in relation to trap formation.

SOURCE ROCKS

The major source rock system in the northern North Sea area is the Upper Jurassic–Upper Cretaceous marine shale of the Viking Group (Doré et al., 1985), which is equivalent to the Humber Group in the United Kingdom sector (Glennie, 1984). These fine-grained clastic rocks were deposited during Bathonian to Ryazanian time and include the highly radioactive Draupne Formation (equivalent of the Kimmeridge Clay in the United Kingdom sector), which rests on

Table 1. Calibrated sterane isomerization kinetic parameters in the study area.

	k_1		k_2	
	E (kcal/mol)	A (cycles/sec)	E (kcal/mol)	A (cycles/sec)
This study	21.55	2.90×10^{11}	21.55	2.075×10^{11}
MacKenzie (1982)	21.75	1.89×10^{11}	21.75	2.22×10^{11}
Zumberge (1984)	21.51	3.33×10^{11}	21.51	3.54×10^{11}

shales of the Heather Formation. The Draupne Formation is of late Oxfordian to Ryazanian age. These shales have been described in terms of their source quality, regional variations, and maturity distributions by several authors (Demaison et al., 1983; Barnard et al., 1981; Barnard and Cooper, 1981; Cooper and Barnard, 1984; Cornford et al., 1983; Thomas et al., 1985).

To calculate the volume of hydrocarbon generation, the Viking Group was subdivided into three time-stratigraphic units. This subdivision was based on lithologic composition, source quality and dating as demonstrated at the reference well 30/6-5 drilled on the Beta South structure (Figure 9). Facies 1 is associated with the Draupne Formation and contains mainly marine type II kerogen with some mixed type II and III material in the lower part. Average total organic carbon (TOC) is 6.8 wt. %, residual source potential (S_2) is 29.6 kg HC/MT rock, and hydrogen index (HI) is 430 kg HC/MT organic carbon. Inspection of kerogen slides suggests that this horizon has a considerable amount of land-derived plant material mixed with amorphous material. The woody and inertinitic components are relatively small. Combined pyrolysis gas chromatography indicates a high liquid potential. Gas–oil ratio (GOR) potential based on pyrolysis gas chromatography is less than 0.2. Facies II corresponds to the upper part of the Heather Formation, which is a marine shale that is slightly siltier than the overlying Draupne Formation. The source quality is poorer than the overlying Draupne shales, comprising mixed type II/III kerogen. Average source quality parameters are TOC = 3.6 wt. %, S_2 = 8.1 kg HC/MT rock, and HI = 228 kg HC/MT organic carbon. Composition of the organic residue, based on visual inspection, suggests a higher content of amorphous, woody, and inertinitic components, while algal material decreases. This amorphous material is probably bacterially reworked terrestrial material, as the chemical data do not suggest any increase in source quality.

The average quality distribution of the total Draupne Formation and the two Heather units in the study area has been mapped in terms of TOC and relative distribution of the amounts of type II, III, and IV kerogens. The resulting quality distribution for the Draupne Formation is shown in Figures 10 and 11. Figure 11 is based on average values from a combination of techniques, such as Rock-Eval pyrolysis, pyrolysis gas chro-

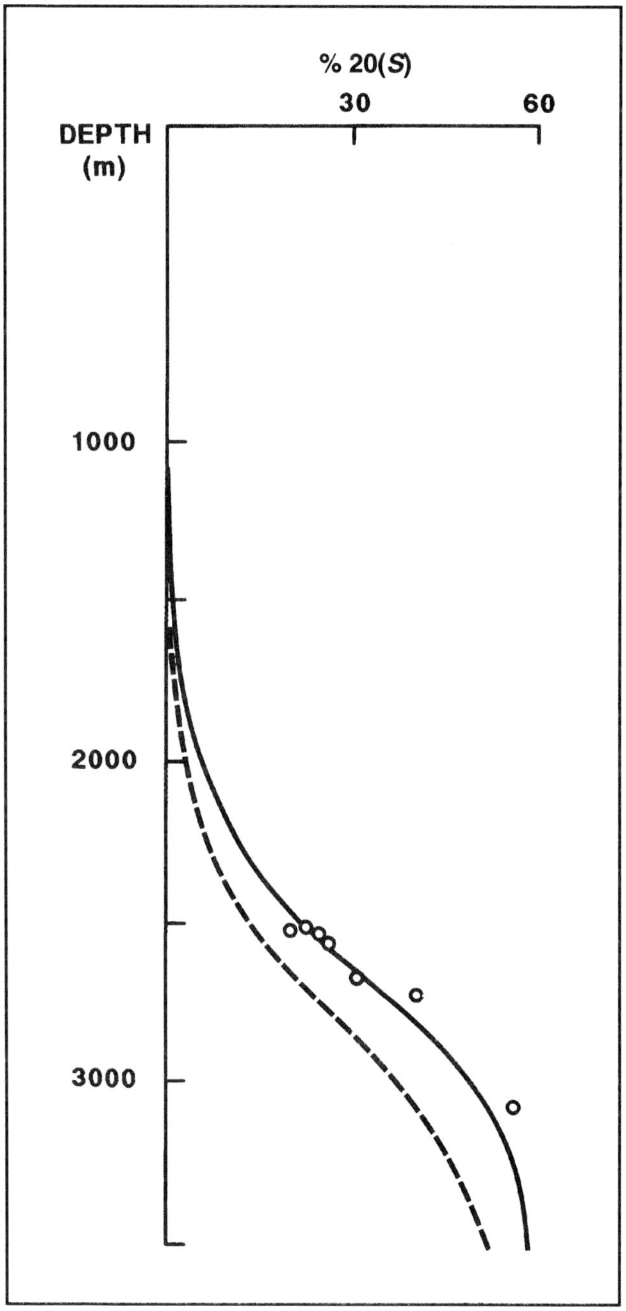

Figure 8. Calculated sterane isomerization ratios using parameters from this study (solid line) and Mackenzie (1982) (dashed line) (see Table 1).

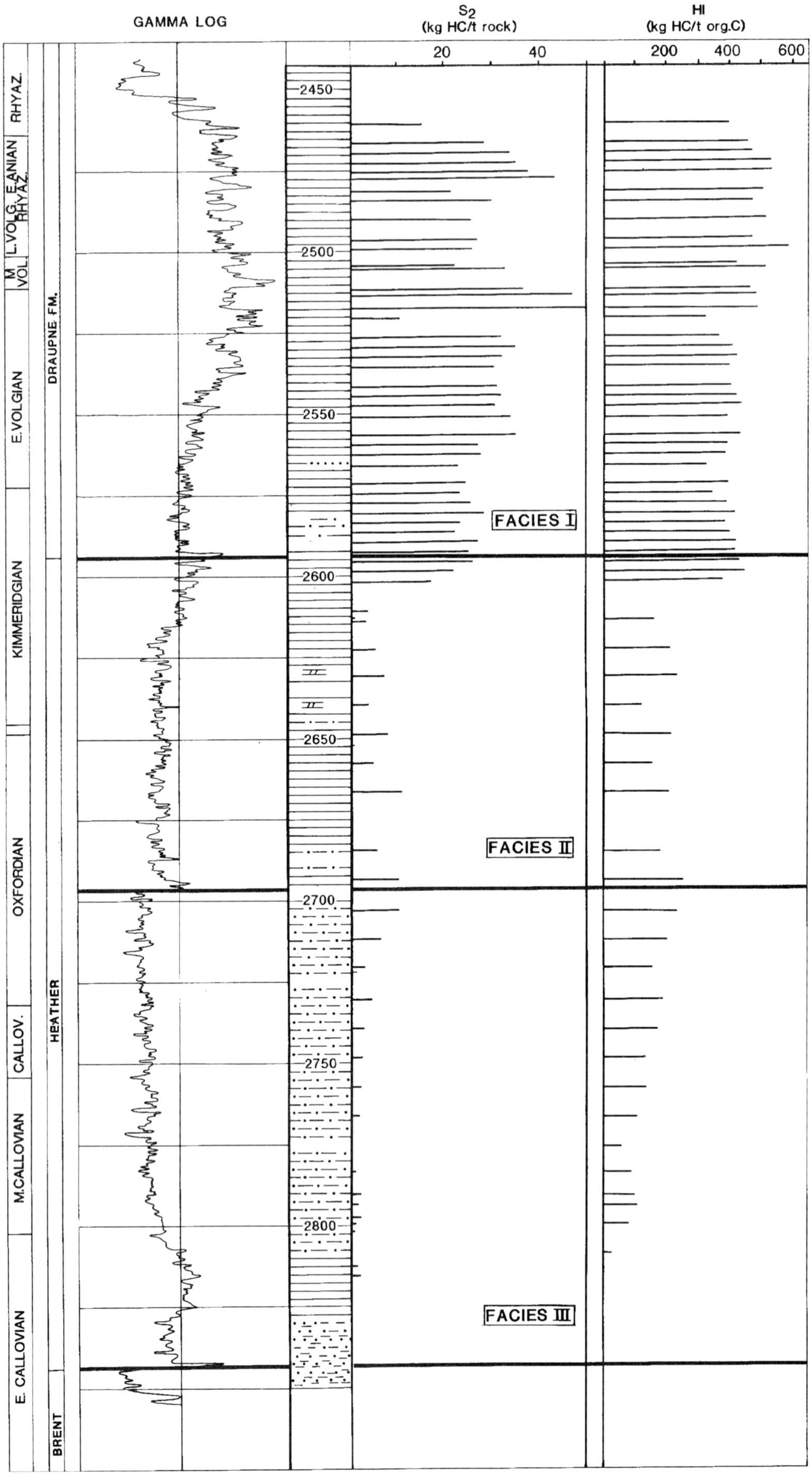
GAMMA LOG
S2 (kg HC/t rock)
HI (kg HC/t org.C)
20
40
200
400
600
RHYAZ.
E.ANIAN RHYAZ.
L.VOLG.
M VOL
E.VOLGIAN
KIMMERIDGIAN
OXFORDIAN
CALLOV.
M.CALLOVIAN
E. CALLOVIAN
DRAUPNE FM.
HEATHER
BRENT
2450
2500
2550
2600
2650
2700
2750
2800
FACIES I
FACIES II
FACIES III

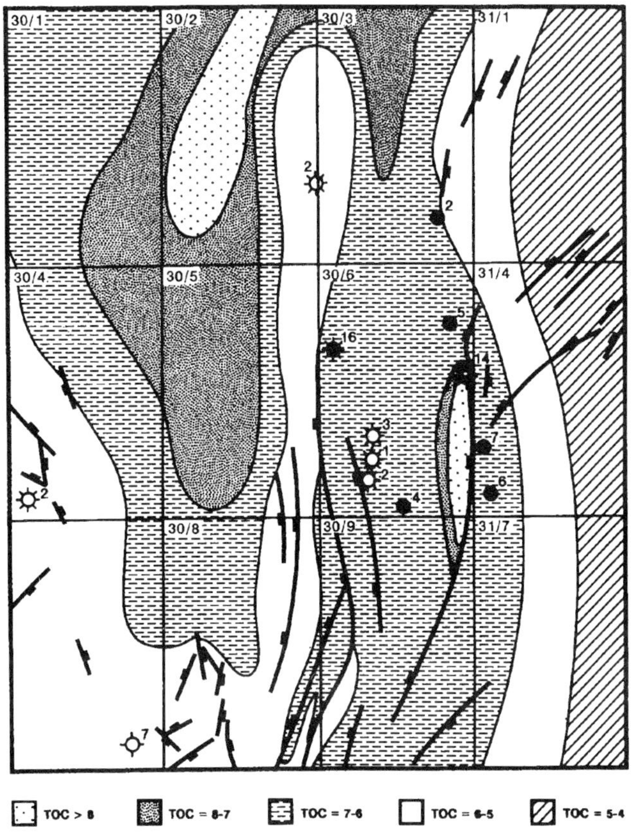

Figure 10. Distribution of the average TOC (%) in the Draupne Formation in the study area.

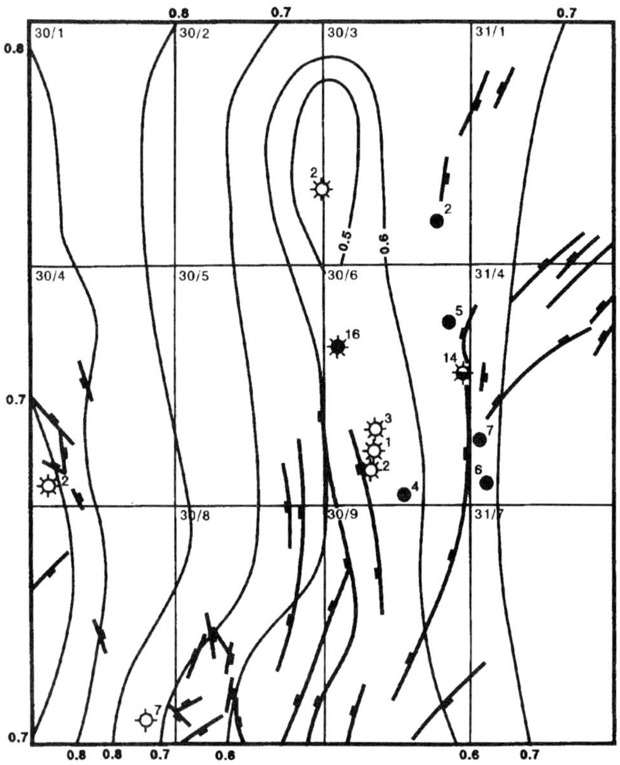

Figure 11. Distribution of the type II (oil-producing) kerogen fraction in the Draupne Formation over the study area.

matography, and visual kerogen typing. To obtain the distributions shown in Figures 10 and 11, extrapolations from areas with well control have been made (e.g., the quality of the Oseberg Back basin has been extrapolated to that part of the Viking Graben believed to be deepest at the time of deposition). Going southward along the axis, the Viking Graben was considered to become shallower (Thomas et al., 1985) and may consequently give rise to lower TOC values, which is in agreement with well data from this area. The maps shown in Figures 10 and 11 disagree with the quality distributions suggested for the Draupne Formation by Barnard et al. (1981) and Cooper and Barnard (1984), in which this area is considered to be dominated by type III kerogen with the possible addition of type IV in the eastern area.

Thickness distributions of Viking Group source rock units (Figure 12) were established by well data (utilizing the subdivision in 30/6-5) and by a number of seismic profiles, as well as the suggestions made by Field (1985). The distribution of the two Heather sections (source facies I and II) are believed to be broadly similar to that of the Draupne Formation, suggesting that the whole Viking Group can be split into three approximately equally thick units in the basinal area. These subdivisions, together with the distribution of the Viking Group siltstones and shales, were later combined with basin modeling results for calculation of hydrocarbon volumes generated in the Oseberg area.

Figure 9. ***(Facing page)*** **The source rock quality trend and the facies subdivision of the Viking Group (Draupne and Heather formations) in well 30/6-5.**

VOLUMES OF GENERATED OIL AND GAS

The volume of oil generated was computed as a function of hydrocarbon generation histories and source rock data using the following formula:

$$HC = [TOC \times F(II) \times TR \times 10 - S_R] \times h \times \gamma \quad (2)$$

where
HC is the generated oil (MT/m^2),

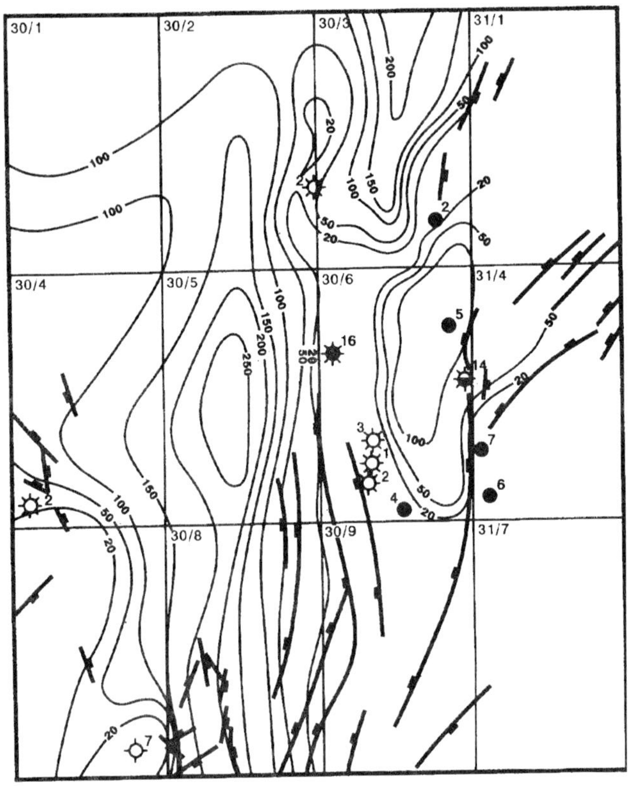

Figure 12. Isopach map (m) of the Draupne Formation in the study area.

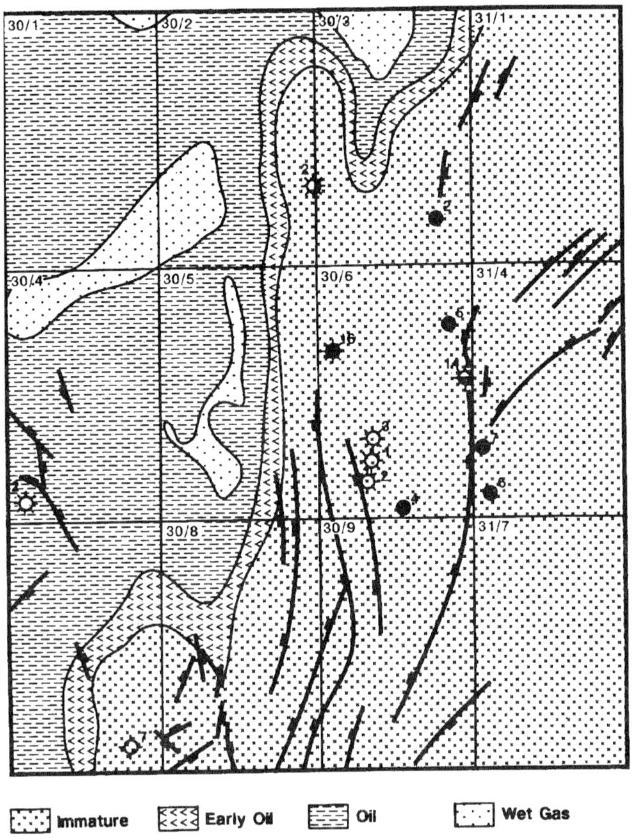

Figure 13. The hydrocarbon generation situation of the Draupne Formation in the study area during Maastrichtian time.

F(II) is the fraction of type II kerogen,
TR is the transformation ratio (kg TOC/MT rock),
h is the source rock thickness (m),
γ is the specific weight of shales (MT/m^3), and
S_R is the retained oil in the source rock (kg/MT).

This relationship is for oil, while a similar one exists for gas. A synthesis to give regional distribution of generated oil and gas volumes required multiplication of a series of maps that represent the regional distribution of the parameters in equation (2) (Figures 10, 11, and 12 showing TOC, F(II), and h, respectively). The transformation ratio (TR) computed by the model output will vary with geological time. Figure 13 illustrates the transformation ratio of oil in the Draupne Formation during the Late Cretaceous, and Figure 14 shows the present hydrocarbon generation. By multiplying the data in these maps, the regional distribution of generated oil volumes in the Draupne Formation was obtained (Figure 15). It can be seen from Figure 15 that particularly large volumes of hydrocarbons have been generated in the Viking Graben close to the Alpha, Gamma, and Alpha North structures and their west flank satellites.

The same exercise has been carried out for the two subunits of the Heather Formation. The total volume of generated oil and gas from the Viking Group shales within the drainage areas of the Oseberg, Brage, Huldra, and Veslefrikk structures (Figure 2b) have been calculated as 9.4 x 10^9 m^3 (3.3 x 10^{11} ft^3) and 1.3 x 10^{12} m^3 (4.5 x 10^{13} ft^3), respectively. Gas has been to be generated from type III kerogen and cracked bitumen retained in the source rock. Maximum retained oil has been set to a fixed value of 5 kg/MT. This value is probably high. Observed values are between 2 and 5 kg/MT and vary with richness and heating rate. The gas contribution from cracked bitumen has been calculated according to the simplified transition of bitumen to methane:

$$C_xH_x \rightarrow CH_4 + C \text{ (residue)}$$

This implies that approximately 50% of the bitumen transfers to gas. Far more hydrocarbons have been generated in the drainage areas than the available trap volumes. The volumes of oil and gas in the various drainage areas have been calculated at different geological times to reveal the generated volumes in relation to

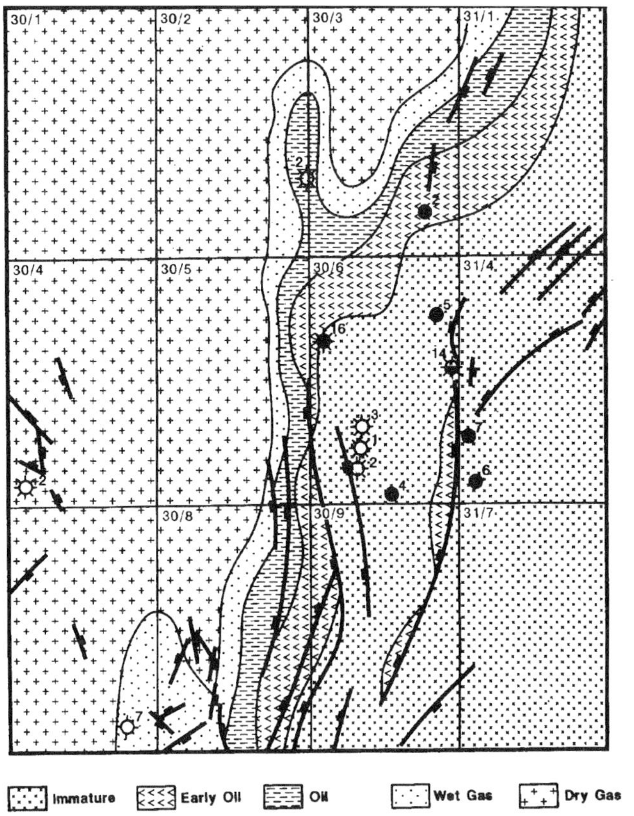

Figure 14. Present-day hydrocarbon generation in the Draupne Formation.

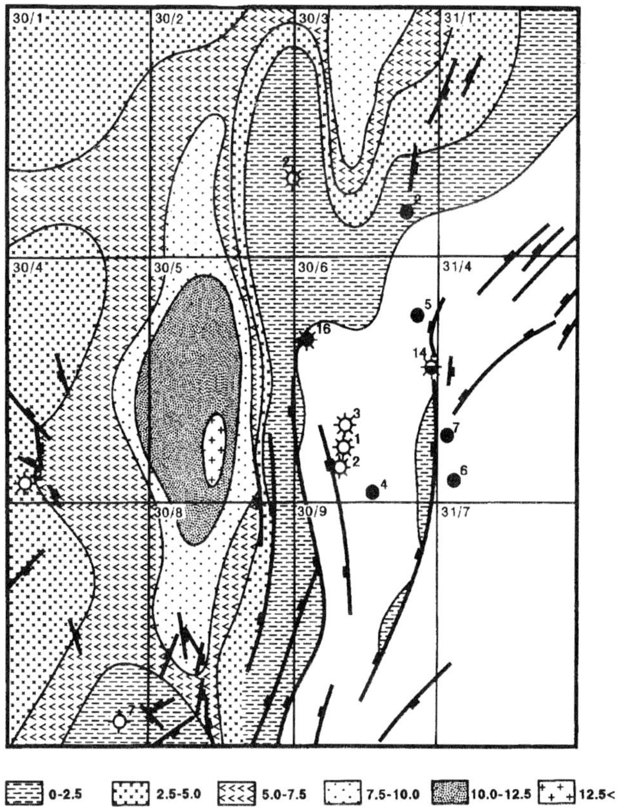

Figure 15. Present-day volumetric distribution (MT/m^2) of total oil generated in the Viking Group.

trap formation and sealing capacity. This is illustrated in Figure 16 for drainage area I and the Oseberg Alpha structure.

Oil generation was very rapid during the Maastrichtian and Paleocene in the Viking Graben. The amounts generated leveled off during the early Eocene due to exhaustion of the source rocks in the deepest part of the Viking Graben. The source area west of the Oseberg Alpha structure and the satellites (Figure 2), however, includes several subbasins that began to generate oil during the Oligocene. This explains the inflection on the oil and the GOR curve in Figure 16. It also indicates that a second phase of oil generation, expulsion, and migration toward the structures together with deeper gas occurred during the Oligocene and early Miocene. Since the mid-Miocene, oil generation has leveled off again, while gas production has started to increase. This situation is not seen in drainage area III (not shown) where the thickest sequences of source rocks reached their major oil production during the Eocene. Although oil has been and still is being generated along the flanks of the basin, no large source rock volumes occur as in drainage area I, which gave rise to a second phase of oil generation.

TIMING OF HYDROCARBON GENERATION

Modeling showed that oil generation and minor expulsion started in the Viking Graben in middle to Late Cretaceous time (Figure 17, grid point 31, 70–75 Ma). The model calculated the conversion of 1 g of ideal type II and type III kerogens with hydrogen indices of 690 and 255, respectively, as a function of geological time for each source rock layer along the simulated profile. Peak oil generation, and even occasional wet gas generation, in the Draupne Formation occurred during late Maastrichtian time throughout the entire graben area (Figure 13). This suggests that oil and particularly wet gas generation occurred in the Viking Graben earlier than proposed by either Goff (1983) or Ungerer et al. (1985). Gas generation from the gas-prone lower Heather Formation started slightly later than oil generation from the Draupne Formation

The model results indicate that during the late Maaastrichtian, there was very little or no cover over a number of the structures in the study area. This is demonstrated for Oseberg Alpha in Figure 16 where top reservoir burial history is shown. Therefore, the

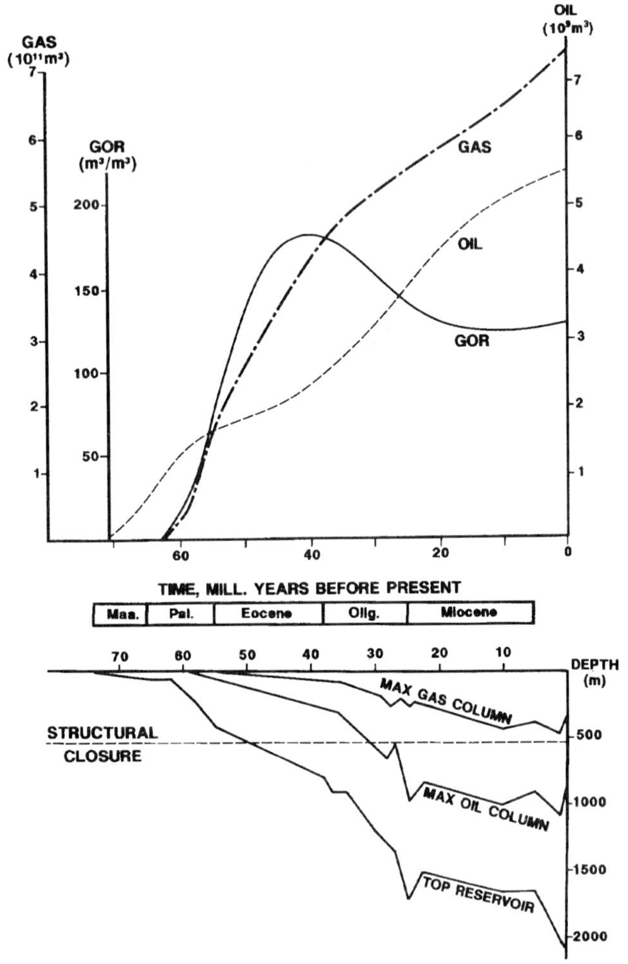

Figure 16. Generated oil and gas volumes, gas–oil ratio (GOR), and cap rock retention capacities for oil and gas as a function of geological time.

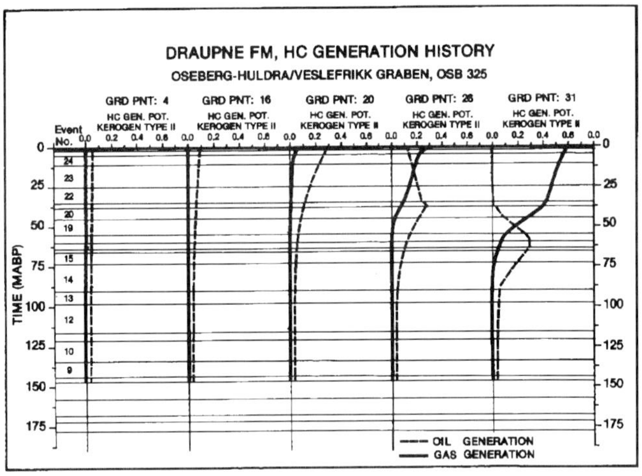

Figure 17. Hydrocarbon generation history of the oil-producing kerogen in the Draupne Formation at five selected grid points of the modeled section OSB325 (Figure 3b), going from the Oseberg Alpha structure and down to the graben between the Huldra and Veslefrikk accumulations.

first phase of migrated hydrocarbons have bypassed uncovered structures, but minor amounts of oil might have been trapped and biodegraded. The Alpha structure was at that time the highest topographical point. The Brage structure to the east was topographically lower and was probably sheltered from hydrocarbon influx from the western drainage area. The only migration fairway to the Brage structure at that time was from the Huldra–Veslefrikk subbasin to the northwest. Along this migration fairway, a number of structures such as Veslefrikk, Beta South, and Theta probably formed obstructions, delaying the influx of hydrocarbons toward the Brage structure. There is little probability that oil charge into the Brage structure occurred during Maastrichtian to Paleocene time.

The computed pressure history shows that abnormal pressures occurred in the Viking Group source rocks simultaneously with oil generation. The pressure was higher in the Viking Group shales than in the underlying Brent sandstones. This would favor migration into these underlying aquifers and further updip along the Brent sandstones where the pressure approaches hydrostatic toward the crest of the structure. However, migration of fluids is a volumetric concept, and consequently, such an event should be simulated by three-dimensional models. Here it is only taken as a brief indication of overpressure build-up as a function of time and depth.

SEALING EFFICIENCY

Two mechanisms for cap rock sealing take place under different geological conditions (Schowalter, 1979; Watts, 1987): membrane sealing and hydraulic sealing. Membrane sealing is controlled by the entry pressure of the largest interconnected pore throat. In hydraulic seals the capillary entry pressure is essentially infinite such that seal failure occurs by development of natural hydraulic fractures. Therefore, seal capacity is thus related to cap rock thickness, minimum effective stress in the cap rock, and degree of overpressure in the system. The two sealing mechanisms fail when the pressure from the hydrocarbon column from buoyancy differences between the hydrocarbons and the formation water exceeds the sealing capacity of the cap rock. In the case of hydraulic seal failure, the resulting heights of oil and gas columns occur according to the following equation (Watts, 1987):

$$T_o(\mathrm{Grad}_w - \mathrm{Grad}_o) + T_g(\mathrm{Grad}_w - \mathrm{Grad}_o) = (\mathrm{Grad}_f - \mathrm{Grad}_o) \times D - P \qquad (3)$$

where
T_o is the height of the oil column,
T_g is the height of the gas column,
$Grad_w$ is the water gradient,
$Grad_o$ is the oil gradient,
$Grad_g$ is the gas gradient,
$Grad_f$ is the cap rock fracture gradient,
D is the depth to the zone of leakage, and
P is the overpressure in the reservoir aquifer.

Field studies have revealed that hydraulic seal failures have been and still are a controlling factor for some of the accumulations in the Oseberg area, e.g., the Oseberg Alpha and Gamma structures. The present-day fracture gradient for the Oseberg Alpha and Gamma structures is 65–70% of the lithostatic pressure. If we assume that this relationship has been constant through geological time, it is possible, together with basin simulation results, to calculate the maximum oil (T_o) and gas (T_g) column heights that the cap rock could seal through geological time. The pressure, temperature, lithostatic pressure, depth to the point of leakage, and gradients for oil and gas (as a function of pressure, temperature, and specific weight) were computed in the model as a function of geological time.

The maximum possible heights of oil and gas columns as a funciton of geological time are plotted in Figure 16 together with the burial curve at the top Oseberg Alpha reservoir. The present-day structural closure (580 m, or 1900 ft) is also shown in the figure. (The structural closure is likely to vary with geological time, and its variations can be quantified with the three-dimensional model which, however, was not used in this study). Figure 16 suggests that gas could never have displaced oil fully beyond the spillpoint and that this structure has probably always contained a considerable oil leg. Furthermore, the first oil generated in the Viking Graben, if it reached the structure, was not trapped. The content of the trap was controlled by the generation, migration, and cap rock leakage until the cap rock was strong enough to hold columns larger than the structural closure. This column was either an oil column or a combined oil and gas column. From then on, the spillpoint together with cap rock leakage were the controlling factors for the trap content. This occurred at 32 Ma for oil. However, considering the large gas volumes generated and focused toward the structure from drainage area I, it is likely that the trap contained free gas and oil when the spillpoint was reached. This is likely to have occurred during the late Oligocene to early Miocene. Similar analysis on structures capped with Upper Jurassic shale (e.g., the Beta South) show a much better trapping potential and indicate that the spillpoints of these structures could have been reached much earlier, basically controlled by the generation and migration.

MIGRATION AND ACCUMULATION

The results of the model and their integration with petroleum geochemical and other geological data are used in a predictive manner to determine migration pathways and location of hydrocarbon accumulations. The results and their implications can be summarized as follows:

1. Because of the large thickness of Viking group shales in the Viking Graben adjacent to the Oseberg structures (drainage area I), significant gas generation and migration from the lower part began shortly after the oil generation commenced in the upper part (Figure 16). Hydrocarbon generation commenced simultaneously with the development of overpressure in the kitchen areas. Overpressure was caused by rapid Late Cretaceous sedimentary loading with subsequent shale undercompaction, combined with hydrocarbon molecular rearrangement and expansion associated with hydrocarbon generation. The thinly covered structures away from the graben remained at normal pressure. During the Late Cretaceous, the Oseberg Alpha, Gamma, and west flank structures had very little trapping potential for hydrocarbons (Figure 16). Gas would have leaked through the thin unconsolidated cap rocks. Oil may have been trapped in the shallow reservoirs or may have reached the surface as seeps. Oil trapped in these reservoirs would have likely been bacterially altered, causing an enrichment of heavy oil components in the reservoirs, such as asphaltenes. Oil and gas have continued to migrate toward the Oseberg structures, which may have disguised the effects of biodegradation.

2. Paleosubsurface topography and pressure distribution suggest that migration may have occurred from several different source areas (Figure 2b). The structures closest to these kitchen areas would have naturally received hydrocarbons first. Structures located away from the graben flanks could probably not acquire hydrocarbons directly, but would have been filled via spill from the flank structures. The Alpha structure and its satellites were probably sourced from the Viking Graben to the west. Spills from these structures may have filled the Brage structures. The Huldra structure could have filled from the Viking Graben to the west and north and from the local depression Huldra–Veslefrikk graben to the east. This small graben would also provide the hydrocarbons for the Veslefrikk structure, which could spill further to other structures (e.g., Beta Sadel and Beta South) on a natural migration fairway, at least to the Brage structure.

3. There was probably a shift in time for filling of the

various structures, which led to a shift in maturity of oils and gases. These postulated migrational features caused certain changes in the structures with respect to reservoir diagenesis and oil and gas properties, which can be observed at the wells drilled. Oils should bear signs of biodegradation, particularly in the west flank satellites and the Oseberg Alpha structure. As a result of the different source areas and differences in the extent of biodegradation, the oil signatures in terms of biomarker patterns and isotope compositions should be different from one part of the study area to the other. Furthermore, oils and gases will be most mature close to the Viking Graben and least mature away from it. The GOR should decrease from the source area to the Horda platform. The time shift in the filling of the various structures should have retarded or stopped diagenesis earlier in the Oseberg Alpha and the west flank satellites than in the Brage reservoirs and other structures that acquired oil via spillover. Therefore, we expect less cementation and more unconsolidated sand in Oseberg Alpha and satellites, while the Brage reservoirs should be more cemented.

4. Hydrocarbon volumes generated in the drainage areas are large enough to fill the structures several times. The relative gas volumes are so high that structures close to the source kitchen are likely to have free gas columns, or their oil may even have been flushed totally by gas. However, a possible cap rock seal failure by hydraulic fracturing may have occurred as the result of overpressure and the buoyancy of vertical extensive gas columns, as suggested in Figure 16 for the Oseberg Alpha structure (which has a vertical closure of 580 m (1900 ft). Gas leakage due to seal failure caused by gas column buoyancy and overpressure will prevent the structures from total loss of oil beyond spillpoint. Overpressuring during the Tertiary is expected in the structures close to the kitchen areas, while the structures to the east (e.g., the Brage structure) have near normal pressure. A gas chimney is observed on seismic profiles over the Alpha and west flank satellites, which may suggest an existing seal failure.

5. The temperature for termination of biodegradation, 60°C, was reached in Oligocene time in the Alpha structure. The second phase of oil generation has provided enough hydrocarbon volume (Figure 16) to disguise early biodegradation. However, some traces of demethylized hopanes should be expected.

MODEL PREDICTIONS AND PROPERTIES OF PRODUCED HYDROCARBONS

The effect on hydrocarbon molecules by the geological framework and related processes (in which they are deposited, buried, generated, migrated, and trapped) can to some extent be deduced from the composition of the trapped petroleum and its distribution in the reservoirs. Because basin modeling simulates the geological framework in terms of its dynamic evolution, predictions can be made of alkane distributions, biomarker maturity, GOR, fluid contact movements, and so on in the entrapped hydrocarbons. Geochemical analyses and interpretation are therefore the most important tool to verify model predictions. Such interpretations are typically obtained through correlation and reservoir petroleum geochemical studies.

Based on the basin modeling and consequent synthesis, the following predictions can be made of molecular compositions and distributions of reservoir fluids in the study area:

- Some structures such as the Oseberg Alpha structure are likely to contain biodegraded oils or oils with imprints of early biodegradation.
- Because of the position of the source rock relative to the paleoshoreline, the nondegraded oils will be paraffinic rich, derived from waxy sapropel (Barnard and Cooper, 1981).
- Because two source areas were active during generation and migration, source differences should be expected in the biomarker and stable isotope contents with respect to structures drained from different source areas. However, these differences can be modified or disturbed by local variations of vertical drainage pattern.
- Because the Oseberg Alpha and its west flank structures most likely have experienced two significant phases of oil generation (Figure 16), the associated oils are expected to be less mature than the oils associated with the Huldra–Veslefrikk Graben (drainage area I, Figure 2a), which has experienced one major oil pulse. A spill-fill situation has likely controlled the filling of several structures in the area. A maturity difference, with the least mature petroleum farther away from the source area, is anticipated.
- Subsidence tilt and overpressure adjustments during Tertiary time may have caused early liquid contacts to be preserved in some of the reservoirs.

n-Alkanes

The oils produced within the study area are paraffinic oils derived from waxy sapropel (Cooper and Barnard, 1984) with low sulfur content and moderate API gravities (~35°) (Dahl and Speers, 1985). They have no immediate appearance of bacterial degradation (Pristane/n-C_{17} is ~2.0 for most oils), but exceptions to this exist. One west flank satellite structure shows clear signs of bacterial alterations. A compartment in this structure produced an oil in which the normal alkanes were completely removed (Dahl and Speers, 1986).

The other oil is depleted in *n*-alkanes. It is believed that the first oil belongs to a compartment that was sealed off from the migration fairway in the early

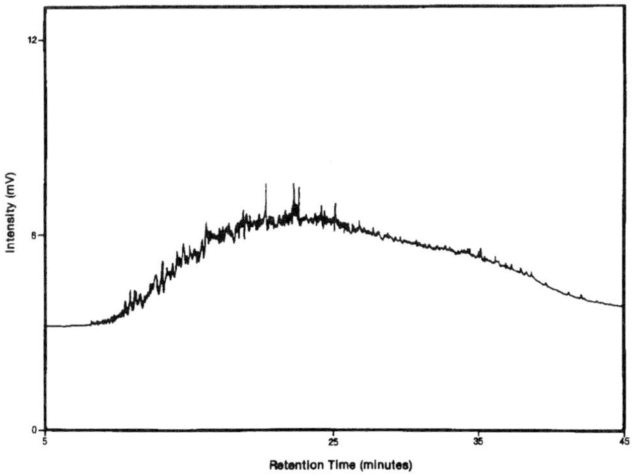

Figure 18. *n*-Alkane distribution of a biodegraded oil produced from the Oseberg cap rock

Tertiary and has since been sealed. The other oil belongs to a compartment in the same structure that was also sealed but was later reopened so that nondegraded oils migrated into the reservoirs and reduced the apparent effect of biodegradation. This explanation is supported by the results of gas analyses, which are discussed later.

From a thin porous layer in the Cretaceous cap rock at the apex of the Alpha structure, a very heavy and severely biodegraded oil was produced (Figure 18) in which demethylated hopanes were found. This suggests that the early migrating oil has not been effectively trapped and much of it bypassed the reservoir. It is most probable that biodegradation occurred in the Alpha reservoir simultaneously with seal leakage, although there is no evidence in *n*-alkane distributions of the Alpha oils today. The presence of minor amounts of demethylated hopanes in the apparent nondegraded oils of the Oseberg Alpha and its satellites support an early biodegradation, as predicted by the basin modeling results.

Biological Markers

The produced oils in the study area have been analyzed to determine the distribution of steranes and triterpanes. Based on interpretation of these analyses, the oils can be grouped into two families. Fragmentograms of the sterane and hopane type compounds representative of the two oil families are shown in Figure 19. Some calculated biomarker parameters from a selection of the oils are shown in Table 2. Although the difference between biomarkers of the two families is minor, the relative concentration of the C_{28}-bisnorhopane (C_{28}/C_{29} and C_{27}/C_{28} in Table 2) and the presence or absence of 25-norhopane (labeled X in Figure 19) seems to give a firm separation of the oils. However, 25-norhopane is likely an indication of remnant biodegradation (J. Connan, pers comm.).

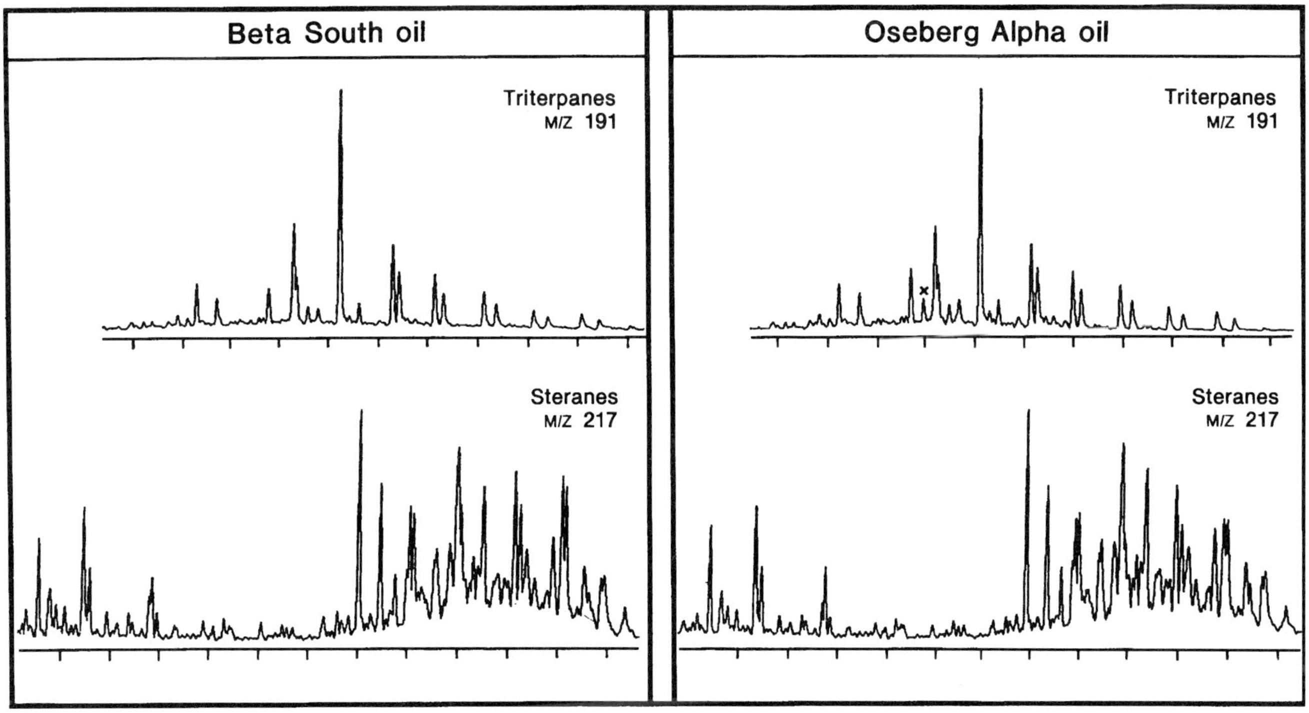

Figure 19. Biomarker distributions from oils of the two oil families.

Table 2. Biomarker parameters of selected oils in the study area.[a]

Oil Family	Structure	T_s (%)	C_{28}/C_{29}	C_{27}/C_{28}	20(*S*) (%)	$\alpha\beta\beta$ (%)
I	Alpha North	56	0.59	1.00	60	59
I	Alpha	55	0.56	0.97	59	58
I	Gamma	53	0.58	0.89	61	57
I	Brage	54	0.61	0.92	57	56
	30/6-16	45	0.60	1.03	50	48
II	Beta South	61	0.35	1.77	56	67
II	Veslefrikk	54	0.39	1.79	51	66
II	Brage Horst	59	0.41	1.87	57	59

[a]Parameters are defined as follows:

$$\% \, T_s = \frac{C_{27}\text{-trisnorneohopane}}{C_{27}\text{-trisnorneohopane} + C_{27}\text{-trisnorhopane}}$$

$$C_{28}/C_{29} = \frac{C_{28}\text{-bisnorhopane}}{C_{29}\text{-norhopane}}$$

$$C_{27}/C_{28} = \frac{C_{27}\text{-trisnorneohopane} + C_{27}\text{-trisnorhopane}}{C_{28}\text{-bisnorhopane}}$$

$$\% \, 20(S) = \frac{C_{29}\text{-}\alpha\alpha\alpha\text{-}20(S)\text{-sterane}}{C_{29}\text{-}\alpha\alpha\alpha\text{-}20(S)\text{-sterane} + C_{29}\text{-}\alpha\alpha\alpha\text{-}20(R)\text{-sterane}} \times 100$$

$$\% \, \alpha\beta\beta = \frac{C_{29}\text{-}\alpha\beta\beta\text{-}20(R,S)}{C_{29}\text{-}\alpha\beta\beta\text{-}20(R,S) + C_{29}\text{-}\alpha\alpha\alpha\text{-}20(R,S)} \times 100$$

The two oil families are related to the postulated drainage areas. Family I consists of the oils from the Oseberg Alpha and its satellites Alpha North and Gamma plus other west flank satellites (Figure 2a). All these oils are from structures with the Cretaceous cap rocks that were postulated to be subjected to early biodegradation. The main Brage structure, east of Oseberg Alpha, also contains oil of this family, believed to be derived from the Viking Graben to the west (drainage area I, Figure 2a). The Brage is capped by Upper Jurassic shales and the potential for biodegradation is low. A likely explanation for the presence of 25-norhopane is due to charge to the Brage structure via spillover from Oseberg Alpha. Family II comprises oils from the Veslefrikk through Beta South structures to the Brage Horst. This can be seen in Figure 20, in which the likely migration routes are outlined. Biomarkers also suggest trends that may be related to the timing migration and trapping histories in the area.

The family I oils are generally less mature than the family II oils with respect to the $\alpha\beta\beta$ parameter. In terms of the 20(*S*) parameter, the situation is opposite. However, the higher 20(*S*) maturities of these oils may be an artifact associated with elusion of the C_{28}-bisnorhopane with the C_{29}-$\alpha\alpha\alpha$-20(*S*)-sterane (Dahl, 1987). [C_{28}-bisnorhopane has a 217 fragment in its mass spectrum, according to Moldowan et al. (1984) and will consequently interfere with the C_{29}-$\alpha\alpha\alpha$-20(*S*) when evolution exists.] Family I oils have higher concentrations of bisnorhopane (C_{28}/C_{29}) than the family II oils (Table 2), which may explain the maturity inconsistencies. (The sterane/triterpane ratio is fairly similar for the two groups: 0.19 ± 0.02.) The T_s ratio, thought to be a maturity parameter (Seifert and Moldowan, 1980), also suggests that family II oils are more mature. Within the families some minor maturity differences [20(*S*) and $\alpha\beta\beta$] occur that can be related to the location of the structures relative to source areas. In both cases, accumulations nearest the mature source are more mature than the Brage structures located farther away from the source areas. One exception exists: the 20(*S*) parameter of the Veslefrikk structure. No immediate explanation for this can be given. However, one possibility is that the Veslefrikk structure has received oils from less mature, nearby sources. This oil has then affected the 20(*S*) parameter more than the $\alpha\beta\beta$ parameter.

The biodegraded 30/6-16 oil, representing the structure that was sealed off earlier from the migration fairway, belongs to family I with respect to both the bisnorhopane content (C_{28}/C_{29} and C_{27}/C_{28}) and its location. The maturity of this oil [20(*S*), $\alpha\beta\beta$, and T_s] is significantly lower than that of other family I oils, which supports the conclusion from the *n*-alkane distribution that this structure was sealed off from the migration fairway, probably in Paleocene time. The content of demethylated hopane type triterpanes has been determined in most of the accumulated oils to assess the degree of biodegradation. As previously mentioned, the oil produced from the cap rock (Figure 18) has been so highly degraded that the regular hopanes were removed completely and demethylated hopanes were formed. The family I oils and the oils in the Oseberg Alpha structure in particular contain traces

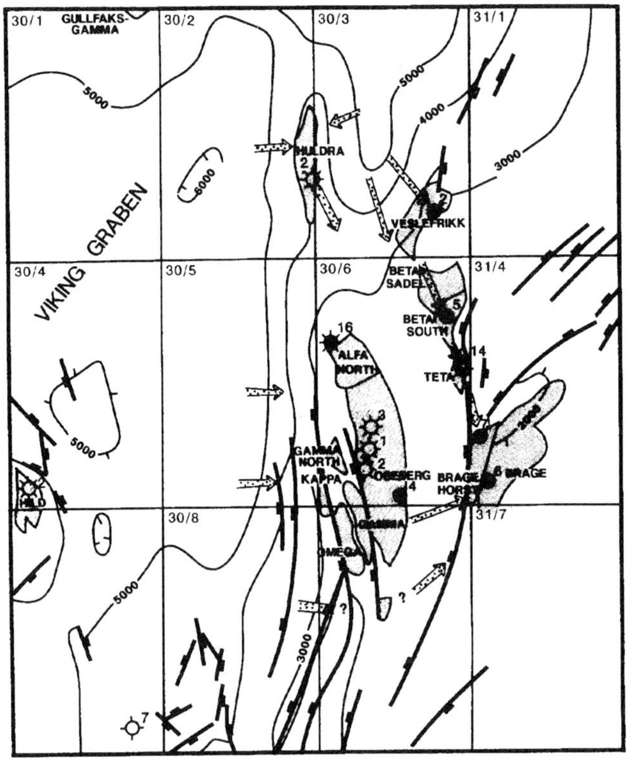

Figure 20. Hydrocarbon migration routes in the study area.

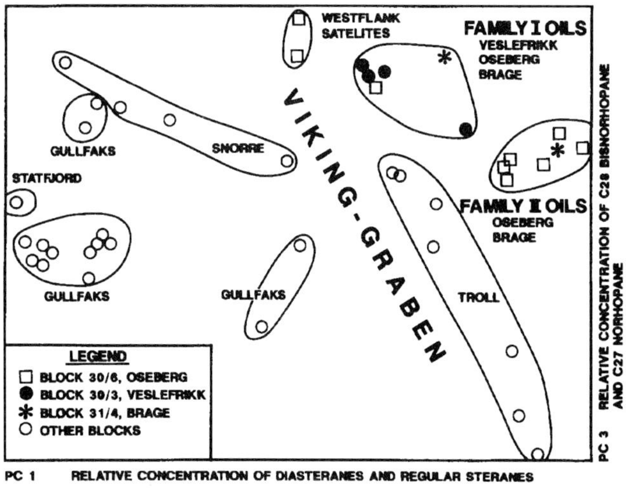

Figure 21. Score plot from factor analysis. Principal component 1 (PC1) is plotted against principal component 3 (PC3), both believed to be source dependent.

of demethylated hopanes that are an indication of early severe biodegradation later mixed with fresh nondegraded oils. Demethylated hopanes have not been encountered in the family II oils.

From this discussion, it is possible to conclude that generation and migration started before the Alpha structure and some of its satellites could trap large amounts of hydrocarbons and that those trapped hydrocarbons were biodegraded. The cap rock thicknesses of the structures associated with family II oils were larger at the onset of migration, and thus the possibility of biodegradation was lower. Although maturity differences within and between the families are minor, consistent trends agree with the modeling results that proximal structures were filled latest.

Multivariate Data Analyses of Oils

Biomarker data from oils in the study area and from fields in the northern Viking Graben region (45 in total) were correlated using multivariate data analyses software (SIMCA) (Wold, 1976). This study has been published in detail by Telnæs and Dahl (1986). Multivariate data analysis yields results similar to those of the few biomarker ratios related to source and maturity dependency, as given in Table 2. Figure 21, a revised score plot from Telnæs and Dahl (1986), demonstrates the scores of principle components 1 and 3 representing the relative concentrations of diasteranes plus steranes and C_{28}-bisnorhopane plus C_{27}-trisnorhopane, respectively. Both these principal components were interpreted as source dependent. Figure 21 shows that families I and II represent separate clusters. The 30/6-16 oils, which are biodegraded and of low maturity, are distinct from the two families. The biodegraded oil encountered in the cap rock (Figure 18) was not included in the analyzed population.

Stable Carbon Isotope Analyses of Oils

Most of the oils from the study area have also been analyzed for their stable carbon isotopes. The analyses have been made on oil fractions such as saturated alkanes, aromatics, NSOs, and asphaltenes plus whole oils. The oils were correlated by comparison of oil fractions in Galimov plots (Galimov, 1973; Stahl, 1979). Stable carbon isotopes for whole oils are within the range reported by Cooper and Barnard (1984) and MacKenzie et al. (1983) for the North Sea: $\delta^{13}C \approx -27.5$ to -31.0‰. The separation of the oils into two groups based on whole oil is fairly vague. However, the Galimov plots separate them into the same two families as their biomarker content (Figure 22). The family I oils (Alpha) are consistently isotopically heavier than the family II oils (Beta South) in all fractions.

Natural Gas Analyses

Several samples of natural gas from the Oseberg and Brage accumulations have been analyzed for hydrocarbon and stable isotope compositions. These data have been interpreted in relation to the publications of Schoell (1983) and James (1983), which present diagrams for the plotting data to reveal genetic relation-

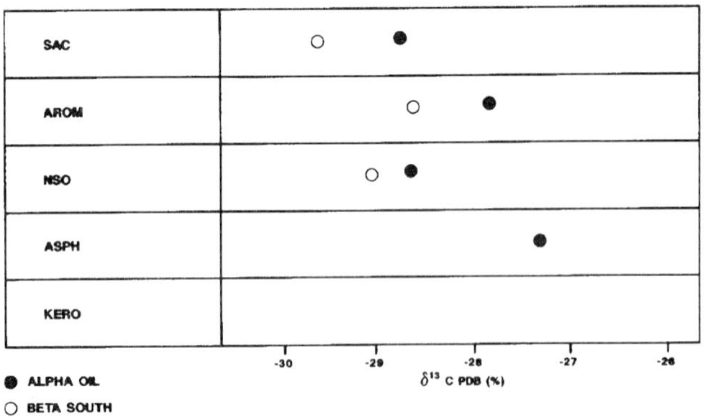

Figure 22. Galimov plot of carbon isotope data from representatives of the two oil families encountered in the study area.

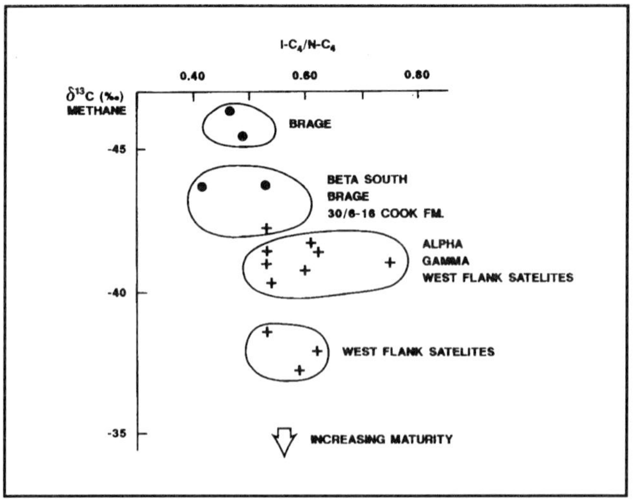

Figure 23. Plot of methane carbon isotopes versus the *i*-C_4/*n*-C_4 ratio for gases and associated gases from accumulations in the study area.

ships and gas maturity. Application of these diagrams suggests that gases in the area have a mixed origin, i.e., they are likely mixtures of two or more gases, one wetter and of fairly low maturity associated with oil or condensate generation and the other more mature and drier. Gas from only two structures was found to be nonmixed.

Gases can be derived from a wide maturity range, from associated gas via cracked oil to late stage kerogen cracking. They are more readily differentiated by their maturity than by their origin. According to Hunt (1979), gases can be only broadly distinguished in terms of their origin as either marine, continental, or biogenic. Source rocks that have produced significant gas in the study area are all dominantly marine. Consequently, such factors as generation history, migration distance, opening and closing of migration fairways as a function of tectonic and diagenetic processes, and cap rock sealing capacities probably control the maturity distribution of gases in the study area. Gas properties related to differences in source area are believed to be disguised by these maturity and migrational effects, therefore gases cannot be related to source areas in the same way as oils.

Figure 23 is a plot of the isotopic composition of methane versus the *i*-C_4/*n*-C_4 ratio, believed to be a gas maturity parameter (Herox et al., 1982). Gases cluster in four different groups, suggesting a general west–east trend of increasingly mature and differentially blended gases. Low maturity and "clean" gases from accumulations that have been sealed off from migration fairways at early stages, such as one of the 30/6-16 samples, normally would have plotted together with the most mature west flank satellites. These features of the gas distribution are compatible with the basin modeling exercise where spillover has been suggested as a significant controlling factor. Differential mixing is also expected as the structures have acquired and continue to acquire a contribution of gas from sources of a wide maturity range.

RESERVOIR STUDIES

In the process of verifying model predictions, reservoir petroleum geochemical studies give extremely valuable information about the predicted timing of migration, the types of petroleum accumulated, and reservoir movements. Reservoir geochemical studies involve dense sampling, extraction, and geochemical analyses of reservoir rocks of preferentially known porosities and permeabilities. Interpretation of the results will reveal vertical (and lateral) oil quality variations, dead oil columns, thin liquid zones in gas legs, and fossil fluid contacts. Such studies can also reveal whether reservoirs have been emptied or are in the process of being filled (Augustson, in press). These data are important for explorationists, but they are also important for field production planning. Fluid inclusion studies on reservoir sediments presented by McLimans (1987) would also be extremely useful during the verification phase. McLimans determined paleotemperature, paleopressure, and paleocomposition of reservoir fluids with respect to burial and migration history. Unfortunately, similar data were not attainable in the Oseberg area. Other valuable information on reservoir paleotemperatures is obtained from general diagenesis studies and isotope analyses of calcite cement.

Tar Mat Formation and Reservoir Processes in the Oseberg Alpha and Gamma Structures

Tar mats are asphaltene-rich heavy oils usually encountered at oil–water contacts or above permeability barriers in reservoirs. This phenomenon is caused by postaccumulation processes within the reservoir that results in enrichment of heavy oils at or above the liquid contact or barrier. Tar mats are

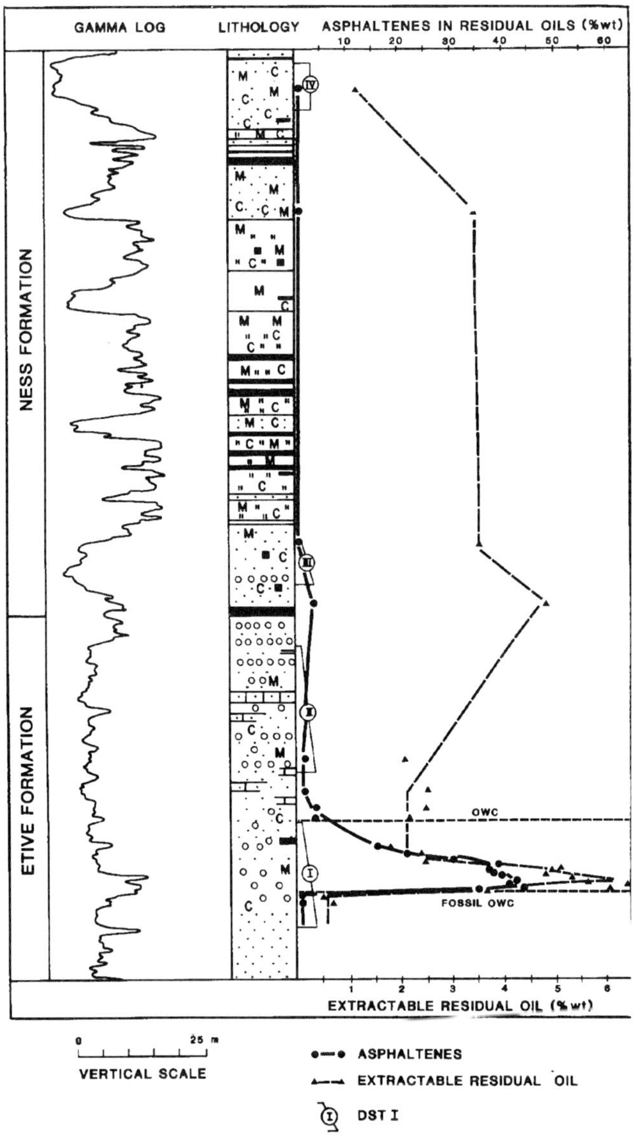

Figure 24. Identification of a tar mat and a fossil oil column in the Oseberg Gamma structure.

attributed to at least three different processes (Miler et al., 1977; Connan, 1984):

1. asphaltene precipitation due to influx of gas,
2. in-reservoir oil maturation and formation of pyrobitumen, or
3. biodegradation.

With this in mind, a scheme for classification of tar mats was suggested by Dahl and Speers (1986). Large tar mats usually occur in extensive sandstone reservoirs with good porosity and permeability conditions. Permeability barriers, which are more common in carbonate reservoirs, create mini-tar mats that are more difficult to detect. Tar mats can cause problems for reservoir production because water drive can be prevented. Oil quality drops as a result of asphaltene enrichment, which can influence field economics.

The tar mat in the Oseberg field was first discovered by production of an asphaltene-rich oil at the oil–water contact. It was also seen as an abnormal response on resistivity logs and as a discrepancy between well test data and porosity and permeabilities measured in the laboratory. The vertical extent of the tar mat was established by measuring the asphaltene contents of residual oils extracted from core samples taken throughout the reservoir as well as below the oil–water contact (Figure 24). Across the tar mat the asphaltene content in residual oils increased from about 5 wt. % to approximately 50 wt. %. Tar mat thicknesses are between 10 and 15 m (32 and 49 ft.). In both wells the lower boundary of the tar mat is shifted relative to the present-day oil–water contact and hence represented abandoned oil–water contacts displaced 8 and 15 m (26 and 49 ft) relative to the present-day contact. These contacts are in line with the present day spillpoints of the Alpha and Gamma structures, which suggests that hydrocarbons have been lost from these structures either through cap rock leakage or tilt of the reservoirs (or a combination of these processes).

Following the tar mat classification scheme of Dahl and Speers (1986), it was concluded that gas deasphaltening was the major process giving rise to the tar mat. This implies that large gas amounts have entered the oil-filled reservoirs and caused the asphaltenes to precipitate and segregate at the oil–water contact. From the GOR curve in Figure 16, the highest generation of gas relative to oil occurred during the Eocene and Oligocene and this process may thus have happened during this period (45–25 Ma). If it happened early during this period, a full hydrocarbon column would not have been established and the tar mat would have been pushed down together with the oil–water contact as the structures were filled.

Residual Oil in the Gas Cap, Oseberg Alpha

A suite of sandstone samples from the gas cap in the Oseberg Alpha structure was analyzed and found to contain 10–20 wt. % asphaltenes in extractable oil residue. As gas and condensates do not contain any significant amounts of asphaltenes, the results support the prediction that the gas cap at some time was occupied by crude oil, which at a later stage was displaced by lighter hydrocarbons such as gas.

Other Reservoir Features

The diagenetic features of the reservoirs in the study area show different sandstone diagenetic histories (H. R. Moe, pers. comm.), which suggests that the Oseberg Alpha and Gamma structures were filled earlier than the Brage main reservoir. A likely explanation for this delay is that oil, and later gas, filled the Oseberg structures to their spillpoints. This oil, saturated with gas, has subsequently filled the Brage reservoirs. Because of the change in temperature and pressure, gas evolution

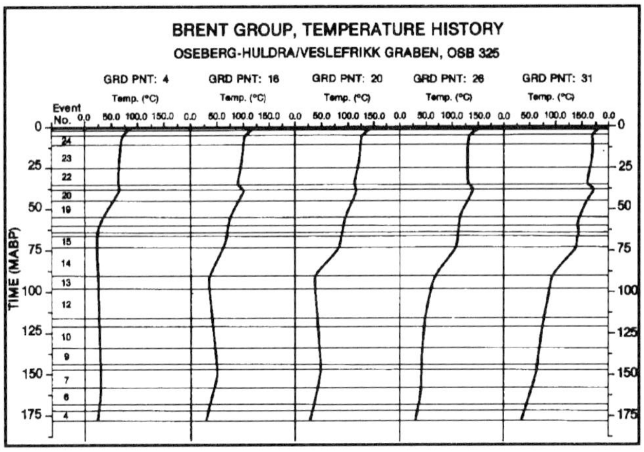

Figure 25. Calculated temperature histories of the Brent Group at selected grid points along the modeled profile OSB325 (Figure 3b) going from Oseberg Alpha to the graben between the Huldra and Veslefrikk accumulations.

created a small, free gas pocket of low-maturity gas in the Brage structure. Oxygen isotopes in calcite cement near the oil–water contact of the Alpha North structure suggests that temperatures were approximately 70°C (160°F) when this cement precipitated. This may indicate that the Alpha, Gamma, and Alpha North structures reached their maximum filling when the reservoir was at 70°C (160°F). (This assumes that calcite cement formation ceases or is significantly retarded when the pore space has reached a threshold oil saturation.) Thermal history modeling (Dahl et al., 1987) indicates that this temperature was first achieved in the early Oligocene (37 Ma) and prevailed to the Pliocene (gridpoint 4 in Figure 25).

ACCUMULATION HISTORY OF THE OSEBERG ALPHA STRUCTURE

By combining the results from this study with data from reservoir diagenesis studies, it is possible to suggest an accumulation history for the Alpha, Alpha North, and Gamma structures as follows:

1. Late Maastrichtian: First phase of oil migration reached the thinly buried Oseberg structures. The oil either escaped and/or was biodegraded (Figure 18). Relatively unconsolidated sand in the Brent reservoir (H. R. Moe, pers. comm.) supports an early retardation of sandstone diagenesis and consequent early filling of the structure.

2. Middle Eocene: A significant oil column has been established. The overlying cap rock is 600–700 m (1970–2300 ft) of marls and shales with a trapping potential of a 200-m (660-ft) oil column (Figure 17). The thickest sequences of oil-prone source rocks in the Viking Graben have exhausted their major liquid potential. However, minor volumes of low maturity oils are still being expelled from the graben margins. The gas may be starting to displace the oil column if the total hydrocarbon columns are overbalanced by the sealing efficiency of the cap rock. A free gas cap is formed or the oil is at least saturated with gassy, asphaltene-rich oil with traces of oil biodegradation disguised completely by influx of more mature paraffin-rich oil. A tar mat is being formed as the oils are undergoing deasphaltening.

3. Late Eocene–early Oligocene: Source rocks in the back basins at the hinge zone west of Oseberg accumulations are getting into the oil-generating zone. This is seen as an incline of the oil volume curve in Figure 16. A new pulse of moderately mature oil is being fed into the Oseberg structures. The hydrocarbon columns are increasing due to better sealing efficiency and to increasing influx of hydrocarbons. However, according to Figure 16, the spill points are not reached during this time and excess gas and oil may have leaked through the cap rock.

4. Middle Oligocene–late Miocene: Oxygen isotope studies of calcite cement in the oil column close to the oil–water contact suggest that temperatures were approximately 70°C (160°F) when this cement was formed. Modeling data suggest such reservoir temperatures are occurring at this time. Oil entrapment prevents additional diagenesis, preserving porosity. Maximum filling of the reservoir may be occurring late in this time (Figure 16), and excess oil is spilling over toward the Brage structure. Tectonic activity probably associated with Atlantic spreading (Glennie, 1984) is occurring with concomitant heating, uplift, erosion, and deposition of sand (Utsira Formation) whose lower boundary is truncated by a mid-Miocene fault. The seals in the Oseberg Alpha and Alpha North structures and most probably the Gamma structure may be weakening or even rupturing and gas is lost from the reservoir. The leakage of gas is balancing the influx of gas, and new oil is leaving the system beyond the spill point.

5. Pliocene–present: A gas chimney seen on seismic data from a reactivated Jurassic–Cretaceous fault suggests that leakage may be occurring throughout this time from the Oseberg Alpha, Alpha North, and most probably the Gamma structures. In periods, influx of gas is still approximately balancing the leakage of gas, while new oil leaves the system beyond the spillpoint. However, recently, due to a pressure increase, the oil is shifted upward away from the spill point, as observed by the tar mat study, and cap rock leakage is currently partially controlling the reservoir content. A better seal over these accumulations might result in total displacement of oil, and Oseberg field may be primarily a gas

field. Because large amounts of gas are probably leaking out of the Oseberg reservoirs into overlying strata, it should be possible to explore for Tertiary gas reservoirs in the area.

SUMMARY

Significant hydrocarbon generation from the Viking Group shales commenced during Maastrichtian time. The first phase of oil flowed into thinly covered structures, bypassing the reservoirs, and small amounts of oil were trapped and biodegraded. The calculated amounts of oil and gas generated and migrated throughout geological time far exceeded the in-place volumes of all structures in the study area. Two source areas for the oils were predicted and confirmed by oil–oil correlations. Predicted cap rock weakness in some of the structures was a significant controlling factor for the hydrocarbon content in these traps. Geochemical analysis, field studies, and gas chimneys on seismic data confirm both an ancient and a recent or present-day reservoir leakage.

Acknowledgments *The authors are grateful to Norsk Hydro A.S. for permission to publish this study. We appreciate the help of Norsk Hydro Research Center's drafting department and O. P. Winge with the diagrams. Special thanks are due to A. S. Steen for GC-MS analysis and to L. Aakvaag, J. B. Olsen, E. Rein, T. Jacobsen, and P. E. Eliassen for the experimental work and geochemical analyses. In addition, we are indebted to R. K. Merrill's editorial handling of this manuscript.*

Part II

Geochemical Methods and Exploration

Chapter 7

Geochemical Exploration Methods

Don Anders
U.S. Geological Survey
Denver, Colorado, U.S.A.

INTRODUCTION

The current understanding of the petroleum system with respect to source rocks and petroleum occurrence is evaluated in this chapter. Particular emphasis is placed on the geochemical methods used in measuring source rock potential, thermal maturation, oil and gas characterization, oil–oil and oil–source rock correlation, and sampling

SOURCE ROCK CHARACTERIZATION

The role of the amount, type, and maturity of organic matter in source rocks with respect to oil and gas generation has been discussed by numerous authors (Dow, 1978; Hunt, 1979; Barker, 1980; Tissot and Welte, 1984; Waples, 1985; Brooks et al., 1987). In general, an *effective source rock* can be defined as any rock that has generated and expelled commercial amounts of oil or gas (Hunt, 1979; Waples, 1985). In contrast, a *potential source rock* is any rock that has the capability of generating oil or gas in commercial quantities but has not done so because of inadequate heating, that is, inadequate thermal maturation (Dow, 1978; Clayton, 1988). Four questions must be answered in characterizing the potential of a source rock: (1) Does the rock have enough organic matter? (2) Is the organic matter of the proper type? (3) Has the organic matter undergone sufficient thermal maturation to generate hydrocarbons? and (4) Have hydrocarbons been expelled?

Source Potential

The *amount* of organic matter in source rocks is important for determining potential for two reasons. First, a minimum concentration of organic matter must be present in a rock before its threshold of liquid or gas saturation is reached and hydrocarbons can be expelled (primary migration). Second, a minimum mass of organic matter must be present within a certain thickness and areal extent of source rock before economic quantities of oil and gas can be expected (Momper, 1978). How much organic matter is sufficient to satisfy the requirements for primary migration is not clearly understood. Several authors have estimated that the minimum value of active (nonrefractory) organic carbon (OC) necessary for carbonates to be source rocks is on the order of 0.3–0.4 wt. % (Gehman, 1962; Ronov and Yaroshevsky, 1969; Dow, 1977; Palacas, 1978; Palacas et al., 1984; Tissot and Welte, 1984). The minimum average value of OC necessary to achieve primary migration in a carbonate evaporite sequence may not be as important as the heterogeneity in the distribution of that organic matter. For example, carbonate source rocks often contain evaporite sequences (e.g., laminate and micrite) or dissolution structures (stylolites) where the concentration of organic matter is much higher (1–30% OC) than the average value for the carbonate sequence (Palacas et al., 1984; Powell, 1984; Spiro et al., 1983; Richardson et al., 1988). These richer organic matter units may be the source of the oil associated with carbonate evaporites. The current consensus concerning the minimum amount of active OC needed for shales to become source rocks is 0.5–1.0% (Ronov, 1958; Barker, 1980; Tissot and Welte, 1984). It should be noted that in basins where oil has been generated, the average OC values are generally much higher than the minimum values just quoted (carbonates 0.7 wt. % and shales >2.0 wt. % OC) (Tissot and Welte, 1984).

The classic method for determining the amount of organic matter in rocks is by combustion of the OC to CO_2 in an oxygen atmosphere after acid leaching the carbonate carbon from the sample. The CO_2 can be chemically trapped in a reaction vessel of ascarite and weighed, or it can be measured by passing the CO_2 through a thermal conductivity cell or by capturing the CO_2 in a cell and measuring the adsorbance with an infrared detector. The major problem with these methods is that they measure the total carbon content and not just the active OC. Active OC can be measured using a pyrolysis device with a programmable oven (Rock-Eval). By programming the pyrolysis temperature, three measurements related to carbon content can be made: (1) free or volatized hydrocarbons (oil) already produced by the organic matter in the rock; (2) pyrolyzable hydrocarbons, that is, oil that can be produced from the remaining organic matter (kerogen) in the rock); and (3) dead carbon (carbon not convertible to oil). The free hydrocarbons give an estimate of the current source potential of the rock, whereas the pyrolyzable hydrocarbons give an indication of the remaining potential in the rock.

The *type* of organic matter in a rock is equally important to understanding the resource potential of a rock as the amount of organic matter is. Different organic matter types have different hydrocarbon producing properties and potentials for oil and gas generation. Kerogen (organic matter insoluble in aqueous alkaline or organic solvents) types I, II, and III are dependent on the relative contributions of algae, bacteria, higher plants, and zooplankton, which are in turn dependent on the paleoecology and depositional environment. The more hydrogen-rich, oxygen-poor kerogens (types I and II) are more prone to produce oil, while the hydrogen-poor, oxygen-rich kerogens (type III) are more prone to produce gas. However, seldom is sedimentary organic matter of a pure kerogen type because the organic matter that is preserved in a sedimentary rock is dependent upon a number of factors, such as paleoecology, depositional environment, and transport mechanisms.

The most common methods (Table 1) for ascertaining the source potential of sedimentary organic matter in rocks involve determining hydrogen to carbon (H/C) and oxygen to carbon (O/C) ratios for the kerogen by pyrolysis, elemental analyses, or petrographic methods. Petrographic approaches are more qualitative, involving visual identification of organic macerals. In contrast, H/C and O/C determinations are more quantitative, involving elemental analyses or pyrolysis. Pyrolysis is the usual method of choice; it is the least labor intensive and can be performed on less than 100 mg of sample (Espitalié et al., 1985). Other useful methods involve pyrolysis fluorescence (Heacock and Hood, 1970) and pyrolysis gas chromatography (Larter and Douglas, 1980; Dembicki et al., 1983; Larter, 1984, 1985)

Maturation

Generation of hydrocarbons (source rock maturation) is dependent primarily upon temperature and secondarily upon time. Regardless of the amount and type of organic matter, hydrocarbons will not be produced if there has been insufficient thermal stress to crack kerogen. Heat sources include heat flow from within the earth's interior or tectonically induced heat flow, such as hot fluid transport, magmatic intrusion, or volcanism. Thermal maturity is assessed not simply to determine if hydrocarbons have been generated but also to determine migration pathways and entrapment mechanisms at the time of hydrocarbon expulsion. Of course, the geological and thermal histories of a region must also be determined (Waples, 1985).

Numerous methods have been developed to help estimate the temperature history of sedimentary rocks (Table 1). Thermal maturity is indicated optically by the amount of carbonization (change in color) of palynomorphs (spores, pollen, or microfossils) in transmitted light and by vitrinite reflectance. Immature palynomorphs are yellow in color, but with increasing maturity, the color changes from yellow to orange or yellowish brown (diagenesis), to brown (catagenesis), and finally to black (metagenesis). The best known version of this technique is the coloration scale developed by Staplin (1969) called the thermal alteration index (TAI). Reflectance of vitrinite (woody) macerals in reflected light, the most often used measure of thermal maturity in sedimentary rocks today, can be correlated to the palynomorph maturation stages as follows: a mean average reflectance (R_m) of <0.5–0.7% indicates diagenesis; 0.7–1.3%, catagenesis or oil window; 1.3–2.0%, late categenesis or main gas zone; and >2%, metagenesis (Dow, 1977; Tissot and Welte, 1984; Bayliss and Magoon, 1988). The limitations of this method are (1) vitrinite may be scarce in types I and II kerogens, (2) transformation rates may vary for different kerogens though they have the same level of vitrinite reflectance, (3) other macerals are often mistaken for vitrinite, and (4) source rocks older than Silurian are believed by many to contain no vitrinite.

A very popular method of measuring the thermal maturity of sedimentary organic matter is the pyrolysis method of Espitalié et al., (1977). In this method, a weighed amount of the crushed rock is heated in an inert atmosphere first at 200°–250°C to quantify the free hydrocarbons in the rocks (S_1) and, second, at a programmed heating rate (10°–50°C/min) from 200°C (or 250°C) to 600°C to determine the amount of active kerogen that can be converted to hydrocarbons (S_2). Where migration has not occurred, the ratio of $S_1/S_1 + S_2$ (transformation ratio) and the measured temperature at maximum conversion of the kerogen to hydrocarbon (T_{max}) have been shown to be directly related to the thermal maturity of the organic matter. Both values increase with increasing maturity. Transformation ratios of <0.1 and T_{max} values of <435°C are generally considered thermally immature; transformation ratios of 0.1–0.4 and T_{max} values of 435°–460°C are considered to be in the catagenesis stage (oil window). Higher values of T_{max} are considered beyond the zone of oil generation. In the early stages of maturation, T_{max} values for type I kerogen are higher than for type II and III kerogens. However, as maturation levels increase, T_{max} differences between kerogen types become less and less and are almost equivalent at peak oil generation

Table 1. Methods used in source rock characterization.[a]

Methods	Abundance of Organic Matter	Quality (Type) of Organic Matter	Maturation of Organic Matter
Optical-microscopic			
Transmitted light (Palynomorphs)	-	+++	+++
Vitrinite reflectance	-	-	+++
Fluorescence (kerogen)	-	+++	++
Pyrolysis			
Rock-Eval	+++	+++	+++
Organic carbon	+++	-	-
Elemental analysis (C, H, O)	-	+++	++
Molecular composition (GC and GCMS)			
n-Alkanes	-	++	++
Steranes	-	++	+++
Terpanes	-	++	++
Aromatic steroids	-	++	+++
Methylphenanthrene index	-	-	+++
Porphyrins	-	++	+++
Miscellaneous			
$\Sigma C_{2-4}/\Sigma C_{1-4}$ (GC)	-	-	+++
C_{5-7}/TOC (GC)	-	-	+++
C_{15+} HCs/TOC (gravimetric)	-	-	+++
Bitumen/TOC (gravimetric)	-	++	++
Electron spin resonance	-	-	++
Stable isotopes (C, H, S) (MS)	-	++	++

[a]Symbols: +++, very useful; ++, often useful; -, not useful.

(Teichmüller and Durand, 1983; Waples, 1985).

Several physicochemical methods have shown promise as thermal maturity indicators. For example, elemental analysis (H/C vs. O/C) of kerogen has been shown to change systematically with increasing thermal maturity (Van Krevelen, 1961; Tissot and Welte, 1984). A major problem with this method is that the ratios vary not only with maturity but also with organic matter type (LaPlante, 1974; Tissot et al., 1974). Electron spin resonance (ESR) has been shown by Pusey (1973), Ho (1977), and Cooper et al., (1975) to be an effective measure of thermal maturity for all forms of organic matter. ESR measures free radicals associated with the large aromatic ring structures of kerogen. The higher the temperature, the more free radicals that are formed. Basic problems associated with ESR include (1) calibration curves are different for each organic matter type, consequently the method cannot be used where the organic matter type is variable; (2) ESR measurements pass through a maximum at an equivalent vitrinite reflectance of 2.0%, making it useful only where metagenesis has not yet occurred; and (3) ESR measurements on whole rocks frequently are small, thereby requiring kerogen concentration.

Various chemical compositions of sedimentary extracts can also be useful in the study of thermal maturation of organic matter. For example, the ratio of the amount of extractable organic matter to OC or C_{15+} hydrocarbons to OC generally increases with increasing maturity to a maximum, then decreases. This is due to the fact that as thermal stress on the organic matter increases, bonds between atoms in the kerogen are cleaved, giving off hydrocarbons. These ratios can be used to establish an evolutionary trend where a series of samples of similar organic matter type can be obtained from various depths. Two additional measures of hydrocarbon change with increasing thermal maturity are the ratios of the abundance of C_{1-4} hydrocarbons to OC or C_{5-7} hydrocarbons to OC. Both measures are usually made on gases released from canned core or cuttings at 100°C (Bailey et al., 1974; Hunt, 1975; Snowdon and Roy, 1975; Claypool et al., 1977; Anders and Gerrild, 1984) and quantified by gas chromatography. In the early stages of maturation, biogenetic methane is the primary gas with low amounts of gasoline-range hydrocarbons. As maturation increases, the abundance of gasoline-range hydrocarbons generally increases until the metagenetic stage where methane (from C–C cleavage) once again dominates. The genetic nature (origin) of the methane can usually be distinguished by their carbon isotope abundances.

Another chemical change often observed with increasing maturity is a change in the *n*-alkane distribution. Immature to moderately mature shale samples, containing even small amounts of terrestrial organic matter, frequently exhibit odd carbon predominance in their C_{25+} hydrocarbon range. In contrast, evaporite and carbonate rocks often exhibit an even carbon predominance where terrestrial input is minimal. These odd or even predominances in the high molecular weight *n*-alkanes gradually decrease and give way to lighter hydrocarbons with increasing thermal maturity (Tissot and Welte, 1984; Palacas et al., 1984; Palacas, 1988).

Pristane/n-C_{17} and phytane/n-C_{18} are two other hydrocarbon ratios that have been shown to decrease with increasing maturity (Tissot and Welte, 1984). These two ratios, however, are often affected by depositional environment, organic matter input, and the thermal dynamics of kerogen degradation and hydrocarbon release.

The ratio of phenanthrene and various methyl isomers of phenanthrene (1-methyl, 2-methyl, 3-methyl, and 9-methyl) have been shown to change in orderly fashion during maturation (Radke et al., 1982). Radke et al. (1982) and Radke and Welte (1983) have shown good correlation between the ratio 1.5 (2-methylphenanthrene + 3-methyl-phenanthrene)/(phenanthrene + 1-methylphenanthrene + 9-methylphenanthrene), called MPI-1, and vitrinite reflectance and have used this correlation to estimate the level of maturation for samples containing little vitrinite. The MPI-1 ratio was first thought to be source independent, but was later shown to vary with changes in organic matter type (Radke et al., 1986).

A relatively new method that has found great application in the past 10 years involves the measure of specific biomarkers (molecules that retain enough of their original chemical structure so that their biological origin is recognizable). The ratio of different biomarkers can indicate different degrees of thermal maturation. Mackenzie (1984) has recognized certain ratios that reflect isomerization, degradation, or aromatization reactions of specific biomarkers with temperature as a function of time. The ratio of reactants to products in a bitumen or oil is usually measured by gas chromatography/mass spectrometry. The ideal parameter is one in which a reactant is converted 100% to product. Seldom in nature, however, does the ideal occur. Generally, there is a reverse reaction that also takes place. For any given temperature, a ratio of reactant to product will be established. At some temperature of reaction, the forward and reverse reactions will become equal, and a reaction equilibrium will be established. Beyond the equilibrium point, the parameter is no longer useful for studying maturation. Several of the more useful ratios (reactant/reactant + product) recognized to exist over a wide range of thermal maturities are as follows:

1. Isomerization at C_{22} in the C_{31-35}-17α(H),21β(H)-hopanes; the ratio of 22(S)/22(S) + 22(R) reaches equilibrium (at about 60%) at the onset of intense hydrocarbon generation (Ensminger et al., 1974; Seifert and Moldowan, 1980; Mackenzie and Maxwell, 1981; Mackenzie, 1984).

2. Isomerization at C_{20} in the C_{29}-steranes has a much wider range of thermal application, reaching equilibrium at a vitrinite reflectance equivalence of about 1.0% (Mackenzie et al., 1980a; Seifert and Moldowan, 1981; Mackenzie and Maxwell, 1981); the ratio of the 20(S)/20(S) + 20(R) [C_{29}-5α(H),14α(H), 17α(H)-steranes] reaches equilibrium at about 54%.

3. Aromatization and/or degradation of C-ring monoaromatic steranes in favor of ABC-ring triaromatic steranes is useful to a vitrinite reflectance equivalence of about 1.0% (Mackenzie et al., 1981b) with an equilibrium (or conversion) of 100% (tri/mono + tri).

Mass-spectral and high-performance liquid chromatographic (HPLC) measured ratios of demetallated deoxophylloerythroetioporphyrins (DPEP) to aetioporphyrins (ETIO) derived from sedimentary extracts have been shown to have a relationship to thermal maturity (Mackenzie and Maxwell, 1981; Barwise and Park, 1983). The ratio of ETIO/ETIO + DPEP can be measured spectrophotometrically or by probed mass spectra on the total porphyrin fraction or on refined demetallated nickel and vanadyl porphyrin fractions separately. The nickel ETIO/ETIO + DPEP ratio was demonstrated by Mackenzie and Maxwell (1981) to be useful to a vitrinite reflectance equivalent of approximately 0.7%, whereas the vanadyl ETIO/ETIO + DPEP ratio appears to be useful over the vitrinite reflectance range of 0.5–1.0%. Factors affecting these ratios include the relative thermal stabilities of ETIO and DPEP porphyrins (Didyk et al., 1975; Barwise and Roberts, 1984; Burkova et al., 1980), the generation of porphyrins from thermal cleavage of kerogen (Mackenzie et al., 1980b; Barwise and Roberts, 1984), the thermal cleavage of the isocyclic ring on DPEP to form ETIO porphyrins (Didyk et al., 1975), and the oxidation cleavage of the isocyclic ring of DPEP to form ETIO at the time of deposition (Baker and Louda, 1980).

The role of kinetic modeling in studying thermal maturation has come into vogue with the advent of inexpensive personal computers and commercially available software programs. All the computer models commercially available calculate thermal maturity based on the theory of time and temperature (TTI) (Platte River Associates), the Rock-Eval kinetics of types I, II, and III kerogen (IFP) (Lawrence Livermore) or the hydrous pyrolysis of vitrinite or kerogen. Some of these programs allow the geologist to use the kinetics of many of the reactions just discussed to model the temperature history (Mackenzie and McKenzie, 1983; Mackenzie 1984; Alexander et al., 1986). The usefulness of these models is limited by how well the geological and thermal histories of a basin are known and how well the kinetic model describes the real behavior of organic matter during burial heating over geological time (Clayton, 1990). Since petroleum and gas generation are kinetically controlled, all the models (Lopatin, 1976; Tissot and Espitalié, 1975; Welte and Yükler, 1981) in widespread use today are based on the Arrhenius equation which takes into account the yield of a product (oil, gas, or specific compound) with respect to time and temperature. Detailed descriptions of laboratory models for determining kinetic parameters can be found in reports by Braun and Burnham (1987), Burnham and Braun (1985), and Burnham et al., (1987, 1988).

OIL CHARACTERIZATION AND CORRELATION

It is generally agreed that conventional oil and gas originates from the thermal transformation of organic matter in sedimentary rocks. Because only a portion of the hydrocarbons generated in sedimentary rocks ever migrates, similar structures to those found in petroleum can be expected to be found in the nonmigrated fluids (bitumen) extracted from source rocks. Unfortunately, numerous things can happen to fluids and the source rock subsequent to migration that can complicate geochemical comparison. Both the source rock and the migrated fluids can undergo further maturation. Fractionation of the migrated fluids can take place during migration, and migrated fluids can undergo biodegradation and water washing. Mixing of hydrocarbons from more than one source can also take place. In spite of these complicating factors, the whole basis for establishing oil–oil, gas–gas, oil–source rock, and gas–source rock correlations is based on finding distinct chemical properties in the migrated fluids and the retained fluids that can be used either to prove or disprove a relationship. The most useful parameters for correlation fall into two categories: (1) bulk parameters, which describe properties of the whole sample and include API gravity, percent sulfur and nitrogen, optical activity, infrared spectra, saturate/aromatic hydrocarbon ratios, and isotopic composition; and (2) molecular parameters that are detailed chemical characteristics that can be measured on a fraction of the sample such as biomarker distributions (steranes, terpanes, benzothiophenes, phenanthrenes, *n*-alkanes, isoprenoids, aromatic steranes, and porphyrins). Whenever possible, several unrelated correlation parameters from both types of analyses should be used to prove or disprove a genetic relationship.

A migrational effect that the analyst should be aware of in comparing oils with bitumen extracts from source rocks is that oils are generally enriched in hydrocarbons and depleted in polar compounds and asphaltenes relative to source rock bitumen (Hunt and Jaieson, 1956; Bray and Evans, 1965). Ternary diagrams of the percent saturate, aromatic, and resins plus asphaltenes are helpful for grouping oil types or families, but are of little use for oil–source rock correlation. Migration, however, does not appear to affect the major hydrocarbon molecular parameters used for comparison.

Many methods for characterizing crude oils have been proposed over the past 30 years: (1) isolation and identification of individual compounds (Rossini and Mair, 1959; Smith, 1968; Erdman and Morris, 1974); (2) distillation (Smith, 1927; Lane and Garton, 1935) followed by measurement of physical properties (e.g., density, viscosity, and refractive index) and chemical properties (e.g., sulfur content and relative amounts of paraffin, naphthenes, and aromatics) of the various distillation cuts; and (3) classification based on bulk compound type distribution, such as asphaltenes, saturated hydrocarbons (normal, branched, and cyclic alkanes), aromatic hydrocarbons (e.g., benzo- and dibenzothiophenes and mono- and triaromatic hydrocarbons), and polar (NSO) compounds (resins) (Tissot and Welte, 1984)). The separation into compound classes can be accomplished by a variety of chromatographic techniques, including liquid, thin-layer, and high-performance liquid chromatography (HPLC). These compound classes can be further analyzed by gas chromatography (GC) and GC–mass spectrometry (GCMS) to obtain carbon number distributions and identification of individual compounds, such as biomarkers. The method of choice depends on the intended end use of the analyses.

Geologists and geochemists generally view oil characterization as a means of correlating oils and relating oils to their source rocks; however, often only oils are available and the source rocks are unknown. In these cases, the nature of the source rocks is often determined through oil characterizations either as marine or nonmarine or as carbonate or shale derived (Peters et al., 1986; Palacas, 1988; Hite and Anders, 1991). Simple classifications of oils by API gravity (condensates, >45° API; medium gravity oils, 45°–20° API; and heavy oils, <20° API) or by boiling point are useful in refinery applications. However, these are of little value to the exploration geologist who needs to know about the rocks from which the oils are derived. The kind of information a petroleum geologist needs is best gleaned from the elements of classification scheme #3 (preceding paragraph). Analyses of oils that yield source-specific information, such as biomarkers, are very helpful in this regard (Seifert and Moldowan, 1986). Stable carbon isotopes can often distinguish marine from nonmarine oils (Silverman and Epstein, 1958; Sofer, 1984). Carbon and hydrogen isotopes have been used successfully to distinguish carbonate- from shale-derived oils (Schoell, 1983b, 1984).

Multivariate analyses, a computerized statistical approach comparing data from large sample sets has been used to differentiate oils derived from a variety of depositional settings and to show genetic relationships among oils (Moldowan et al., 1985; Zumberge, 1987; Mello et al., 1988). Characterization of condensates derived from mature source rocks is more difficult because most of the source-specific parameters have been lost or severely altered.

The existence and recognition of distinct physical and chemical similarities between hydrocarbons in a reservoir and the insoluble (kerogen) and soluble (bitumen) organic constituents still retained in the source rock is the basis for geochemical correlation (Anders, 1988). Correlations (oil–oil, oil–source rock, gas–gas, and gas–source rock) are generally attempted to answer specific questions. How many families of oils, condensates, or gases are present in a basin? Do the reservoired hydrocarbons have more than one source? Can the source rock(s) for the hydrocarbons be substantiated so as to establish migration pathways and timing of generation?

Many of the early physical and chemical techniques for correlation (e.g., API gravity, color, pour point, and distillation ranges) were found to be only marginally successful in showing genetic relationships. These physical methods have gradually given way to more useful detailed chemical analyses. Gross chemical properties, such as percent sulfur and nitrogen, optical rotation, vanadium/nickel ratios, and carbon and sulfur isotope ratios, often have limited usefulness because of overlapping ranges (McIver, 1962), but all have been useful at times in breaking out or distinguishing families of oils in basins with multiple sources. For example, McIver (1962) used vanadium/nickel ratios to distinguish Cretaceous oils from Jurassic, Triassic, and Paleozoic oils of the Wind River basin. Williams (1974) used carbon isotopes and optical rotation to break out two families of oils in the Williston basin, and Thode (1981) distinguished three oil families in the same basin using sulfur isotopes. Koons et al. (1974) utilized a multiple parameter approach to study the oils of the lower part of the Tuscaloosa Formation (Upper Cretaceous) of the central Gulf and was able to recognize two families of oils—one oil type influenced by terrestrial organic matter and the other derived primarily from marine organic matter.

In choosing correlation parameters, compound types should be chosen that have distinguishing characteristics and are little affected by migration, maturation, or other alteration processes (van Eggelpoel, 1964; Gransch and Eisman, 1970; Koons et al., 1974; Powell and McKirdy, 1975; Seifert and Moldowan, 1978, 1981, 1986). During the past two decades, much emphasis has been put on biomarkers in combination with bulk properties to attain correlations. For a summary of techniques in geochemical correlation, see reports by Barker (1975, 1980), Tissot and Welte (1984), and Anders (1988). Table 2 lists and evaluates methods in geochemical correlation. Correlation of altered oils (biodegradation, water washing, and maturation) requires the use of chemical properties least acted upon by the alteration processes. For example, water washing has the tendency to reduce the light aromatic content, whereas thermal alteration increases the light aromatic content. Paraffins and API gravity increase with thermal maturation but are decreased by biodegradation. Organic sulfur and nitrogen, asphaltenes, and optical activity of hydrocarbons are decreased by greater thermal maturation, but are increased during biodegradation. Isoprenoids survive mild biodegradation; multiring naphthenes, carbon isotope values, vanadium/nickel ratios, benzothiophenes, and aromatic steroids are still useful parameters with more severe biodegradation (Seifert and Moldowan, 1979; Moldowan and Seifert, 1980; Welte et al., 1982; Palacas et al., 1986, 1988; Anders and Magoon, 1986; Anders et al., 1987). However, even the more resistant biomarkers are eventually altered if biodegradation proceeds far enough. The resistance of hydrocarbons to biodegradation proceeds in the following relative order: polyaromatic hydrocarbons and thiophenes > triterpanes > steranes > isoprenoids > *n*-alkanes.

Generally, the methods or analyses used in oil–oil correlation are applicable to oil–source rock correlations as well (Table 2). It is advisable to use both bulk and biomarker parameters from more than one molecular weight range to evaluate oil–source rock genetic relationships. It is important in oil–source rock correlations to establish that the hydrocarbons associated with the source rock are indigenous and not oil stained by migrated fluids. This can usually be accomplished by Rock-Eval transformation ratios or by hydrocarbon/TOC ratios (Espitalié et al., 1985). By far the most successful examples of oil–source rock correlations generally involve two or more measurements of biomarker distributions as well as bulk properties (Seifert and Moldowan, 1978; Seifert, 1978; Seifert et al., 1980; Palacas et al., 1984; Magoon and Claypool, 1985; Anders and Magoon, 1986).

GAS CHARACTERIZATION AND CORRELATION

Gas characterization is less complex than oil characterization simply because of the narrow range of hydrocarbon (C_{1-4}) composition. Nonhydrocarbon gases (CO_2, H_2S, N_2, H_2, Ar, and He) from air, volcanic, geothermal, and radioactive sources are also usually present with hydrocarbon gases. Parameters frequently used to characterize natural gases include methane/total gas, nonhydrocarbon gas content, and carbon and hydrogen stable isotopes (generally of methane). Gases are more difficult to correlate with source rocks because the more complex biomarkers are absent. Methane isotopic composition is often very useful for determining the source of the gas (Claypool and Kaplan, 1974). Values of $\delta^{13}C$ (‰) relative to the Peedee belemnite (PDB) standard for biogenic methane range from -90 to -55‰; for catagenetic methane, -55 to -30‰; and for metagenetic methane, -40 to -20‰. Values of $\delta^{2}D$ (‰) relative to standard mean ocean water (SMOW) for biogenic methane are <-180‰; for catagenetic methane, <-140‰; and for metagenetic methane, -150 to -130‰ (Stahl, 1975; Stahl and Carey, 1975; Galimov, 1980; Schoell, 1980, 1983b; Rice and

Table 2. Methods used in geochemical correlation.[a]

Methods	Application				
	Oil–Oil	Oil–Source	Gas–Gas	Gas–Oil	Gas–Source
Elemental analysis					
Sulfur (%)	+++	+++	-	-	-
Nitrogen (%)	+++	+++	-	-	-
V/N	+++	+++	-	-	-
Stable isotopes (MS)					
Carbon	+++	++	+++	+++	++
Sulfur	+++	++	-	-	-
Hydrogen	+++	++	+++	+++	++
Bulk parameters					
Saturates/aromatics	+++	+	-	-	-
Hydrocarbons/nonhydrocarbons	+++	+	-	-	-
Infrared spectra	++	++	-	-	-
Hydrous pyrolysis—kerogen	-	++	-	-	-
Hydrous pyrolysis—asphaltenes	-	++	-	-	-
Optical activity	++	+	-	-	-
API gravity	++	-	-	-	-
Detailed composition (GC)					
CO_2, N_2, H_2S, He	-	-	++	-	-
$C_1/\Sigma C_{1-4}$ (wetness)	-	-	++	-	-
C_{5-7} (normals/branched/cyclic)	+++	-	+	+	-
Molecular parameters (GC or GCMS)					
n-Paraffins	+++	+++	-	-	-
Naphthenes	+++	+++	-	-	-
Isoprenoids	+++	+++	-	-	-
Aromatics	++	++	-	-	-
Thiophenes	+	+	-	-	-
Diterpenoids	+++	+++	-	-	-
Triterpenoids	++	++	-	-	-
Steranes	++	++	-	-	-
Aromatic steroids	++	++	-	-	-
Porphyrins	++	+	-	-	-
Miscellaneous					
Correlation index (USBM)[b]	++	++	-	-	-
PNAA[c]	++	++	-	-	-
GOR/API gravity	+	-	-	-	-

[a]Symbols: +++, very useful; ++, often useful; +, sometimes useful; -, not useful.
[b]U.S. Bureau of Mines distillation technique (LeTran, 1971).
[c]PNAA is paraffins, naphthenes, aromatics, and asphaltene + resins.

Claypool, 1981; Schoell et al., 1983). Isotopic composition of individual homologs combined with their relative distributions and inorganic gas compositions can provide a means for correlation (Colombo et al., 1966; Menendez, 1973; Stahl, 1978; Schoell, 1977, 1980; Héritier, 1979; Mattavelli et al., 1983).

SAMPLING

Proper sampling of rocks and oils for geochemical analyses is critical for oil–oil and oil–source rock correlations and exploration programs in frontier and poorly defined regions. The availability of samples from such areas is often limited to outcrops, seeps, and surface-stained rocks. In such areas, it is important to obtain samples of the most important stratigraphic and lithologic units represented in the basin (Tissot and Welte, 1984). Care should be taken to obtain samples that are as fresh as possible, such as from active seeps and from unoxidized and unweathered rocks. The samples should be stored in glass containers or sealed metal cans, avoiding all possible contamination from waxed cardboard containers, bags, or plastics. If cuttings are available, they should be sealed in water in metal cans or glass bottles with a bactcriacide, while core samples should be sealed in cans or wrapped in foil and frozen in dry ice at the well site for later analyses. Samples of drilling mud, pipe dope, and any other ingredients that may have been used in the drilling process should be obtained for analysis. Collection of crude oil at well sites should be obtained under pressure in the reservoir if at all possible. Sampling at the well head at atmospheric pressure results in light hydrocarbon losses. However, if no analyses of light hydrocarbons (boiling point <175°C) are deemed necessary, well head sampling is sufficient. Pipeline and storage tank collections are less desirable because of potential contamination between the well head and the collection site.

Chapter 8

Geochemical Techniques in Relation to Organic Matter

R. A. Noble
ARCO Oil and Gas Company
Plano, Texas, U.S.A.

GAS CHROMATOGRAPHY

Airspace Gas

One of the simplest screening methods for evaluating the hydrocarbon potential of a sedimentary section is to measure the composition and concentration of light hydrocarbons (C_{1-5+}) that are released from drill cuttings at the surface. This is readily achieved by gas chromatography (GC), but because of the volatility of the light hydrocarbons, special consideration must be given to the methods of sample collection and storage. The most widely used collection technique is to transfer 250–500 mL (0.5–1 pint) of cuttings from the shale shaker to a 1-L capacity stainless steel can. Water is added to bring the level to within 2 cm (~1 in.) from the top of the can, and a few drops of a bactericide, such as zephrin chloride, are added to prevent bacterial activity. The can is sealed, labeled, inverted so as to place the manufacturers seal on top, and transported to the laboratory for analysis. Canned cuttings are usually collected at 30–100 ft (10–30 m) intervals, although the number of samples ultimately analyzed will depend upon the scope of the intended study and the type of information desired (Hunt, 1979).

The gas that accumulates in the void above the surface of the liquid is usually referred to as airspace or headspace gas. Several reports also refer to these light hydrocarbons (C_{1-4}) as cuttings gas (see Hunt, 1979, and references therein). However, the term *cuttings gas* is somewhat ambiguous since some companies have chosen to use it to describe the light hydrocarbons that are released from the washed cuttings after mechanical pulverization and heating *in situ*. To avoid any confusion in terminology, the latter will be referred to as disaggregate cuttings gas. The terms *airspace gas, headspace gas*, and *cuttings gas* are therefore synonymous, although only the first of these will be used in this chapter.

The laboratory procedure used for the analysis of the airspace gas requires placing a silicone rubber plug on the upper surface of the can and then puncturing the can through the silicone plug. A hypodermic needle is inserted through the silicone seal to obtain a measured volume of gas, which is injected into a gas chromatograph fitted with a column designed specifically for gas analysis (packed or capillary). The analysis is usually conducted isothermally using helium carrier gas and a flame ionization detector. Each analysis typically takes 5–10 minutes, and individual peaks are recorded for methane (C_1), ethane (C_2), propane (C_3), isobutane (i-C_4), and normal butane (n-C_4). After elution of n-C_4, the column flow is reversed (backflush mode), and a single peak is recorded for pentanes plus heavier hydrocarbons (C_{5+}). Detector responses for each analyte are compared with those of standards, and after obtaining appropriate volumetric measurements of the can and its contents, the light hydrocarbon gas concentrations (C_{1-5+}) are calculated in parts per million (ppm) by volume of sediment.

Airspace gas concentrations provide information on the hydrocarbon potential of a sedimentary profile. However, due to the effects of pressure depletion and outgassing at the surface, not all rock types are equally suited to this type of analysis. Course-grained reservoir rocks, which are typically more porous and permeable than fine-grained nonreservoir units, lose most of their gaseous hydrocarbons prior to canning at the well site. Therefore, unless the reservoir unit is unusually tight, limited information is available from the analysis of airspace gases. Fine-grained rocks, in contrast, retain a significant amount of their light hydrocarbons after

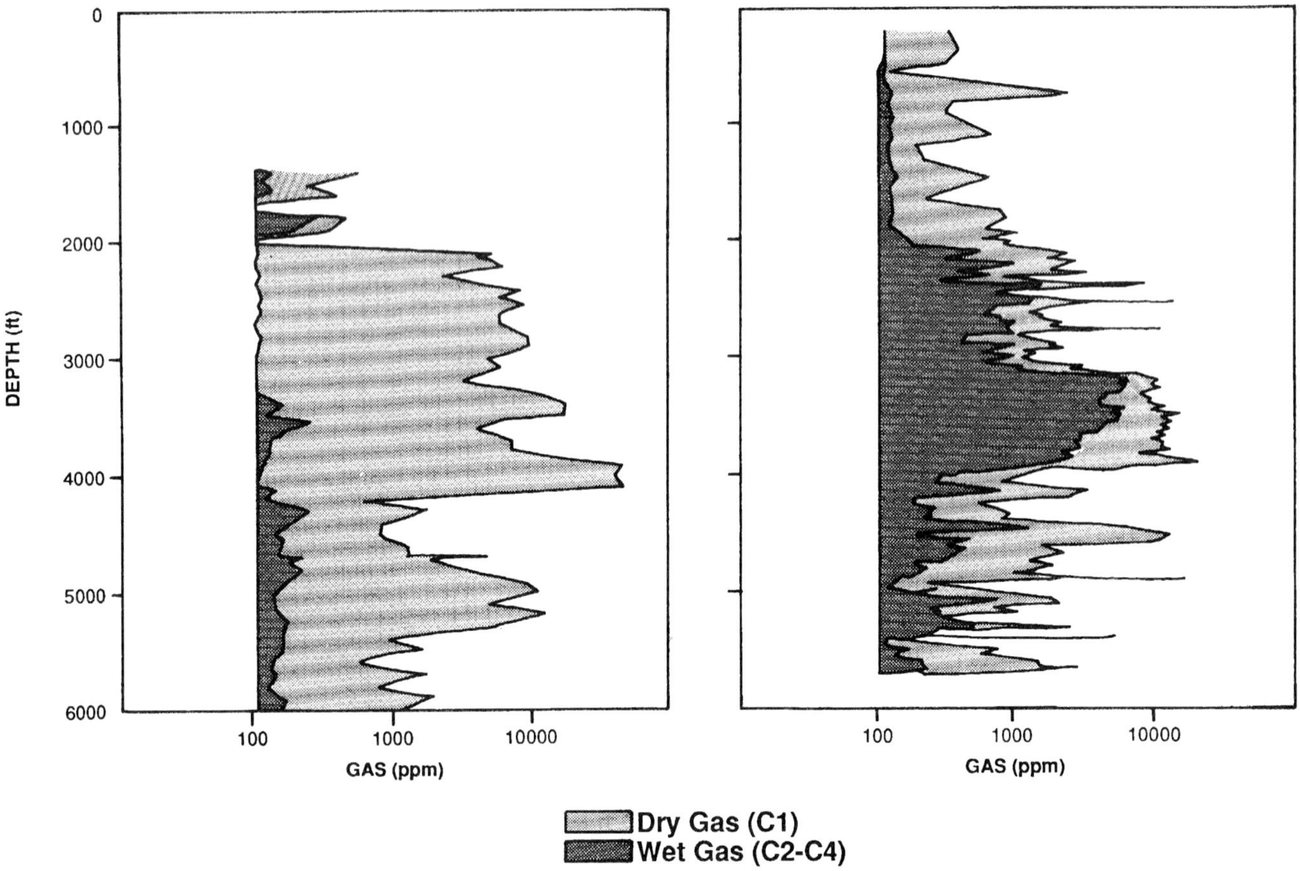

Figure 1. Airspace gas logs for two wells in the Western Canada basin. (After Evans and Staplin, 1971.)

reaching the surface and are more amenable to airspace gas analysis. Fortunately, a correlation often exists between the occurrence of producible hydrocarbons in porous reservoir units and high concentrations of airspace gas in shale cuttings from the same borehole. Correspondingly, Hunt (1979) has stated that as a general observation, hydrocarbon accumulations are usually found in reservoir rocks of stratigraphic sections where the nonreservoir rocks contain relatively large amounts of gas.

From an exploration standpoint, the concentration and composition of the airspace gases are important in evaluating the types of reservoired hydrocarbons that can occur within a given sedimentary section. To evaluate the data, airspace gas concentrations (in ppm by volume) are usually plotted versus depth on a semilog diagram. Figure 1 shows an example of typical plots obtained from published data for two wildcat wells drilled in the Western Canada basin (Evans and Staplin, 1971). Total gas concentrations in shaley intervals that exceed 100 ppm are considered to be above the "background" level and those above 1000 ppm and 10,000 ppm are considered to be rich and very rich, respectively. The concentration of wet gas components (C_{2-4}) are shown in the darker shaded regions of the log. Cuttings from the well on the left in Figure 1 recorded mainly dry airspace gas (predominantly methane) throughout the entire section, which defines this as a gas-prone area. The cuttings from the other well (on the right) afforded a high yield of wet gas, particularly around the 3000–4000 ft (900–1200 m) interval. This defines a more oil-prone section, and suitable traps and reservoirs associated with these intervals could contain liquid hydrocarbons.

Disaggregate Cuttings Gas

As previously discussed, the light hydrocarbons released through mechanical pulverization and heating of the cuttings *in situ* are referred to as disaggregate cuttings gas. Typically, 5–10 g of washed cuttings are placed in a stainless steel cell fitted with a side arm and septum. Two stainless steel balls and 20 mL of water are added, bringing the airspace in the cell to about 10 mL. The cell is placed in a hot water bath (90–95°C) to promote the transfer of lightly sorbed hydrocarbons into the gaseous phase and is then mechanically agitated for 10 min. An aliquot of gas is withdrawn through the septum and analyzed by gas chromatography using the same conditions as previously described for airspace gas. The concentrations of gaseous hydrocarbons are reported in parts per million by volume of sediment.

The absolute concentration of disaggregate cuttings gas from a particular stratigraphic interval is often lower than the corresponding airspace gas concentration. Furthermore, the composition of the gases vary, with the disaggregate cuttings gas generally being depleted in methane relative to the airspace gas. Therefore, for the purpose of ranking the hydrocarbon potential of different horizons, only gas concentrations and compositions obtained from the same analytical technique should be compared. Some companies prefer to analyze for both airspace gas and disaggregate cuttings gas and then sum the values to derive a total light hydrocarbon concentration. This provides the most reliable means of evaluating hydrocarbon potential, but it is more time consuming and expensive than a single analysis. Moreover, the condition of samples received from the well site often dictates the type of analytical method that can be used. For instance, if only bagged cuttings are available, then disaggregate cuttings gas analysis is the only option. However, if the samples are canned at the well site, then the analysis of airspace gas is generally preferred to disaggregate cuttings gas. In either case, data can be obtained that will allow a preliminary assessment of the petroleum potential of a sedimentary sequence, and the light hydrocarbon analyses provide an effective method for screening for the most productive zones.

Bitumen Extract

Petroleum geochemists use the term *bitumen* to describe the organic matter of fine-grained sedimentary rocks that is soluble in common organic solvents (e.g., methylene chloride, chloroform, and methanol). It represents the indigenous organic fraction that has not been expelled from the source rock during subsidence and thermal maturation. Other frequently used synonyms for bitumen include extractable organic matter (EOM), soluble organic matter (SOM), and total soluble extract (TSE). Based on this definition, it is clear that source rock bitumen is different from the residual oil fraction that may occur in reservoir rocks, either at depth or exposed at the surface. The latter represent nonindigenous hydrocarbons that have undergone secondary migration from their source beds. However, they are also sometimes referred to as *solid bitumens* or *heavy bitumens*, particularly when the deposit has been altered by in-reservoir processes such as biodegradation and water washing. The distinction between these native substances and the indigenous bitumen extracted from fine-grained source rocks is therefore an important one for a thorough understanding of geochemical terminology.

Bitumen analyses are normally performed on prospective source rock intervals, where total organic carbon values generally exceed 1–2% by weight. Two or three analyses per source rock interval are usually enough to characterize the soluble organic matter, although in extremely heterogeneous sections, a higher sampling frequency may be required. The bitumen can be isolated from the sediments by a number of solvent extraction techniques, which include Soxhlet reflux, high-speed flow blending, and ultrasonic agitation. After extraction, the bulk of the solvent is removed by distillation, and the bitumen is dried to constant weight under a stream of nitrogen gas at 40°C. This method of isolation removes most of the volatile components with less than 15 carbon atoms, and the yield of bitumen is therefore reported as parts per million by weight of C_{15+} compounds.

Once isolated in this manner, the C_{15+} bitumen extract has chemical and physical properties that resemble a "topped" crude oil, and both are analyzed in a similar way. The remainder of this discussion on analytical techniques is therefore equally applicable to bitumen extracts and crude oils.

Asphaltenes are removed from the C_{15+} bitumen extract or crude oil by precipitation with *n*-pentane. The remaining soluble material is separated by liquid chromatography into subfractions containing organic compounds with similar chemical functionalities, namely, saturates (or aliphatics), aromatics, and polar compounds (NSO fraction). Of these fractions, the saturates and aromatics are the most amenable to analysis by gas chromatography and can be used to provide the following information.

Saturate fraction gas chromatograms are used most frequently as a means of "fingerprinting" source rocks and crude oils during the preliminary stages of an oil–source rock correlation study. Tentative relationships established in this way are then further investigated using more detailed methods, such as GCMS (see next section). The saturate gas chromatograms are also used as an initial guide to characterizing the environment of deposition, type of organic matter, and thermal maturity of a source rock. Figure 2 shows the saturate gas chromatograms for two carbonaceous shales that have very different geological origins. Sample 1 is a lacustrine shale that contains organic matter derived predominantly from freshwater algae. It was deposited under suboxic conditions and has presently attained a thermal maturity level corresponding to the onset of oil generation. Sample 2, in contrast, contains organic matter derived predominantly from land plants and was deposited in a more oxidizing deltaic environment. It has presently not attained the maturity level required for hydrocarbon generation, and at higher maturity, it would be expected to act as a source for waxy crude oils. These inferences are based on the distributions of specific compound classes, such as normal alkanes and

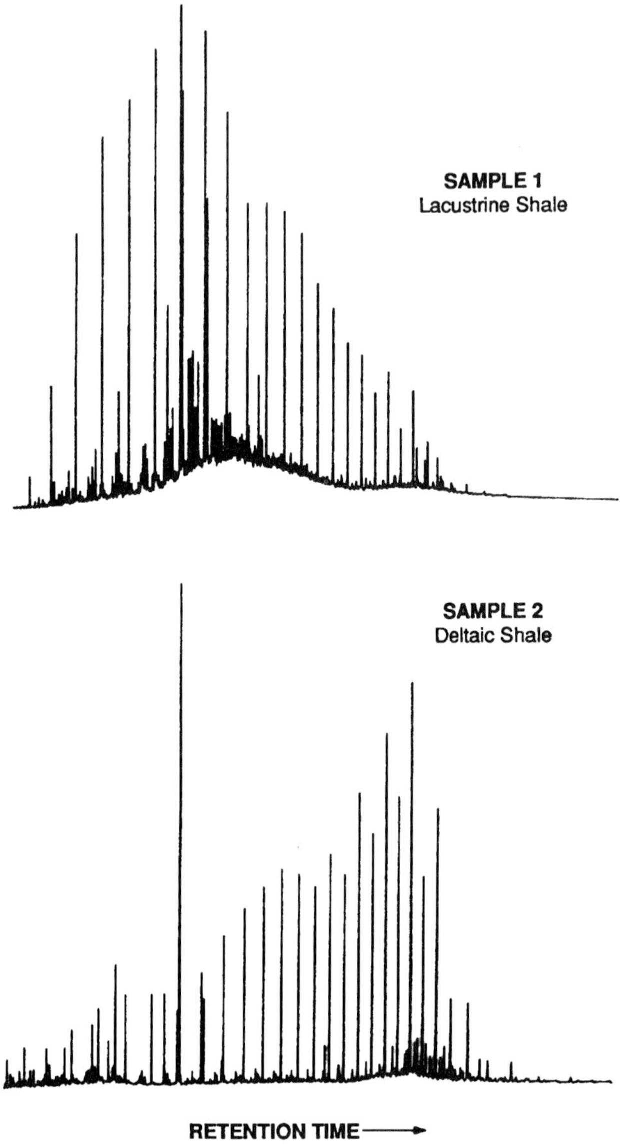

Figure 2. Gas chromatograms for the saturated fraction of two carbonaceous shales.

isoprenoid alkanes, and can be readily achieved by geochemists and/or geochemically oriented geologists who are familiar with the origin and evolution of sedimentary organic matter.

Analysis of aromatic hydrocarbon fractions by gas chromatography is a more recent development in petroleum geochemistry (reviewed by Radke, 1987). This method shows particular promise in the area of thermal maturity assessment, where molecular parameters have been formulated that correlate with conventional maturity indicators such as vitrinite reflectance. This is particularly attractive in the case of crude oils since the molecular parameters can be used to estimate the thermal maturity of source rocks from which a crude oil was expelled. Figure 3 shows a gas chromatogram of a typical aromatic hydrocarbon fraction. The major peaks correspond to different isomers of alkylnaphthalenes and phenanthrenes. In addition to the maturity applications mentioned previously, parameters have also been developed that provide information on the provenance of the organic matter and the lithology of the source rock (i.e., carbonate versus siliclastic). Continued research in this area should result in additional applications in petroleum exploration.

GAS CHROMATOGRAPHY–MASS SPECTROMETRY (GCMS)

Gas chromatography–mass spectrometry (GCMS) is an analytical technique used to study the occurrence and distribution of specific organic compounds in complex hydrocarbon mixtures. As its name suggests, it is a combination of two analytical methods, gas chromatography and mass spectrometry, each of which is a powerful tool in its own right. By combining the two techniques, very small amounts of sedimentary organic matter can be accurately analyzed for compounds of geochemical interest. The types of samples required for this purpose are the same as those used for gas chromatography analysis: C_{15+} bitumen extracts of fine-grained sediments (outcrops, drill cuttings, or cores); produced crude oils and DST samples; oil shows and seeps; and asphalts and other heavily degraded substances. One of the main attractions of the GCMS technique is that very small amounts of material are required for analysis. Another benefit is that the technique is not sensitive to variables such as length of warehouse storage or physical condition of the sample, i.e., wet versus dry cuttings or preserved versus unpreserved core. This permits analysis of just about any type of source rock or petroleum sample, provided that it has not been subjected to chemical alteration or high-temperature drying (above 100°C).

GCMS analysis of petroleum extracts is used primarily to characterize the distributions of trace organic substances refered to as biological marker compounds (biomarkers). These biomarkers are organic compounds whose carbon skeletons can be related directly to precursor molecules that were incorporated into the sediments from the remains of living organisms. Hence, the term *chemical fossil* has also been used to describe them, due to the obvious precursor–product relationship they show with specific biological substances (for a thorough review, see Mackenzie, 1984; Philp, 1985b).

The biomarker patterns found in crude oils and source rocks depend both on the type of organic matter incorporated during sedimentation and the subsequent chemical changes that occur during burial in the subsurface. Since these are complex processes that vary considerably from one depositional setting to another,

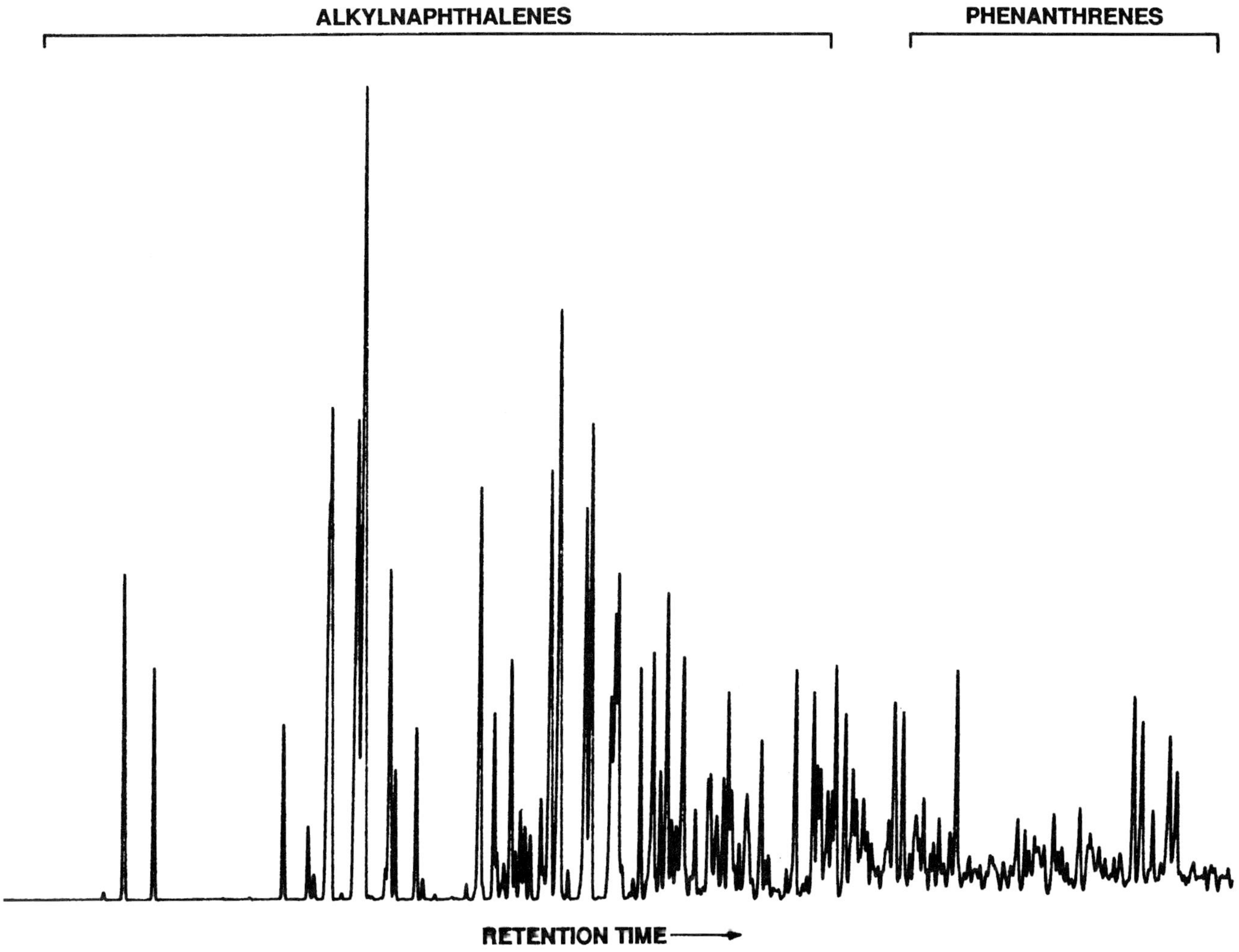

Figure 3. Gas chromatograms for the partial aromatic fraction of a carbonaceous shale.

the resulting biomarker distributions are often very different and provide a useful means for fingerprinting geological samples. Accordingly, biomarker patterns can be used to correlate crude oils with their source formations, numerous examples of which have been presented in the geological literature (e.g., Seifert et al., 1980). Certain biomarker species are also very resistant to weathering processes such as biodegradation, which permits correlation of samples otherwise devoid of any source indicators (e.g., surface seeps and tar sands). Nevertheless, despite the obvious attraction of using biomarkers for correlation purposes, it must be remembered that these compounds occur in very small concentrations in crude oils and ancient sediments, and thus care must be taken to avoid cross-contamination of samples. It is therefore strongly recommended that whenever possible, biomarker analyses should be supplemented with other geochemical data, such as gas chromatography profiles and stable isotope ratios, before any conclusions are drawn on oil–source rock relationships.

There are many different types of biomarkers that occur in sedimentary organic matter. The most frequently encountered species are usually classified according to their basic carbon skeleton. For example, hopanes, steranes, cheilanthanes, drimanes, and porphyrins all represent different classes of biomarkers. Each class has several different members, all of which have the same basic carbon skeleton but which differ slightly in their orientation and number of carbon atoms. The biomarker groups can often be related to specific types of precursor organisms, which in turn can be used to derive information about the type of organic matter and the environment in which it was deposited. Some of the more notable examples are the association of hopanes with procaryotic microorganisms (bacteria), oleananes and lupanes with higher plants (angiosperms), phyllocladanes and pimaranes with conifers, and certain steranes with dinoflagellate algae. Based on biomarker associations of this type, GCMS analysis of petroleum extracts is now routinely applied to the source characterization of sedimentary organic matter.

Biomarker distributions are also sensitive to the time–temperature history of sedimentary rocks. With increasing burial in the subsurface, biomarker patterns often change in a systematic way, which can be used to infer maturation trends. The most commonly used compounds for this purpose are hopanes, steranes, and aromatic steroids, all of which are readily detected by GCMS. The biomarker maturity parameters provide a useful supplement to conventional maturity indicators such as vitrinite reflectance and spore coloration and also have the advantage of being applicable to the thermal history assessment of crude oils.

Chapter 9

Stable Isotopes in Petroleum Exploration

Zvi Sofer
Amoco Production Company
Tulsa, Oklahoma, U.S.A.

INTRODUCTION

The importance of determining the abundance of stable isotopes of various elements in organic matter to help understand different geochemical processes has been recognized since the early stages of the development of the field of organic geochemistry. In gas and petroleum exploration, stable isotopes have been useful for correlation, maturation, and alteration studies and in evaluation of source rock depositional environment. Because organic matter is composed of carbon and hydrogen, as well as some nitrogen, oxygen, and sulfur, isotopic compositions for these elements in organic matter are reported in the literature.

The isotopic composition of an element is determined by first converting it quantitatively to a gaseous form (CO_2, H_2, N_2, SO_2, and CO_2 for carbon, hydrogen, nitrogen, sulfur, and oxygen, respectively) and then analyzing the gas on a mass spectrometer. Isotopic compositions are reported as δ values per mil (‰), which are defined as

$$\delta(‰) = [R_{sample}/(R_{standard} - 1)] \times 1000 \qquad (1)$$

where R_{sample} is the ratio of the heavy-to-light isotopes of the element of interest ($^{13}CO_2/^{12}CO_2$, HD/H_2, $^{15}N^{14}N/^{14}N_2$, $^{34}SO_2/^{32}SO_2$, and $C^{18}O^{16}O/C^{16}O_2$ for the previously listed elements, respectively) and $R_{standard}$ is the respective ratio in an internationally accepted standard (PDB for carbon, SMOW for hydrogen and oxygen, atmospheric N_2 for nitrogen, and troilite from the Cañon Diablo meteorite for sulfur). By definition, all standards have δ values of 0.0‰. Samples with R larger than that of the standard have positive δ values. Negative δ values are observed in samples where R is smaller than the ratio in the standard.

Because δ values can be negative, such terms as *isotopically more negative, isotopically more positive, isotopically heavier, isotopically lighter, heavy isotopic composition, light isotopic composition*, and so on are commonly used. These all describe a relative situation pertaining to the amount of heavy isotope in a sample. Isotopically more positive or more negative describe which δ values (or groups of δ values) represent samples that have more or less of the heavy isotope, respectively. Isotopically heavier or lighter are equivalent to isotopically more positive or negative, respectively.

Due to the analytical difficulties in quantitatively converting organic oxygen and nitrogen to CO_2 and N_2, isotopic compositions of these two elements are rarely used in petroleum exploration. Hydrogen isotopes are used occasionally in petroleum exploration and are commonly used in gas exploration, while sulfur isotopes are used occasionally in petroleum exploration. Carbon isotopes, however, are by far the most frequently used in gas and petroleum exploration. The following sections briefly describe the uses of the isotopic compositions of carbon, hydrogen, and sulfur in petroleum and gas exploration.

CARBON

The isotopic composition of the insoluble organic matter (kerogen) in rocks is commonly analyzed. For rock extracts and crude oils, bulk isotopic compositions are commonly determined. These analyses included whole-rock extracts and whole oils, as well as saturate, aromatic, heteroatomic (NSO), and asphaltic fractions obtained by various precipitation and liquid chromatographic techniques. Recently, it has become possible to determine the isotopic composition of individual hydrocarbons in crude oils and rock extracts. However, the usefulness of these measurements for petroleum exploration is still under investigation and will not be discussed here. Natural gas can be easily separated into individual compounds using a gas chromatograph (GC) or cryogenic techniques. Therefore, the isotopic composition of methane and sometimes ethane, propane, and the two butane isomers can be determined.

The isotopic composition of insoluble sedimentary organic matter (kerogen) varies from –15 to –35‰, but most kerogens show a narrower range of –23 to –32‰. In general, it seems that there is a trend of increasing δ values in kerogens from Paleozoic to Recent times. However, examples of extreme δ values are abundant throughout the geological record. Therefore, attempts to associate certain ranges of δ values in individual samples with different geological ages have failed. In Recent deltaic and estuarine depositional environments (with a significant nearshore land-derived organic input), the isotopic composition of the organic matter becomes more positive seaward (Scalan and Morgan, 1970). This led to the idea that land-derived organic matter is isotopically more negative than marine-derived organic matter. Later studies (Deines 1980) and indirect evidence based on the isotopic composition of oils (Yeh and Epstein, 1981; Sofer, 1984) have shown that the isotopic composition of kerogen is not indicative of any particular type of depositional environment. The isotopic composition of kerogen is probably the most important factor in determining the isotopic composition of the oil generated from it. Sthal (1977) has shown, in what he termed as *type curves,* the relationship between the isotopic composition of kerogen and the isotopic composition of various oil and rock extract fractions (saturates, aromatics, NSOs, and asphaltenes). Normally, the isotopic composition of the various fractions becomes less negative with increasing polarity from saturates to kerogen. These type curves can therefore be used to estimate the isotopic composition of an oil generated from a given kerogen (and vice versa) and hence become helpful in oil–source rock correlation studies.

Isotopic compositions of rock extracts and oil fractions generally range between –17 and –36‰, with most samples falling between –23 and –32‰. They are important in establishing genetic relationships between source rocks and oils and among various oils. Fuex (1977) suggested that in oil–source rock and oil–oil correlations, a genetic relationship is usually established when the isotopic composition of the saturate and/or aromatic fraction differs by less than 1‰. Sofer (1984) has shown that this is true for correlations of marine oils. He also showed that nonmarine oils can be isotopically different by as much as 4‰ and still be genetically related. Genetically related marine oils or oils and source rocks can have δ values that differ by as much as 2‰ if significant differences in maturation, migration, or biodegradation are involved. Both maturation and migration tend to increase the δ value of the saturate and aromatic fractions as well as the δ value of the whole oil or extract. Biodegradation processes do not cause systematic changes in the isotopic composition of genetically different oils (Sofer, 1984). In some oils, both the saturate and aromatic fraction become isotopically more positive, while in others, no change is observed. In a few cases, one of the fractions does not change, while the other becomes isotopically more positive or more negative. In genetically related oils, however, biodegradation causes similar changes.

Carbon isotopes are often used to distinguish between oils derived from marine and nonmarine organic matter. The differences between marine and nonmarine oils are manifested in the relationships between the isotopic composition of the saturate and aromatic fractions (Sofer, 1984). The canonical variable (CV), derived from multivariant discriminant analysis is one parameter that formulates this relationship:

$$CV = -2.53\delta^{13}C_{sat} + 2.22\delta^{13}C_{aro} - 11.65 \qquad (2)$$

Values of CV larger than 0.47 indicate a mostly nonmarine source for an oil, whereas CV values smaller than 0.47 indicate a mostly marine source. Alternatively (and practically more useful), on a crossplot of $\delta^{13}C_{aro}$ vs. $\delta^{13}C_{sat}$ (the isotopic composition of the C_{15+} aromatic vs. saturate fractions), the line that best separates marine oils from nonmarine oils has this equation:

$$\delta^{13}C_{aro} = 1.14\delta^{13}C_{sat} - 5.46 \qquad (3)$$

Nonmarine oils plot above this line, while marine oils plot below this line (Figure 1). Since this work was published in 1984, it has been recognized that lacustrine and marine oils follow these relationships in about 90% of the cases. Deltaic oils, however, which were expected to plot above the best separation line, often plot below the line. The relationships shown in equations (2) and (3) were derived from an extensive data base consisting only of crude oils. For an unknown reason, rock extracts seem to follow this relationship in less than 90% of the cases. Some modifications of these relationships may be required in the future to better characterize source rock extracts. Still, the type of diagram shown in Figure 1 is very useful for displaying isotopic data and for illustrating genetic relationships among oils and between oils and source rocks.

Isotopic compositions of methane and other natural gas components have been used to determine genetic characterization (e.g., sapropelic–liptinitic vs. humic organic matter) and maturity of the source rock of the gases. It is difficult to use similarities in isotopic compositions to establish genetic relationships (which are different from genetic characterizations) of natural gases to specific source rocks or other natural gases in a basin. This is because natural gases easily migrate and mix and because there is a multitude of active sources of different depositional environments and levels of thermal maturity that could contribute to the same

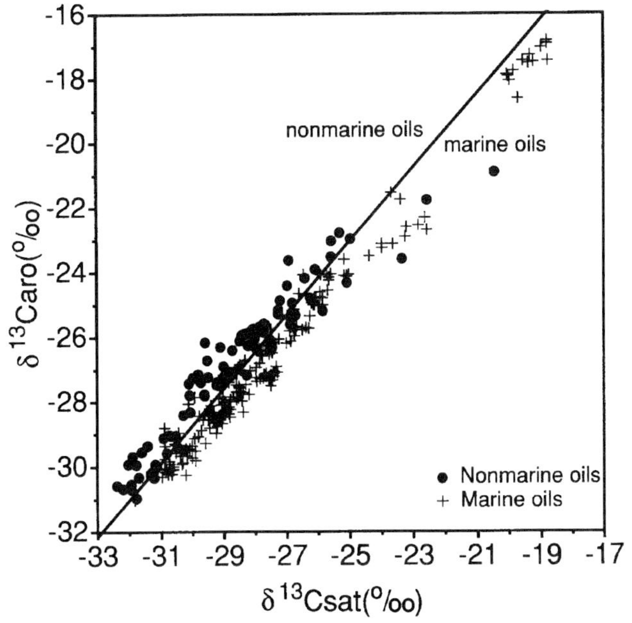

Figure 1. Isotopic composition of saturate and aromatic fractions of crude oils. The best separation line between nonmarine and marine oils ($\delta^{13}C_{aro} = 1.14\delta^{13}C_{sat} + 5.46$) is drawn.

reservoir. Schoell (1983) has analyzed the isotopic compositions of a large number of natural gas samples of various known origins and has observed that on a $\delta^{13}C$ vs. δD diagram, several groups can be identified. Biogenic methane from marine source rocks can be distinguished from biogenic methane of continental origin. Also, gases associated with petroleum can be distinguished from gases associated with condensates and from nonassociated gases. On the methane $\delta^{13}C$ vs. % C_{2+} (gas wetness) diagram, Schoell (1983) has shown that dry biogenic gas has very negative δ values (<-55‰); with increasing maturity of the source, gas wetness increases and the methane isotopic composition becomes less negative. Highly mature gases become dry (due to cracking of the C_{2+} compounds) and, depending on the source, have δ values of –40 to –30‰ (for associated gas and nonassociated gas of sapropelic–liptinitic origin) and of –30 to –20‰ (for nonassociated gas of humic origin).

James (1983, 1990) has used the carbon isotopic differences among the methane, ethane, propane, *n*-butane, and *n*-pentane in natural gases to assess the maturity of the gas source on the level of organic maturity (LOM) scale. Although the absolute isotopic compositions of the individual gas components are source dependent, the differences seem to be source independent. The differences systematically decrease with increasing maturity. The procedure to estimate the maturity of the gas source involves plotting on a moving scale the $\delta^{13}C$ of the various gas components vs. a fixed LOM scale. At a point where the measured differences match with a set of quasi-thermodynamically calculated differences, the level of maturity is read. It should be noted that this procedure cannot be applied to coal gases.

HYDROGEN

Hydrogen isotopes have been analyzed in kerogen, rock extracts, crude oils, and natural gases. Similar to carbon isotopes, they are used for correlation and genetic characterization. Schoell (1984) has reviewed the hydrogen isotopic compositions of various sedimentary organic materials. Kerogens of various origins show a range of compositions from -60 to -170‰, whereas oils show a range of -70 to -200‰. Similar to carbon isotopes, it seems that no relationships can be established between the hydrogen isotopic composition and any particular depositional environment.

However, on a plot of δD_{sat} vs. δD_{aro}, carbonate-sourced oils do seem to be isotopically different from shale-sourced oils (marine and lacustrine). Both fractions of carbonate-sourced oils seem to have δ values more positive than –120‰, whereas shale-sourced oils seem to have δ values more negative than –120‰. As for carbon isotopes, similar isotopic compositions (less than 10‰) between similar rock extracts and oil fractions and among similar oil fractions suggest genetic relationships. Various δD vs. δD and δD vs. $\delta^{13}C$ plots can be used to illustrate such relationships. The effect of maturation, migration, and biodegradation on the hydrogen isotopic composition of crude oils is not well documented.

Hydrogen isotopic compositions of methane in natural gases show a range of –50 to –300‰. Biogenic methane of marine and continental origin shows a range of δD values from –150 to –200‰ and –200 to –250‰, respectively. Low-maturity thermal methane shows δD values of –300‰ that become less negative with increasing maturity (up to –50‰). A plot of $\delta^{13}C$ vs. δD for methane has been used by Schoell (1984) to genetically characterize natural gases and assess their thermal maturity.

SULFUR

Sulfur isotopes have been determined in kerogens, crude oils, and hydrogen sulfide occurring in natural gases. As with other stable isotopes, some workers have attempted to use them for oil–source rock and oil–oil correlations. Sulfur isotopes have also been used to determine the origin of sulfur in crude oils by assuming that primary sulfur would be biogenic in origin and would therefore have $\delta^{34}S$ values much lighter than +22‰ (which is the isotopic composition of marine evaporitic sulfates). Secondary sulfur would have an inorganic origin (from thermal interaction of crude oil and sulfates) and would therefore have $\delta^{34}S$ values near +22‰.

Sulfur isotopes in oils have been notoriously difficult to interpret in terms of geochemical processes. The reason may be that sulfur isotopes have been determined on bulk oil samples that contain some elemental sulfur in addition to organosulfur compounds. The elemental sulfur is probably of secondary inorganic origin (which is isotopically positive), whereas the organosulfur compounds could be both primary biogenic (which is isotopically negative) and secondary inorganic in origin. Fractionating sulfur species in oils and analyzing them separately may resolve some of the problems. Orr and Sinninghe-Damste (1990) have written an extensive review on sulfur and sulfur isotopes in petroleum systems.

Chapter 10

Characterization of Organic Matter by Various Pyrolysis Techniques

R. P. Philp and A. Galvez-Sinibaldi
School of Geology and Geophysics
University of Oklahoma
Norman, Oklahoma, U.S.A.

INTRODUCTION

A major goal in the application of organic geochemistry to petroleum exploration is to provide the best possible input data for mathematical models that integrate the geological and geochemical information to predict the nature, amount, and time required for oil and gas generation (Tissot and Welte, 1984). During the past decade, pyrolysis has become a common analytical method for the analysis of kerogens and whole rocks because it provides the analyst with information about a number of parameters that can give a measure of the quality, quantity, and petroleum generative potential of prospective source rocks (Peters, 1986; Arnosti and Müller, 1987; Tarafa et al., 1988). The various pyrolysis techniques that are commonly used in geochemical laboratories are summarized diagrammatically in Figure 1.

Rock-Eval pyrolysis of whole rock or kerogens provides information on the quantity, quality, type, and thermal maturity of associated organic matter in these materials (Peters, 1986). Pyrolysis methods combined with a high-performance separation technique (capillary gas chromatography) take advantage of the ability of the separation technique to analyze, and identify at the molecular level, the composition of the pyrolysate. The pyrolysate composition provides information that can be used to evaluate, with a high degree of accuracy, the source of the organic material and the maturity level of the samples being analyzed (Larter, 1984). Hydrous pyrolysis of whole rocks is used in the laboratory to simulate the evolution of petroleum-like compounds from the heating of the geological samples, providing a tool to perform oil–source rock correlation studies (Winters et al., 1983; Lewan, 1985).

ROCK-EVAL PYROLYSIS

The Rock-Eval pyrolysis method, which falls under the category of bulk flow pyrolysis (Larter, 1984), generally requires a pyrolysis chamber (Pyrolyzer), a conduit, and a splitter for the pyrolysis products to flow to a series of traps and detectors. The detectors most commonly used are the flame ionization detector (FID) and the thermal conductivity detector (TCD). The pyrolysate is swept by an oxygen-free carrier gas (He or N_2) and split, such that one aliquot of the effluent is directed to the FID and the other to a CO_2 trap and then to the TCD (Figure 2). Various Rock-Eval pyroanalyzers are commercially available, most of which are variants of the instrument developed at the Institute Français du Petrole (Espitalié et al., 1977). Some incorporate additional features, such as the Rock-Eval II system from Delsi Inc., which permits the simultaneous determination of total organic carbon (TOC). Other systems are simpler and lack the means to trap and analyze the CO_2 that is produced during the pyrolytic procedure and consist only of the pyrolyzer and an FID detector assemblage (e.g., the Pyran system from Ruska Laboratories).

When a sample of whole rock is analyzed, the recorded output from the typical Rock-Eval apparatus consists of three peaks: the first two peaks correspond to signals from the FID and the third from the TCD. The first peak represents the products evolved at pyrolysis temperatures up to about 300°C. The area under this peak (S_1) can be related to the yield of free organic matter (bitumen) distilled from the rock. The area under the second peak (S_2) represents the amount of hydrocarbon-like products evolved during the thermal degradation of kerogen. The second peak also provides information about the maturity level of the sample by

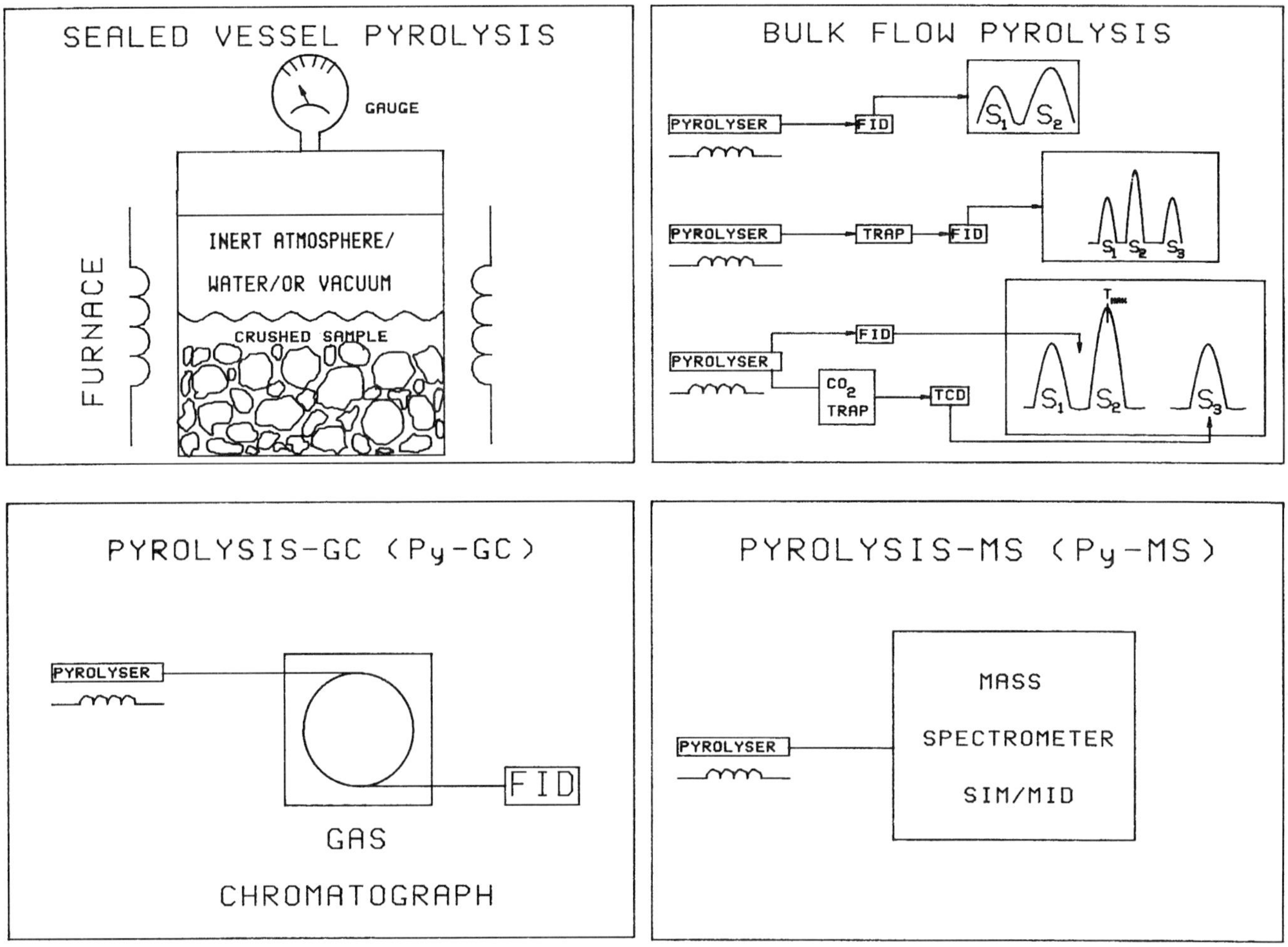

Figure 1. Simplified diagrams showing the various analytical pyrolysis methods currently used in organic geochemistry.

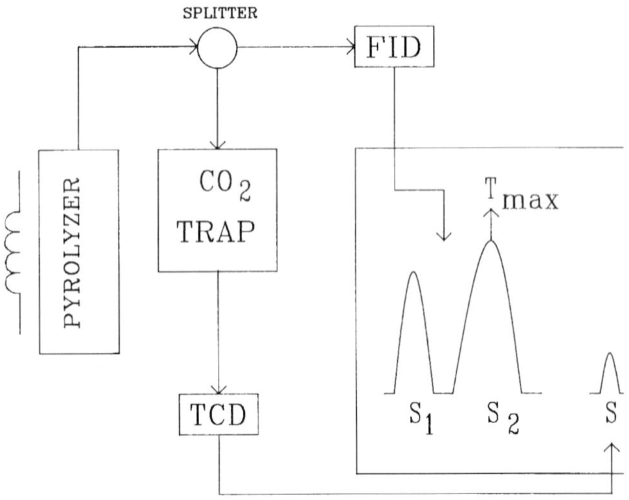

Figure 2. Diagram of a typical Rock-Eval apparatus assemblage and its recorded output.

means of the T_{max} parameter (the temperature at which this peak reaches its maxima). The area of the third peak (S_3) corresponds to the yield of CO_2 that is generated during the pyrolysis from oxygen-containing functional groups in the kerogen (Figure 2).

The three peaks have been traditionally labeled S_1, S_2, and S_3 following the lead of the Institut Français de Petrole, which developed the first Rock-Eval system. If no CO_2 and TCD are provided, the results will only show two peaks (from the FID signal). Most U.S. publications now refer to these peaks as P_1 and P_2, which are analogous to the S_1 and S_2 peaks of the French apparatus. (In this chapter, S_1 and S_2 will be used to designate these peaks.) The results are commonly expressed in milligrams of hydrocarbons (HC) per gram of rock or milligrams of HC per gram of organic carbon (C_{org}), from which several parameters can be derived (Peters, 1986).

Several bulk parameters have been defined to allow

Table 1. Rock-Eval source rock evaluation parameters.[a]

Generation Potential	TOC (wt. %)	S_1	S_2
Poor	0.0–0.5	0.0–0.5	0.0–2.5
Fair	0.5–1.0	0.5–1.0	2.5–5.0
Good	1.0–2.0	1.0–2.0	5.0–10.0
Very good	>2.0	>2.0	>10.0

Type	HI[b] (mg HC/g C_{org})	S_2/S_3
Gas	0–150	0–3
Gas and oil	0–150	3–5
Oil	>300	>5

[a]After Peters (1986).
[b]Assumes a level of thermal maturation equivalent to $R_o = 0.6\%$.

for the characterization of the source rocks being analyzed in terms of productivity and maturation level. The ratio of S_1 to the sum of the S_1 and S_2 peaks defines the *production index (PI)*, a parameter that is used to determine the extent of petroleum generation based on the fact that S_1 will increase and S_2 will decrease when petroleum is generated from the organic matter. The *hydrogen index (HI)* corresponds to the quantity of organic compounds or "hydrocarbons" from S_2 relative to the TOC (S_2/C_{org}). Conversely, the *oxygen index (OI)* corresponds to the quantity of CO_2 relative to the TOC. These two indices are analogous to the atomic H/C and O/C ratios used in Van Krevelen diagrams to determine kerogen type and therefore source quality (Horsfield et al., 1983). Since pyrolysis does not require any prior sample preparation or kerogen isolation procedure, it has proven to be a more rapid and hence less expensive method than the conventional elemental analysis. Table 1 summarizes some of the assessments that can be made using the Rock-Eval technique.

The T_{max} parameter mentioned earlier has been shown to increase as the maturity of the sample increases, therefore it can be used together with PI to estimate maturity levels. T_{max} is measured from the pyrograms as the pyrolysis temperature at which the S_2 peak reaches its maxima, but it is strongly dependent on the nature of the organic matter (source) (Peters, 1986). As a general rule, it can be said that the top of the oil window is reached when PI approximates 0.1 and T_{max} is in the range of 435°–445°C ($R_o = 0.6\%$). The bottom of the oil window (wet gas zone) can be roughly delineated by a PI greater than 0.4 and a T_{max} greater than 470°C (Peters, 1986).

The mineral matrix in which the kerogen is contained has been shown to affect the measurements obtained from Rock-Eval analyses. Both HI and OI are strongly affected by the mineralogy and carbon content of the rock samples (Katz, 1983). Furthermore, pyrolysate composition (S_1 and S_2 yields) can also be affected by surface and interlayer absorption phenomena caused by some clays and other sedimentary minerals (Dembicki et al., 1983). Espitalié et al. (1980) studied the effect of various minerals upon the yields of kerogen pyrolysates and found that clay minerals can reduce the yields by 30–85%, whereas gypsum or calcite can reduce the yields by as much as 15% for samples with TOC values of less than 0.9%. It was also shown that if the organic carbon content exceeds 10%, then the mineral effect is negligible, due to saturation of the active mineral sites by the large amounts of pyrolysate produced (Espitalié et al., 1980).

In conclusion, Rock-Eval analysis is a technique well suited for the rapid determination of source quality, maturity, and petroleum potential of rock samples. Its versatility and ease of operation, as well as the high turnaround rate of analysis, permits its use at the well site for the screening of a large number of samples for more detailed analysis (Larter, 1984).

PYROLYSIS–GAS CHROMATOGRAPHY

Pyrolysis–gas chromatography (PY-GC) takes advantage of the ability of the chromatographic system to resolve complex mixtures into individual components. The fingerprints obtained in this way can be used to classify the different kerogen types in terms of the chemical composition of the pyrolysates (Larter and Douglas, 1980; Larter, 1984). PY-GC methods are also employed to elucidate the structure of kerogens and to monitor structural changes caused by the natural (catagenesis) or artificial maturation of the organic matter present in sediments (Monthioux et al., 1986).

Early studies using PY-GC showed that hydrogen-rich kerogens produced pyrolysates dominated by aliphatic compounds (*n*-alkanes + alkenes) with minor contribution of aromatic species, whereas hydrogen-poor kerogens produced pyrolysates with minor contributions of aliphatics and an abundance of aromatic compounds (Larter, 1984). As a result of this observation, Larter (1984) developed a general classification scheme by measuring the ratio of aliphatic vs. aromatic components in kerogens, humic materials, and coal macerals (Figure 3). Specific use was made of the *m,p*-xylene vs. *n*-octane ratio in combination with the carbon percent (dry ash-free basis) rank.

In 1985, Larter and Sentfle described an improved kerogen typing scheme that not only used quantitative PY-GC data but also incorporated petrographic data. This procedure allows for the characterization of kerogens at all levels of maturity and results in the classification of eight basic kerogen types that can be correlated to their respective depositional environment. In summary, these basic kerogen types can be described as follows:

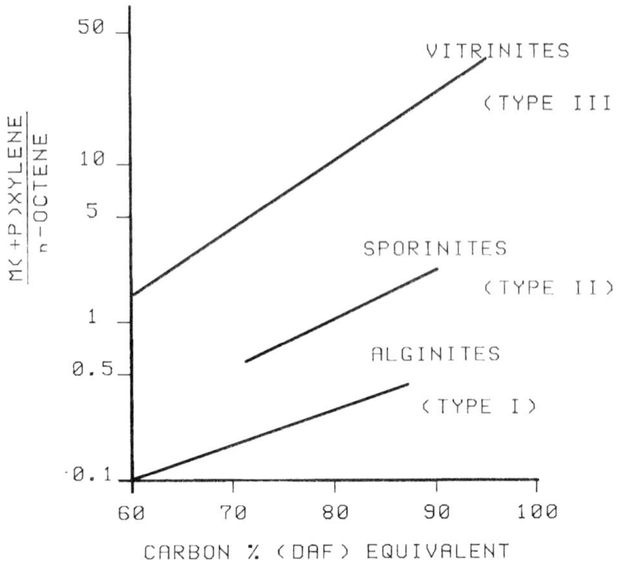

Figure 3. A plot of the *m,p*-xylene/*n*-octane ratio vs. carbon percent (DAF) used to classify the distinct kerogen maceral groups. (After Larter, 1984.)

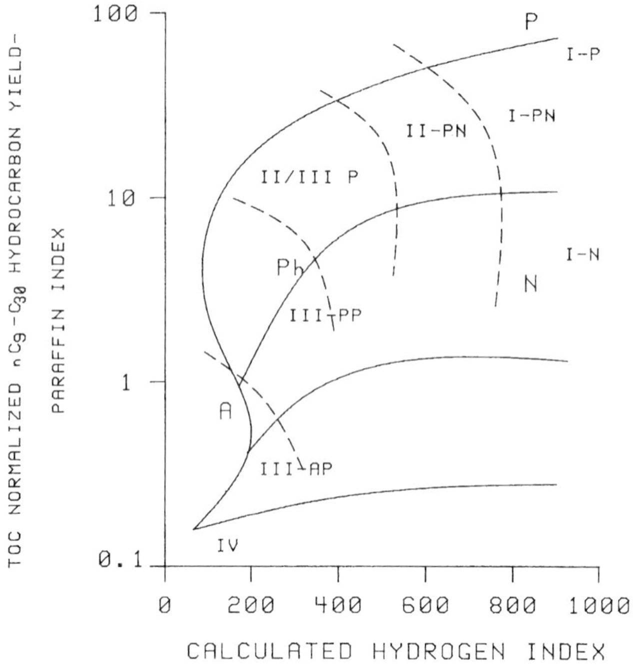

Figure 4. Classification of the basic kerogen types in terms of pyrolysate composition. P = paraffinic, N = naphthenic, Ph = phenolic, A = aromatic, I = kerogen type I, II = kerogen type II, III = kerogen type III, and IV = kerogen type IV. (After Larter and Sentlfe, 1985.)

1. *Type I* kerogens are characterized at low-maturity levels by high HI, but the thermal degradation products may be highly paraffinic (Ip), mixed paraffinic/naphthenic (Ipn), or naphthenic (In), depending on the source facies. Examples of these kerogen subtypes include torbanites, tasmanites, and algal kerogens rich in resinite, respectively.

2. Marine source beds containing amorphous kerogen from organic debris of algal and bacterial sources characterize the *type II* kerogens, which are characterized by intermediate HI and relatively low paraffin content (IIpn). Examples of this type would include kerogens from sedimentary deposits of the Alaskan North Slope basins, the North Sea (Kimmeridge Clay), the southern California basins (Miocene Monterey Shale), and the Paris basin. Type II kerogens that are derived from land plants and are rich in exinite with intermediate HI indices generally produce high-paraffin content pyrolysates (IIp).

3. Depending on the depositional environment, immature *type III* kerogens may have paraffin contents ranging from high (IIIp) to almost zero (IIIap). As maturation proceeds, type III kerogen pyrolysates will become dominated by paraffins and phenolic compounds (IIIpp). Higher maturity levels will produce an increase in the pyrolysate of the concentration of aromatic and phenolic compounds (IIIap), with minor contributions, if any, from paraffins.

4. *Type IV* kerogens are characterized by pyrolysates dominated only by aromatic hydrocarbons.

This information is summarized in Figure 4, which is taken from the work of Larter and Sentfle (1985).

Maturity levels, as just described, can be determined by measuring T_{max} from the Rock-Eval data. However, this method provides only an approximation and has been shown to be affected by the mineral matrix of the sediments (Espitalié et al., 1980; Horsfield and Douglas, 1980; Katz, 1983), by the source materials (kerogen type), and by the maturity level of the organic matter in the samples (Dembicki et al., 1983). To circumvent these effects and to get a better estimate of the maturity level of the organic matter in sediments, a more accurate index of maturity is necessary. Such an estimate can be obtained by measuring the extent of a particular chemical reaction that can reflect the relationship between the temperature history of a sample and its maturity. Such an approach is widely used in petroleum geochemistry in oil–oil and oil–source rock correlation studies (Mackenzie et al., 1981b; Mackenzie et al., 1982). Goossens et al. (1988) has introduced a new molecular parameter for the estimation of maturity in unextracted samples by using PY-GC, namely, the pristane formation index (PFI). Previous work by Van Grass et al. (1981) had shown that the increase in the pristane concentration with increasing depth was accompanied by a decrease of the pyrolysis product prist-1-ene. Assuming a monomolecular reaction and first-order kinetics, a direct relationship between the maturation product (pristane) and the precursor (pristene) can be made. Kinetic parameters can then be calculated that permit reconstruction of the temperature history of the sediments. It has been reported that the PFI shows a good correlation with other more general maturity parameters such as carbon preference index (CPI), T_{max}, vitrinite reflectance (R_o), and depth (Goossens et al., 1988).

PY-GC in combination with mass spectrometry–multiple ion detection (PY-CG-MS-MID, or PY-MID) can also be used to determine source and maturity by selectively monitoring specific ions. Philp and Gilbert (1985b) demonstrated the suitability of this approach by studying a series of vitrinite concentrates isolated from Sydney basin (Australia) coals and using variations in distributions and concentrations of the hopane and sterane biomarkers. It was found in particular that the hopene and 17β(H)-trisnorhopane concentrations were affected by increasing maturity. The changes in the sterane distributions were shown to follow the same trend as those found in extractable fractions that have been discussed in many papers (Seifert and Moldowan, 1981; Philp and Gilbert, 1985b).

HYDROUS PYROLYSIS

Artificial thermal maturation experiments take advantage of the fact that the maturation of organic matter can be considered as a kinetic phenomenon and decribed in terms of the Arrhenius equation (Monthioux et al., 1985). When using the kinetic approach, one can reduce the reaction time by increasing the temperature at which a reaction takes place. It is commonly assumed that the reaction rate doubles for every 10°C (18°F) increase in temperature. Therefore, this approach allows for the simulation of petroleum generation (which is thought to take place on the order of millions of years) at time intervals ranging from hours to days or months, depending upon reaction temperatures (Winters et al., 1983; Monthioux et al., 1985). Many analytical pyrolysis experiments have been designed to fulfill the purposes of artificial thermal maturation experiments, but up to now, all of them can be classified as belonging to one of four basic types: (1) open system pyrolysis, (2) sealed tube pyrolysis, (3) pyrolysis in an autoclave, and (4) confined pyrolysis. The last three of these procedures can be modified so that the reaction can take place under vacuum in the presence of an inert gas (He or N_2) (anhydrous pyrolysis) or in the presence of water (hydrous pyrolysis). The nature of the pyrolysis products is affected by the type of pyrolytic procedure used. However, it has been found that if the pyrolysis reaction takes place in the presence of water, the

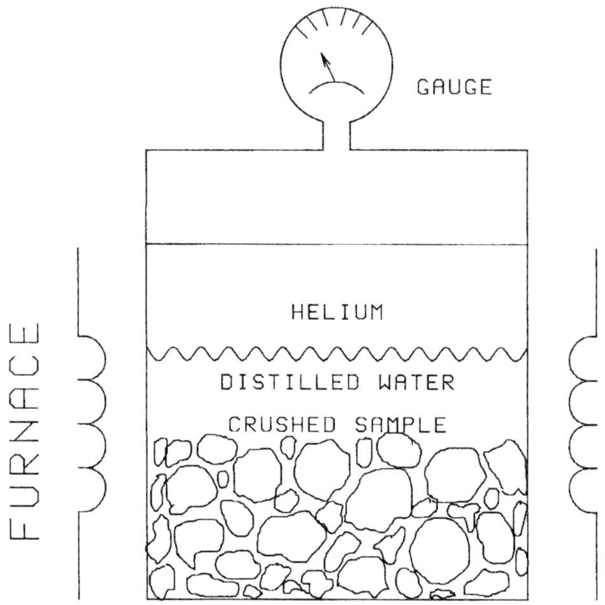

Figure 5. Diagram of a hydrous pyrolysis reaction vessel.

pyrolysate composition closely resembles that of oils found in natural systems (Winters et al., 1983). The procedure takes into consideration the *in situ* generation of petroleum-like products. It also allows for the study of the changes that occur on the kerogen structure as a result of maturation and facilitates monitoring of the changes in biomarker distribution of the "oils" so produced (Winters et al., 1983; Lewan, 1985; Monthioux et al., 1985, 1986). Stable isotope studies have also been performed on the pyrolysates obtained by this method (Lewan, 1983).

Hydrous pyrolysis, as described by Winters et al. (1983), involves heating a weighed amount of broken pieces of shale in a stainless steel autoclave to which a certain volume of distilled water is added (Figure 5). Oxygen is removed from the reaction vessel by pressuring the system several times with helium. The temperatures and reaction times are defined by the maturity of the kerogen to be analyzed and the final objectives of the experiment. Typical heating temperatures are in the range of 290–360°C (an upper temperature limit is imposed by the critical temperature of water, 374°C). As a result of water expansion, a buildup of pressure is expected to be in the range of 3000–5000 psi. The length of the experiments normally takes from 1 to 5 days. At the end of the experiment, the system is allowed to cool to ambient temperature, the final pressure is measured, and a sample of gas is taken for analysis. On the top of the water, an oil layer forms that can be further analyzed to provide information about maturity and biomarker distributions.

Pyrolysis of kerogen aliquots at different tempera-

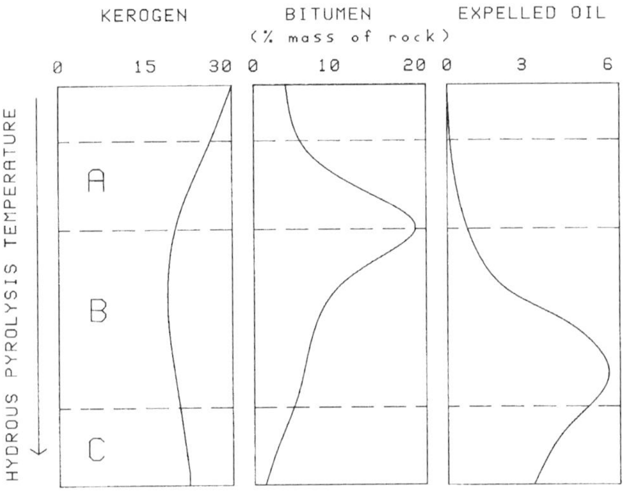

Figure 6. Evolution paths for kerogen, bitumen and expelled oil during a hydrous pyrolyis experiment. (a) Bitumen generation stage. (b) Oil generation stage. (c) Oil cracking stage.

tures shows that the process of hydrocarbon generation can be divided into three main stages. The first stage is bitumen generation, which results from the cracking of the kerogen matrix and is accompanied by production of small amounts of petroleum-like products (Lewan, 1985). Next is the oil generation stage, which is reached at temperatures high enough to produce cracking of the kerogen and bitumen. The third and final stage is oil cracking, which is a stage comparable to that of metagenesis when the petroleum-like pyrolysate and the residual bitumen suffer severe cracking and produce gas and pyrobitumen (Lewan, 1985; Huizinga et al., 1988). If the concentrations of kerogen, bitumen, and expelled pyrolysate are plotted against pyrolysis temperature, a profile of the process can be made that allows for the graphic representation of the oil generation process and the identification of these stages (Figure 6). It can be observed in this profile that, contrary to the general assumption, the peak of bitumen production does not coincide with that of expelled pyrolysate (Lewan, 1985; Huizinga et al., 1988), hence the necessity for involving a step-by-step process.

The temperature (and therefore time in the natural system) at which a rock will commence to generate hydrocarbons depends to a large extent upon kerogen composition. When pyrolyzing type II kerogens of similar maturities but different sulfur contents, Lewan (1985) observed that kerogens with higher sulfur content produce hydrocarbons at a lower temperature than those with lower organosulfur content. This phenomenon has important economic implications, as it is now accepted that source rocks with a significant sulfur content can produce oil at a lower level of maturity (Lewan, 1985; Orr, 1986).

Hydrous pyrolysis data can also be used for kinetic studies of the oil generation process (Lewan, 1985; Huizinga et al., 1988). By considering the overall reaction from kerogen to oil and assuming a first-order reaction mechanism, a good correlation is found that allows for the calculation of kinetic parameters such as frequency factors, activation energies, and rate constants. These parameters can be incorporated into basin models, along with other geological data, to predict timing and quantity of oil generation (Lewan, 1985).

SUMMARY

A large variety of pyrolysis techniques are available for the characterization of geochemical samples. The nature of information obtained will depend to a large extent on the type of pyrolysis technique used. In brief, the major pyrolysis approaches are used either to fingerprint the pyrolysis products and relate these back to the original source material or to simulate natural maturation processes. With increasing development and sophistication of available analytical techniques, our ability to analyze and evaluate the products of these pyrolysis experiments has greatly increased. With this added knowledge, it is becoming more and more evident that it is possible to utilize much of the pyrolysis data in a fashion similar to that of the biomarkers in the soluble fraction. Furthermore, derivation of kinetic parameters, which are subsequently used as input for basin modeling studies, is becoming another important role for pyrolysis experiments. Finally, the simulation of petroleum generation by hydrous pyrolysis experiments in the past few years has become a major experimental technique in many geochemical laboratories. Data are used for modeling studies as well as for obtaining information on the nature and type of products that are generated by various source rocks.

Chapter 11

Total Organic Carbon (TOC) Analysis

Daniel M. Jarvie
Humble Instruments & Services
Humble, Texas, U.S.A.

INTRODUCTION

Total organic carbon (TOC) is a measurement of the organic richness of sedimentary rocks. Its utility for screening potential petroleum source rocks is based on the source of known petroleum reserves. This source is organic matter derived from a variety of biological origins which has been deposited and buried through geological time (Tissot and Welte, 1984; Hunt, 1979). While some constituents of petroleum are directly incorporated into petroleum (biological markers or biomarkers), the bulk of petroleum is formed from the thermal decomposition of organic matter through time. The reactions resulting from increased thermal stress yield the large variety of carbon-containing compounds of which oil and gas are composed. Since the oil or gas potential of a formation is related to its carbon content, TOC analysis is typically the first screening analysis completed to evaluate the general potential of a formation to generate hydrocarbons. Additional analyses confirm the source potential and elucidate source characteristics such as the level of thermal exposure (maturity) and the likelihood of oil, gas, or a mixture of the two being generated (kerogen type).

Organic carbon is distinguished from inorganic carbon by its derivation. Organic carbon is derived from biogenic matter, whereas inorganic carbon is derived from mineral matter. Inorganic carbon is an oxidized form of carbon often combined with calcium, magnesium, or other elements commonly found in carbonate rocks.

Carbon constitutes 75–95 wt. % of hydrocarbons by molecular weight, with an average value near 83 wt. %. The nonhydrocarbon organic constituents of petroleum (nitrogen, sulfur, and oxygen containing compounds) and the bulk of the asphaltene fraction as well as the bitumen and kerogen of petroleum source rocks also contain high percentages of organic carbon.

MODEL OF ORGANIC CARBON

To help in understanding the derivation of the organic carbon content of a sediment sample, a descriptive model is presented in Figure 1, which illustrates the distribution of carbon in organic carbon (Espitalié et al., 1984). In a given volume of sediment sample, only a very small percentage of the volume is from organic carbon. Because TOC values are reported in weight percent carbon, a TOC value of 1 wt. % means that in 100 g of sediment sample, there is 1 g of organic carbon. This is illustrated in Figure 1 by the small square in the lower right corner of the larger square representing an aliquot of sediment.

The TOC value consists of three components: the carbon in extractable organic matter (*EOM carbon*), *convertible carbon*, and a *residual carbon* fraction. The latter two are incorporated into the kerogen of a sediment.

The EOM carbon is composed of the carbon contained in the oil and gas already formed. It is derived from the products of thermal cracking of kerogen and the directly incorporated biological markers. The EOM carbon is generally a very small fraction of the TOC. In a typical shale or carbonate sample, the EOM carbon fraction is less than 1% of the TOC itself. However, in reservoir rocks, this percentage may be much higher.

The second component of the TOC content is the convertible carbon contained in the kerogen. Kerogen is the complex structure into which the originally deposited organic matter is transformed through diagenesis (Tissot and Welte, 1984; Hunt, 1979). The convertible carbon represents the *remaining potential* of a sediment sample to generate oil or gas.

The residual carbon component is also part of the kerogen and represents the organic carbon present in a form that has *no potential* to generate oil or gas due to its chemical structure and composition. This is because its highly condensed chemical structure has very little hydrogen per unit of organic carbon.

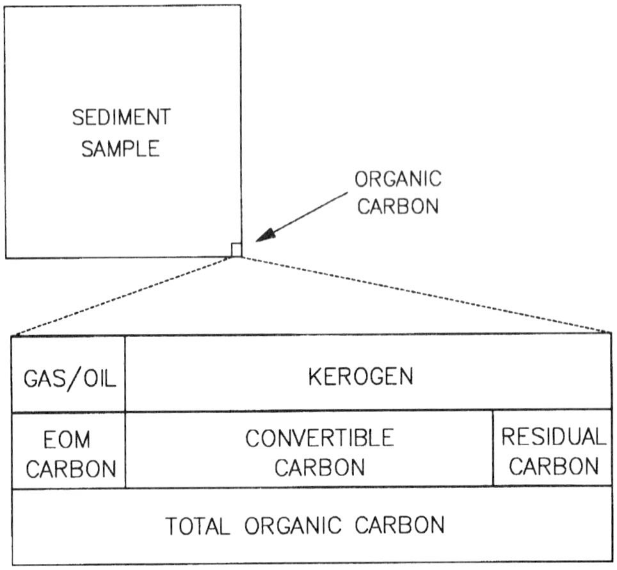

Figure 1. Model of organic carbon distribution in a sediment sample. TOC is the weight percent total organic carbon in a sediment sample. The carbon constituents of the TOC are the extractable organic matter (EOM), the remaining potential to generate hydrocarbons (convertible), and the nonreactive carbon having no potential to yield hydrocarbons (residual). (Adapted and modified from Espitalié et al., 1982.)

In summary, the EOM carbon is that portion of the organic carbon representing what has been generated but not expelled, whereas the convertible carbon represents the remaining hydrocarbon generation potential. Not all organic carbon can yield petroleum-like products, and the residual carbon describes that fraction. Whether a sedimentary deposit has oil, gas, or limited potential is largely due to the chemical composition of the convertible carbon fraction.

ANALYTICAL METHODS

There are several analytical methods for determining TOC, but the two most common are the Leco combustion method and the combined pyrolysis--oxidation method of Rock-Eval.

Leco Method

The Leco method requires approximately 1 g (~0.5 teaspoon) of pulverized rock. To remove inorganic carbon in the form of carbonates, chemical treatment of the sample is required prior to analysis. To achieve this, the sample is carefully soaked in hydrochloric acid (HCl) for 12–16 hours with intermittent stirring. At the end of this time or when the dissolution of carbonates is observed to be complete (no effervescence with stirring or additional acid), the sample is rinsed free of the HCl solution by using water, filter paper, and a filtering flask. The sample on the filter paper is then allowed to dry.

Calibration of the Leco Carbon Analyzer is achieved using a steel standard of known carbon content. The instrument is blanked to subtract any effect of the combustion crucible, filter paper, and catalyst. The carbon in the sample is oxidized to carbon dioxide, which is detected by either an infrared (IR) detector or a thermal conductivity detector (TCD). An IR detector is specific for carbon dioxide, whereas a TCD will respond to other compounds, such as sulfur dioxide and water, which may inflate the TOC value if not properly removed.

The operation of the Leco is basically straightforward, but there are some important chemical and operational facts to be aware of. First, in Recent sediments, part of the organic carbon may be hydrolyzed and lost in the water rinse (Peters and Simoneit, 1982). Second, a high-carbon switch must be activated if a high response is recorded across the detector from high TOC samples analyzed at too high a weight (often this is only caught by observing the real-time response of the instrument). Third, ineffective trapping of sulfur compounds and water inflates the response of instruments with TCDs.

Carbonate carbon can be determined by the Leco method by completing two analyses. Both the total carbon content and total organic carbon content are determined, and the difference is the total carbonate carbon.

Rock-Eval/TOC Method

The Rock-Eval/TOC method combines Rock-Eval pyrolysis and an oxidation cycle. Approximately 100 mg (about the volume of a pencil eraser) is required per analysis. No prior chemical treatment is required, as the processing steps eliminate carbonate carbon interference. The first portion of the TOC is obtained from the normalized carbon content of the Rock-Eval S_1 peak (free hydrocarbons or thermal extract) and S_2 peak (kerogen cracking or pyrolyzate peak). The normalized S_1 and S_2 values are equivalent to the EOM carbon and the convertible carbon, respectively. (These are jointly referred to as pyrolysis carbon by Espitalié et al., 1984.) The S_1 and S_2 values are normalized by multiplying by 0.083, which is derived from the average weight percent of carbon in hydrocarbons and by conversion from milligrams of hydrocarbon per gram of rock or parts per thousand (‰) to parts per hundred (%).

The residual portion of the TOC is obtained from the oxidation of the sample in a second oven following pyrolysis. This portion of the TOC yield is referred to as residual carbon (the S_4 value converted to %, i.e., times 0.10) (Espitalié et al., 1984). There is no carbonate interference since the sample was essentially decarbonated by thermal decomposition in the pyrolysis oven. The EOM and convertible carbon are summed

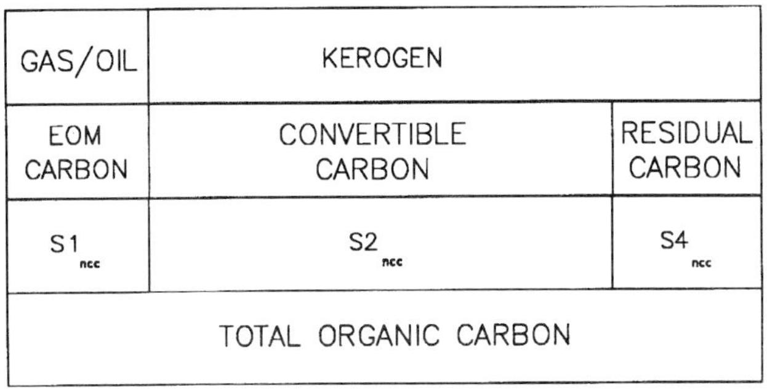

Figure 2. Relationship of Rock-Eval data relative to the model of TOC. S_1 and S_2 are derived from the Rock-Eval pyrolysis process, whereas S_4 is derived from the Rock-Eval oxidation process. (ncc = normalized carbon content; 0.083 wt. % for S_1 and S_2; 0.1 wt. % for S_4.) (Adapted and modified from Espitalié et al., 1982.)

with the residual carbon to obtain the TOC value based on instrument calibration against a known rock standard. The Rock-Eval/TOC relationships are summarized in Figure 2, which illustrates the relationship of various Rock-Eval measurements to the TOC model.

A key advantage of the combined Rock-Eval/TOC method is the availability of both TOC and pyrolysis data nearly simultaneously, which together permit a screening assessment of maturity, kerogen type, and organic richness.

Care must be exercised when operating the Rock-Eval/TOC system to ensure that (1) the instrument is accurately calibrated for both S_2 (pyrolysis) and S_4 (oxidation) because two separate calibration curves are required for high TOC (>10%) samples, and that (2) the sample was not dropped prior to oxidation by the automatic sampler. Nonlinear responses for the oxidation oven may result in low TOC values and high hydrogen indices (S_2 x 100/TOC). Samples with TOC values greater than 10% should be evaluated for TOC at reduced sample weights and longer oxidation times. Dropped samples due to instrumentation malfunction can be readily discerned because the EOM plus convertible carbon will equal the TOC, i.e., the residual carbon is not included in the computation.

Overall, variability in TOC analysis largely results from sample preparation techniques and variations in instrumentation as well as from operational and calibration differences. To cross-check accuracy, known rock standards of varying organic carbon content and maturity should be included with unknown samples.

In addition to these chemical methods, there has been some success determining potential source intervals by utilizing the data obtained from conventional well logging techniques (see Herron, Chapter 13, this volume). Schmoker (1981) described a technique to predict TOC from gamma ray logs. However, the technique was not universal and required local calibration. Meyer and Nederlof (1984) used gamma ray, density, and resistivity log responses to compute organic rich intervals mathematically. Passey et al. (1990) have used an overlay of the sonic transit time curve on a resistivity curve to predict TOC in the Alberta basin. The curves overlay one another in organic-lean intervals, but separate in organic-rich or reservoir intervals. By first calibrating the separation against TOC and maturity, they were able to obtain a TOC versus depth profile from well logs only. Reservoir intervals were distinguished by a gamma ray curve and were eliminated from the analysis.

The principal limitation of these methods at the present time is the requirement to calibrate each field or basin with laboratory-derived TOC values. This technology will grow in accuracy and use in the coming years.

INTERPRETATION

The convertible carbon of any kerogen ultimately yields hydrocarbons, and an inert carbon residue as hydrogen is depleted. Even the leanest TOC sediment will generate a minute quantity of hydrocarbons as a result of thermal maturation. The key question is whether a sediment contains *sufficient* organic carbon of the *right type* (oil or gas prone) and *maturity* to generate and expel hydrocarbons resulting in commercial accumulations. Of course, this requires a definition of what is commercial, which may vary considerably depending on location and operator.

TOC data only answer the sufficiency question, whereas other analyses are required to determine kerogen type and maturity. Other factors affecting realization of an effective source rock are the chemical and physical characteristics of the mineral matrix during generation and expulsion, its position relative to carrier beds, and the size of economic fields (Jones, 1981). Therefore, samples with adequate TOC content should be evaluated in greater detail by pyrolysis, microscopy, pyrolysis–gas chromatography, extraction–gas chromatography, and so forth as necessary to determine their true source potential.

There is a tendency to choose a lower TOC value for carbonates than for shales in qualifying potential source rocks. This is based primarily on the interpretation of data from Gehman (1962), which showed that carbonates have a higher concentration of hydrocarbons per unit TOC than shales. However, Jones (1984) interpreted this data differently and stated that Gehman's data may simply infer that potential source rocks have a higher concentration of hydrocarbons per unit TOC than nonsource rocks. In assessing published TOC data for potential carbonate and shale source rocks, Jones (1984) indicated that the minimum TOC values are approximately the same for carbonates and shales.

A highly descriptive table of TOC values for shales and carbonates is often used, as shown in Table 1a (Peters, 1986). However, a simple qualitative approach may also be appropriate because the TOC is only the initial assessment of the petroleum-generating capability of a potential source rock. This approach is summarized in Table 1b.

When evaluating source potential, it is important to consider the TOC value and the TOC composition. The quantity (TOC) along with the kerogen type (oil or gas proneness) more accurately describe source potential. For example, a 4 wt. % TOC sample with 50 wt. % oil-prone kerogen effectively has a 2 wt. % oil-prone organic carbon content (Cornford, 1986).

In addition, the geologist needs to consider the volume of source rock available to supply the hydrocarbons necessary to fill a trap. A volumetric model of hydrocarbon generation and expulsion can be used, as shown by Cooles et al. (1986).

Table 1. Classic interpretations of TOC content in sediments (based on early oil window maturity)

(a)

Generation Potential	TOC in Shales (wt. %)	TOC in Carbonates (wt. %)
Poor	0.0–0.5	0.0–0.2
Fair	0.5–1.0	0.2–0.5
Good	1.0–2.0	0.5–1.0
Very good	2.0–5.0	1.0–2.0
Excellent	>5.0	>2.0

(b)

	TOC (wt. %)
Inadequate	0.0–0.5
Marginal	0.5–1.0
Adequate	>1.0

Consideration of Kerogen Type

The derivation of the originally deposited organic matter affects both the quantity and quality of the hydrocarbons generated. This is due to the concentration of the originally deposited organic matter, its chemical composition, and its preservation as a result of sedimentation rates and environmental conditions during deposition (Tissot and Welte, 1984; Dow, 1978; Demaison et al., 1984; Cornford, 1986).

Kerogen type is often defined by petroleum geochemists using the elemental composition of kerogen in terms of hydrogen and oxygen content relative to carbon content. Type I kerogens have a high atomic ratio of hydrogen to carbon and a low atomic ratio of oxygen to carbon. Type III kerogens are just the opposite, and type II kerogens are intermediate between types I and III. Palynologists describe kerogen types I, II, and III as algal-amorphous, herbaceous, and woody, respectively (Hunt, 1979).

Elemental analysis of kerogens to determine atomic ratios has largely been replaced by a common screening technique used to describe kerogen type derived from Rock-Eval and TOC data. The Rock-Eval S_2 peak is normalized against TOC to derive the hydrogen index (HI) (S_2 x 100/TOC). The hydrogen index has been correlated to atomic hydrogen/carbon ratios by Espitalié et al. (1977). Hydrogen index is simply an indication of the hydrogen content per unit of organic carbon. It is essentially the availability of hydrogen per unit TOC that reveals whether a source rock has oil, mixed oil/gas, gas, or limited hydrocarbon potential. For example, coals have a very high organic carbon content but yield primarily gas upon thermal decomposition. This is due to the structure and composition of the organic matter and, concomitantly, to the lack of hydrogen, i.e., coals are hydrogen poor. Oil shales, which are also organic rich, yield primarily oil because of their structure and composition, i.e., they are hydrogen enriched.

In a type I source rock, the percentage of oil-prone convertible carbon in TOC is high (generally greater than 70 wt. %), illustrating its high potential to generate oil (Figure 3a). Such a source rock is hydrogen rich (HI > 700) and yields a higher concentration of paraffinic (waxy) hydrocarbons than do kerogen types II or III (Hunt, 1979; Tissot and Vandenbroucke, 1983).

Type II source rocks contain between 30 and 70 wt. % convertible carbon in TOC (Figure 3b). Those with a higher percentage of convertible carbon have more oil than gas potential. They have a lower hydrogen content than type I source rocks (HI = 300–700) and yield a complex mixture of liquid hydrocarbons (Hunt, 1979).

Type III source rocks have a lower potential to generate hydrocarbons than either type I or II source rocks. Being hydrogen poor (HI < 300), type III source rocks yield primarily gaseous hydrocarbons. The percentage of convertible carbon contained in the TOC is less than 30 wt. % (Figure 3c). The TOC content is primarily residual carbon, which has no potential for hydrocarbons.

Thus, although TOC content is important, it must be evaluated in the context of kerogen type and is only one indication of the potential to generate hydrocarbons.

Effect of Thermal Maturation on TOC Value

With increasing thermal maturation, the free hydrocarbon content (EOM carbon) increases while the kerogen (convertible carbon) content decreases due to conversion to oil or gas (Figure 4a and b). Oil-prone convertible carbon (hydrogen rich) cracks more completely under lower thermal stress than gas-prone convertible carbon (hydrogen poor) due to differences in cracking kinetics. Ultimately, kerogen is cracked

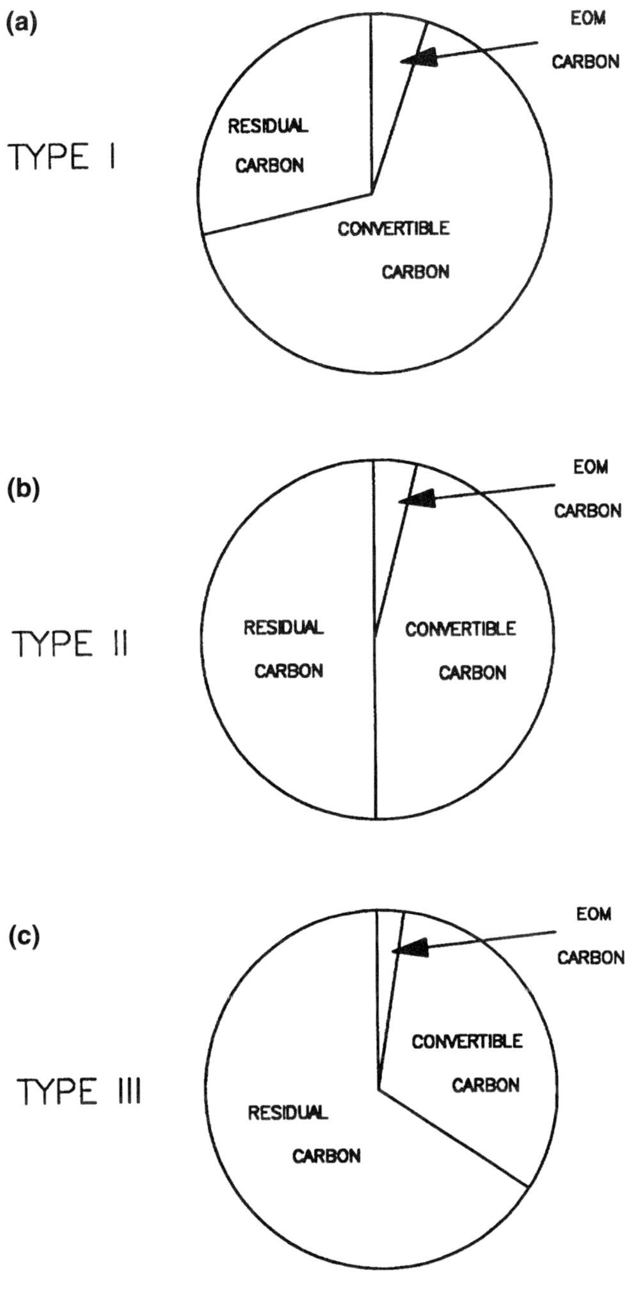

Figure 3. Distribution of organic carbon in different kerogen types. (a) In a type I (oil-prone) source rock, the convertible carbon predominates. (b) In a type II (oil/gas-prone) source rock, the convertible and residual portions are approximately equal although variable depending on oil or gas predominance. (c) In a type III (gas-prone) source rock, the residual carbon predominates. (Note: circle size would vary depending on TOC content.)

completely to hydrocarbons and a carbon-rich, hydrogen-poor residue. Because of the conversion of convertible carbon to EOM carbon and the expulsion of gaseous and liquid hydrocarbons, the TOC value of a sediment decreases (Figure 4c).

TOC values without maturity data (such as Rock-Eval T_{max} or vitrinite reflectance) are insufficient to screen potential source rocks. If the rock has a very high potential (i.e., it has a high TOC and contains oil-prone organic matter), but it is immature, few hydrocarbons will have been generated, none expelled, and no commercial reserves accumulated. Of course, downdip or hotter locations may be more mature and could supply traps if a migration pathway is available. On the other hand, if a very high TOC value is determined in a very mature rock, the TOC is almost entirely residual carbon, the convertible carbon having been exhausted in the generation process. In such a case, the timing of generation and trap formation, the availability of carrier beds, and hydrocarbon preservation are the important questions to be addressed.

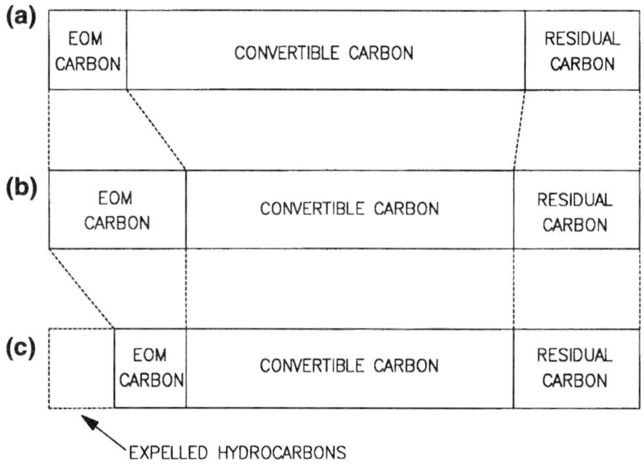

Figure 4. Effect of thermal maturation on a given volume of organic carbon. (a) With increasing thermal maturity, the initial EOM and convertible carbon amounts are changed as a result of generation. (b) The EOM carbon increases at the expense of the convertible carbon. (c) With expulsion, the original TOC value is decreased by the amount of carbon contained in the expelled EOM carbon fraction. The residual carbon increases slightly with maturation.

Organic Richness Contour Mapping

A contour map of TOC values of potential source units is useful in the first phase of analyzing the geochemical potential of a prospect to yield hydrocarbons. Acquisition of TOC data from available well samples or outcrops can be used to construct organic richness maps (e.g., see Webster, 1984). Ideally, samples from possible source horizons gathered at a variety of horizontal and vertical grid points should be analyzed. Organic richness maps of various horizons will delineate prospective source rocks. When overlaid with additional screening data such as Rock-Eval pyrolysis data, the best oil or gas prospects may be assessed. For example, isomaturation (T_{max}), isopotential (TOC and S_2), isogeneration (transformation ratio), and organic facies (HI) contour maps with geological maps may delineate the best targets for oil or gas.

Mass Balance Model

Cooles et al. (1986) have proposed an algebraic scheme for computing the amounts of petroleum generated and expelled which is based on a mathematical model of TOC. Their model is similar to the simple model of this paper. However, the terminology is different, as the convertible carbon is called reactive carbon, which is further subdivided into oil- and gas-prone fractions (labile and refractory carbon, respectively). Their model can be used to assess the petroleum generation and expulsion efficiencies of potential source rocks. In addition, computation of original TOC and HI can be achieved for input into quantitative basin modeling programs.

A key assumption in their model is that the residual carbon (their inert carbon), i.e., the carbon having no potential to generate hydrocarbons, remains constant during the main phase of petroleum generation. Burnham (1989) has stated that while this assumption is good for type II and III sources (less than 10–20% error), it is poor for type I (50% error). In fact, most exploration targets involve kerogen types II and III.

CONCLUSION

Measurement of TOC is the first step toward assessing the potential of a sediment to generate hydrocarbons. Sufficient organic carbon is required to yield commercial quantities of petroleum. However, this carbon must be present in a form that can be converted to petroleum. Thus, TOC data should be utilized to choose prospective source rocks for additional screening analysis, such as kerogen type and maturity assessments. These data together can then be used to evaluate the effective source potential of a sediment and the utility of additional detailed analyses of samples.

Chapter 12

Vitrinite Reflectance as a Tool To Assess Thermal Maturity

J. T. Senftle and Charles R. Landis
ARCO Oil and Gas Company
Plano, Texas, U.S.A.

INTRODUCTION

Thermal evolution (maturation) of kerogen is chemically and physically similar to coalification. When progressively buried under sedimentary deposits, peat undergoes chemical and physical changes passing from lignite to bituminous coal, then to anthracite. During this thermal evolution, the chemical composition of plant substances is irreversibly altered. The carbon content of the deposited plant material gradually increases, hydrogen is depleted, and various chemical compounds are decomposed. These progressive changes are known as coalification. The degree of coalification is referred to as the rank of a coal.

Coal petrographers have used vitrinite reflectance to study coalification in detail for many decades. Coal rank is determined by the mean maximum reflectance of vitrinite. Vitrinite reflectance has been successfully demonstrated as a reliable indicator of organic maturation in sedimentary rocks (Castano and Sparks, 1974) and is widely used in the oil industry for the evaluation of kerogens. Early work suggested that vitrinite reflectance is a maturity measurement independent of kerogen composition, organic facies, and depositional processes (Stach et al., 1982). More recent work (Jones and Edison, 1979; Bostick, 1979; Price and Barker, 1985) has shown that this is not always correct; thus, vitrinite is identified cautiously in shales and other clastic sedimentary rocks.

DISCUSSION

Among the optical properties used to characterize organic matter, reflectance is the most common property measured. Reflectance is the proportion of normally incident light reflected by a plane and polished surface of the substance under consideration. Reflectance is expressed as a percentage of the light incident on the surface. For vitrinite reflectance, only the amount of light reflected in the green visible wavelengths (546 nm) is measured. Thus, the amount of reflected light is a small fraction of the incident light, less than 4% in most cases.

The standard method to determine reflectance employed for coals is outlined by the American Society for Testing and Materials (ASTM) (1976). A stabilized light source, oil immersion of known refractive index (1.518), and a microscope photometer are used to compare the reflected intensity of macerals to that of a glass or synthetic mineral standard. Commonly, a tungsten filament lamp provides incident polychromatic or "white" light upon the sample.

Reflectance analysis is done on the vitrinite macerals for several reasons: (1) vitrinite is the predominant maceral, (2) individual vitrinite often appears homogeneous, (3) the particle size is large enough to permit measurements, and (4) the uses of coal in the steel industry focus on the behavior of vitrinite because it plays a major role in coking properties (McCartney and Teichmüller, 1972; Davis, 1978). Capitalizing on many of the same principles, the petroleum industry has embraced vitrinite reflectance as a method of determining organic maturation of kerogen in sedimentary rocks. As with coals, vitrinite in fine-grained clastic rocks often appears homogeneous and is sufficiently large to permit measurement. Furthermore, vitrinite reflectance offers a means to evaluate organic maturity over temperatures ranging from early diagenesis through catagenesis to metamorphism (Figure 1). This thermal range spans the entire sequence of hydrocarbon generation, preservation, and destruction in rocks.

There are, however, many technical and geological

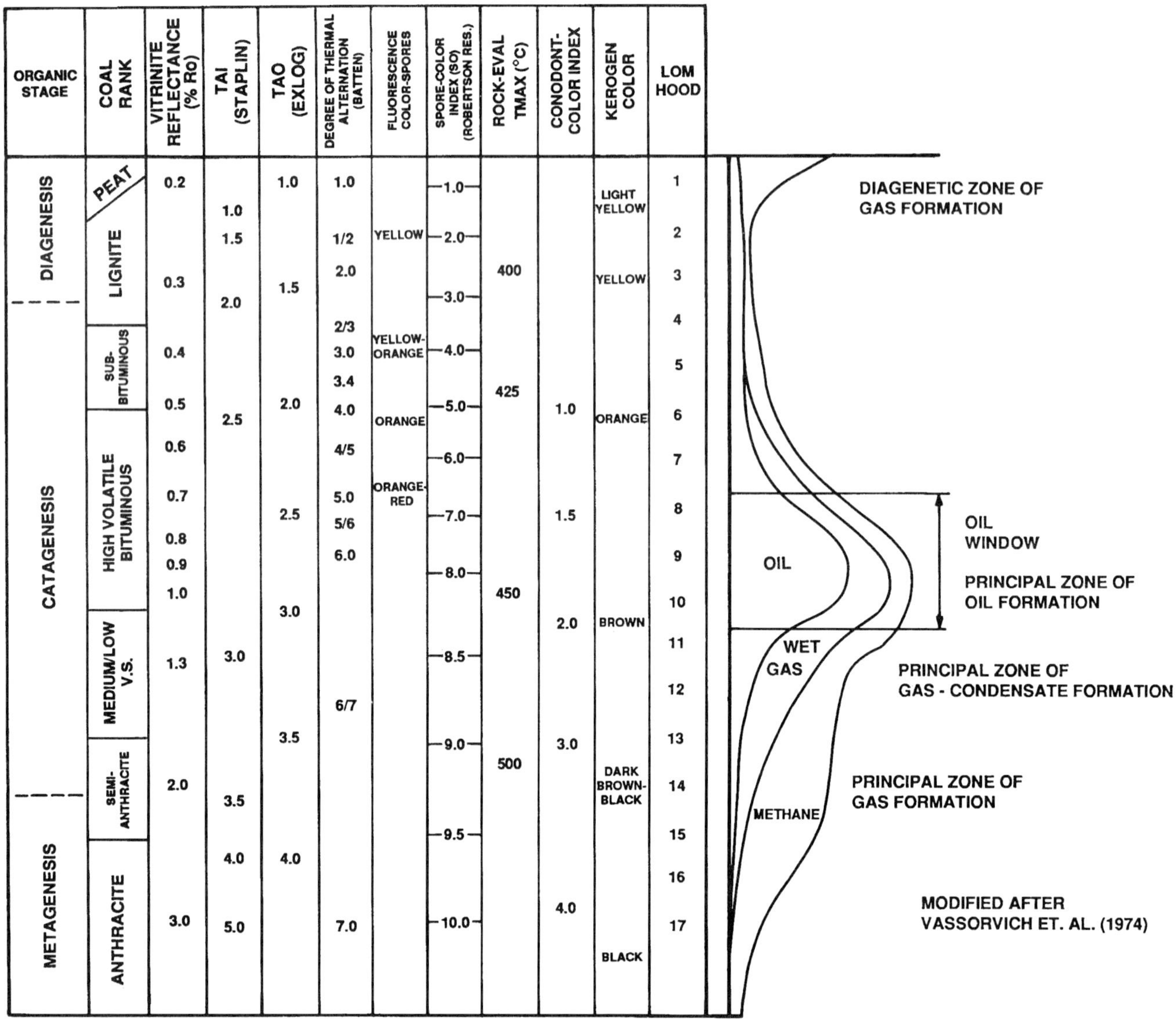

Figure 1. Correlation table of common organic maturation parameters with coal rank and hydrocarbon generation.

problems associated with vitrinite reflectance as a thermal indicator of the various stages of hydrocarbon evolution in sedimentary rocks. First, sample handling and drilling procedures can yield misleading measurements, particularly in older well cutting samples. Second, the ability to measure vitrinite reflectance is often controlled by the occurrence of vitrinite in rock. Third, many sedimentological processes which govern the deposition of shales are also deleterious to the preservation of petrographically distinct, indigenous vitrinite. Consequently, fragments of other macerals such as inertinite and allochthonous vitrinite, oxidized vitrinite, and solid hydrocarbon must be excluded from the reflectance measurements.

Vitrinite reflectance can be measured from whole-rock mounts or from a sample concentrated with kerogen after the removal of the mineral matrix. Whole-rock mounts are polished rock slabs or chips embedded and polished in epoxy pellets or on microscope slides. The pellets are ground and polished on rotating laps with sequentially finer silicon carbide and aluminum oxide powders. Typically, buffing culminates with a submicron slurry. Kerogens are often concentrated from a rock with such procedures as handpicking of indigenous lithology from ditch samples, crushing, sequential HCl and HF digestion, and heavy liquid separation. Kerogen concentrates are mounted in plastic pellets and polished, similar to whole-rock mounts.

The advantage of whole-rock vitrinite reflectance analysis is the distinction among the vitrinite population (i.e., primary vs. recycled or oxidized vitrinites) and between bitumen and inertinite macerals. The actual selection of indigenous vitrinite particles is subjective, drawing from the principles of coal petrology. The decision is based not only on the appearance of the vitrinite, its form, relief, association with pyrite, and the staining of the associated clays, but also on the appearance of the sediments and color of the bitumen in the mineral groundmass. From the entire sample, the petrographer can select appropriate particles for analysis. One limitation of whole-rock petrography is locating enough acceptable vitrinite particles for analysis. The organic matter in sedimentary rocks frequently amounts to no more than 1–5%. Since vitrinite may be a minor element, determination may not be possible or results may be limited. Jones and Edison (1979) note that in many samples, not enough individual measurements are available to calculate a statistically significant average. Generally, 20 measurements are needed to evaluate a sample. In all cases, the petrographer should report the number of particles measured, standard deviation, and average value as the results.

Concentrated kerogen preparations are widely used within the petroleum industry to enrich pellets in vitrinite. This permits the acquisition of a large number of measurements (50–100 readings) from humic organic matter, which can be summarized into histograms and average reflectance (Figure 2). (The statistical averaging of values implies that a single population of vitrinite is being sampled.) Yet, the determination of thermal maturity is difficult using reflectance data alone due to the complex mixture of constituents within a kerogen sample. Thus, kerogen concentrates analysis can produce less reliable distinctions between primary and recycled or oxidized vitrinites, and between bitumen and inertinite macerals. The problem is further complicated by variations in kerogen preparation procedures used among analytical labs. Typically, preferential recovery of organic constituents within the kerogen concentrate samples is the most significant problem.

A range of reflectance histograms may be encountered in well sections (Figure 2). Most reflectance histograms represent multiple populations of vitrinite due to incorporation of caved overlying rocks during drilling, vitrinite and inertinite identification problems, and reworked vitrinites of higher reflectance deposited into the sedimentary section. The identification of the indigenous vitrinite population can be assisted through study of samples from the well taken above and below the sample interval and the comparison with additional maturity indicators, such as fluorescence color and thermal alteration index (TAI) of indigenous spores.

Coal seams within well sections serve as reference points in maturity profiles. Vitrinite is abundant in coals and, in most cases, forms *in situ*. The vitrinite population in coals is relatively less complicated compared with clastic rocks, thereby permitting measurements of vitrinite of common origin and diagenetic history. However, when dealing with ditch cuttings, care must be taken to sample indigenous lithologies. This problem is amplified for coals since they are prone to cave and washout.

Vitrinite reflectance profiles, i.e., plots of reflectance (% R_o) vs. depth, are effective tools for evaluating the thermal history of sedimentary basins. Vitrinite reflectance varies systematically due to temperature over geological time; thus, it is dependent upon subsidence history, depth of burial, and geothermal gradient. By convention, vitrinite reflectance profiles are log-linear plots to assist exploration geologists in direct correlations with increasing depth.

Vitrinite reflectance profiles can contribute to studies of common geological problems. Reflectance profiles can be used to study geological history and hydrocarbon generation with respect to sedimentary thermal conditions. The latter case is an important application of long-standing coal petrographic techniques. Numerous workers have illustrated the utility of this approach (Castano and Sparks, 1974; Bostick, 1979; Robert, 1980; Dow, 1977; Dow and O'Connor, 1982). Careful scrutiny of the profiles yields important thermal information from a sedimentary succession not as readily available from other geological techniques, if at all. Common vitrinite reflectance profiles encountered from thick sections reflect a range of basin-scale geological processes (Figure 3), which include

1. continuous sedimentation (constant geothermal gradients),
2. tectonic activity, interrupted sedimentation, and erosional events (deviant profiles), and
3. localized thermal events (strong positive anomalies).

A fourth variety of reflectance profile indicative of the troublesome phenomenon of vitrinite reflectance suppression is less documented in the literature. Superimposed upon common geological processes is a poorly understood problem describing the downward bias of reflectance values. Consequently, the expected rate of increase in reflectance with depth is retarded in thick sections, or the profile exhibits a strong negative anomaly in particular horizons (Figure 4). Many studies have linked anomalously low vitrinite reflectance values to significant concentrations of exinite, including amorphous organic matter in coals and rocks (Ting, 1977; Jones and Edison, 1979; Hutton and Cook, 1980; Kalkreuth, 1982).

Other studies indicate that anomalous vitrinite reflectance data may relate to depositional environment, plant communities, and hydrogen-enriched "saprovitrinites" in some coals and oil shales (Sitler, 1979; Newman and Newman, 1982; Toxopeus, 1982). Low-reflecting, hydrogen-enriched vitrinites are common in type I and type II kerogen assemblages (Price and Barker, 1985). These authors believe hydrogen enrichment of vitrinite occurs in source rocks

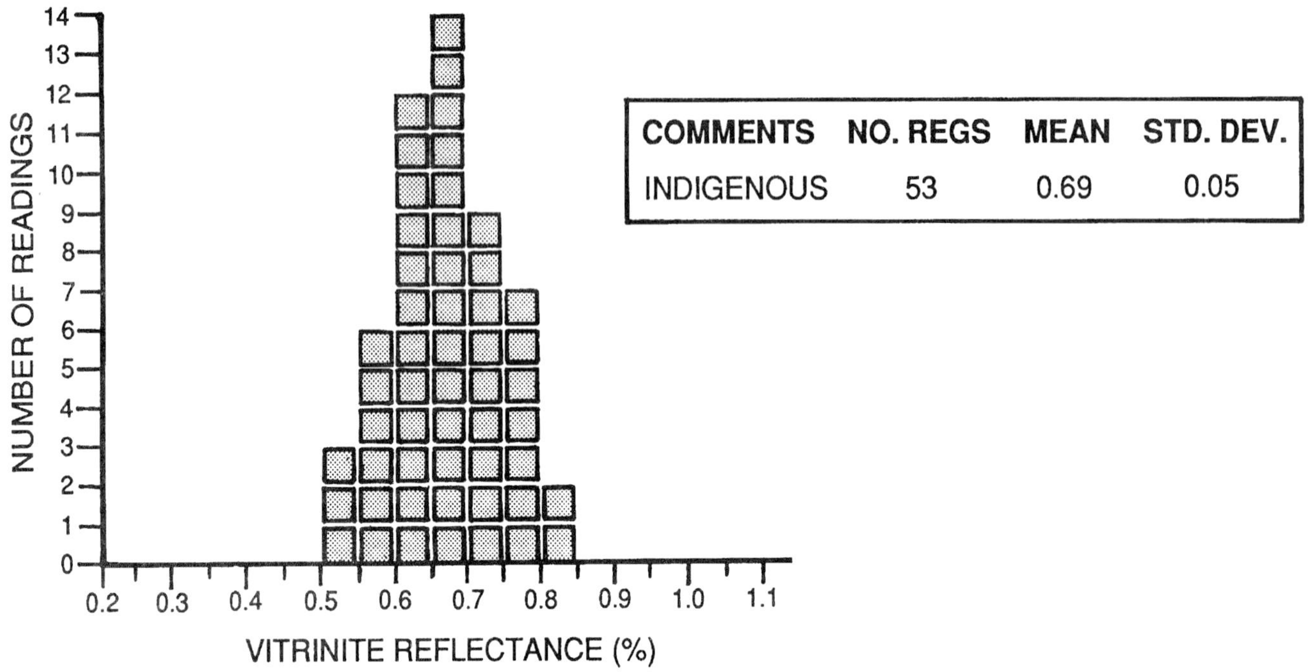

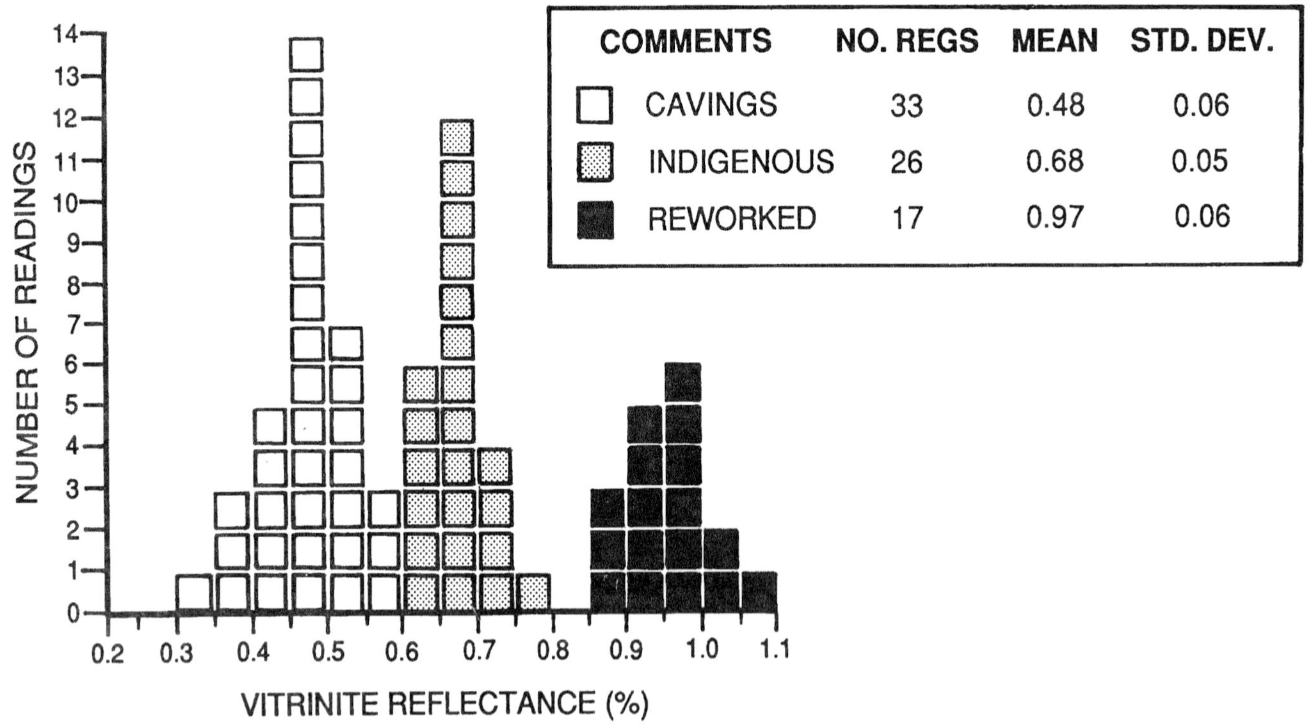

Figure 2. (Top) Ideal reflectance histogram of a normal population of indigenous vitrinite and corresponding descriptive statistics. (Bottom) Typical reflectance histogram exhibiting overlapping populations of caved, indigenous, and reworked reflective organic matter.

Figure 3. *(Next page)* Representative reflectance profiles of geological sequences encountered downhole.

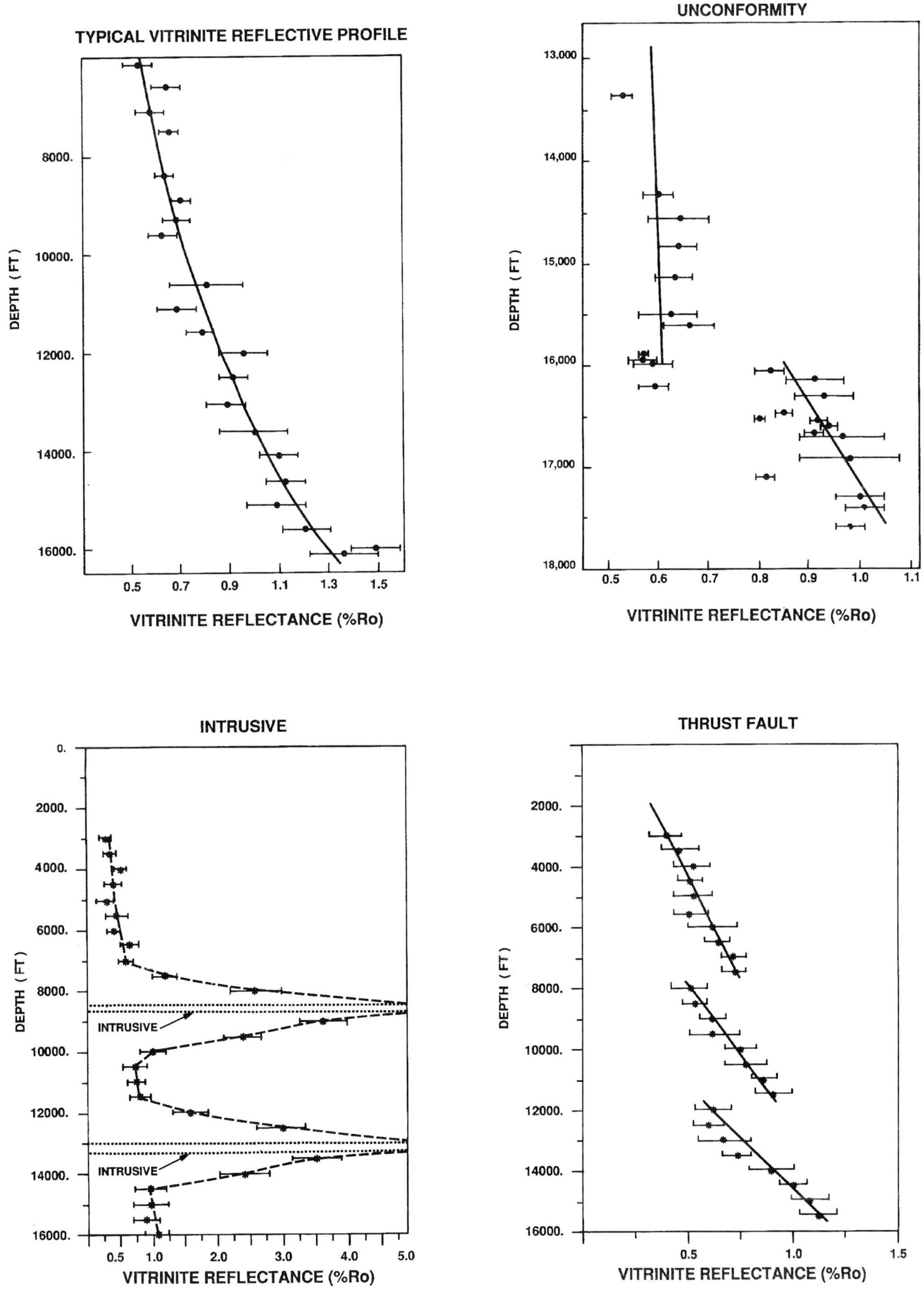
TYPICAL VITRINITE REFLECTIVE PROFILE
DEPTH (FT)
8000.
10000.
12000.
14000.
16000.
0.5
0.7
0.9
1.1
1.3
1.5
VITRINITE REFLECTANCE (%Ro)
UNCONFORMITY
DEPTH (FT)
13,000
14,000
15,000
16,000
17,000
18,000
0.5
0.6
0.7
0.8
0.9
1.0
1.1
VITRINITE REFLECTANCE (%Ro)
INTRUSIVE
DEPTH (FT)
0.
2000.
4000.
6000.
8000.
10000.
12000.
14000.
16000.
INTRUSIVE
INTRUSIVE
0.5
1.0
2.0
3.0
4.0
5.0
VITRINITE REFLECTANCE (%Ro)
THRUST FAULT
DEPTH (FT)
2000.
4000.
6000.
8000.
10000.
12000.
14000.
16000.
0.5
1.0
1.5
VITRINITE REFLECTANCE (%Ro)

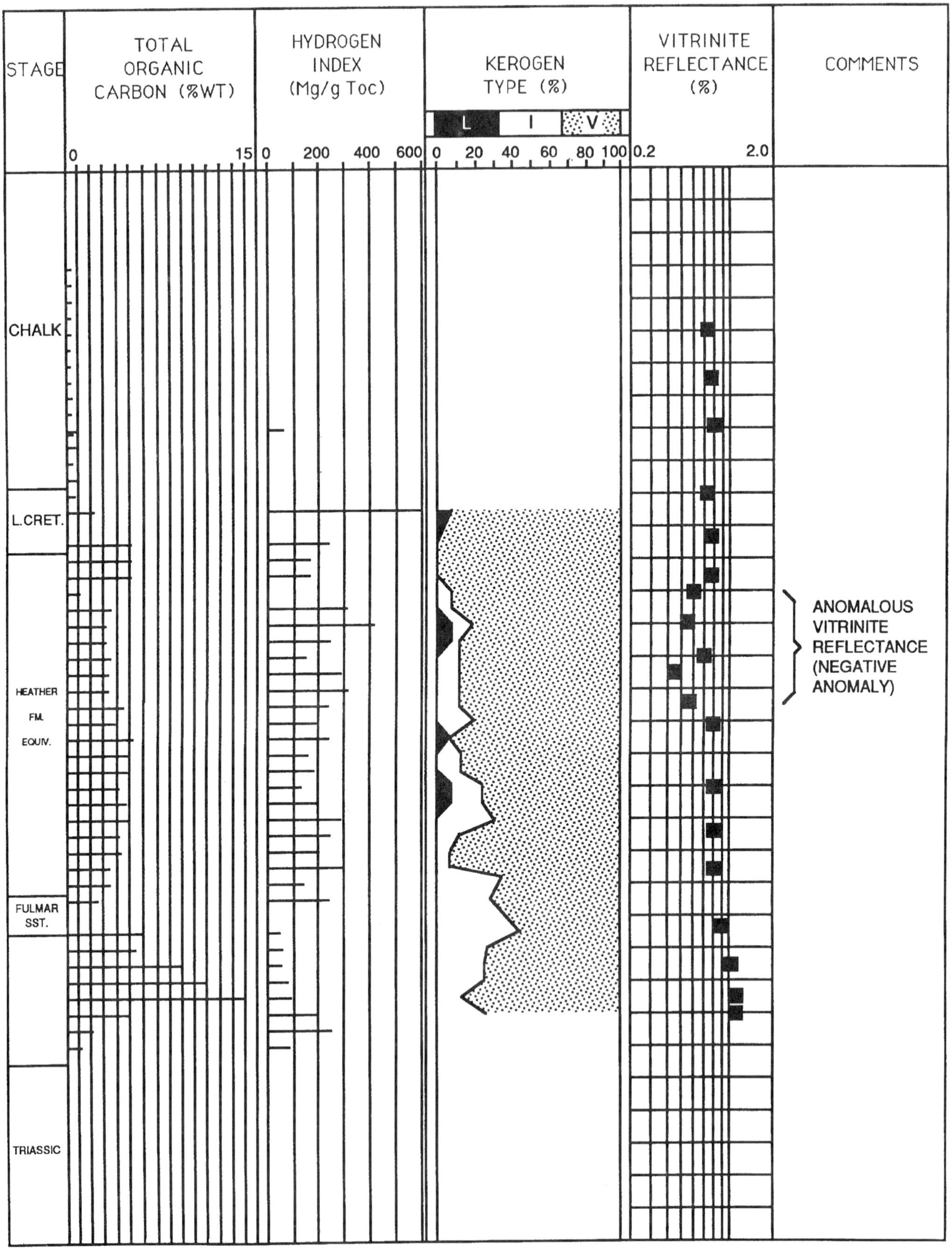

Figure 4. Reflectance profile exhibiting vitrinite reflectance suppression (negative anomaly). Note association of negative anomaly with hydrogen index (HI) increase.

during deposition in anoxic environments and early diagenesis. Other workers conclude that vitrinite is impregnated by bitumen or mobile hydrocarbons (Jones and Edison, 1979; Hutton and Cook, 1980; Hutton et al., 1980; Kalkreuth, 1982).

CONCLUSIONS

Vitrinite reflectance analysis is a powerful tool for the evaluation of organic matter maturation. A major application has been the relationship between coalification and maturation processes responsible for petroleum formation along with the evaluation of variations in the thermal histories of sedimentary basins. As research into hydrocarbon generation and kerogen reaction kinetics continues, vitrinite reflectance analysis will provide a means to study the generation of petroleum.

Due to the limitations of vitrinite reflectance in fine-grained, oil-prone rocks, we recommend that vitrinite reflectance microscopy be used in conjunction with additional optical parameters, such as fluorescence, transmitted microscopy, and chemical maturation indicators. One method cannot guarantee sufficient reliability and accuracy. Furthermore, since vitrinite is only marginally prone to oil generation, it is an indirect parameter by which liquid hydrocarbon generation can be monitored, unlike its role during the coking process. Thus, extrinsic deleterious effects may mask systematics so often apparent in coal petrology. These problems can be managed in most cases. Vitrinite reflectance, however, has been standardized, and this measurement has been compared with other petrographic and chemical parameters to define the stages of hydrocarbon generation.

Chapter 13

In Situ Evaluation of Potential Source Rocks by Wireline Logs

Susan L. Herron
Schlumberger–Doll Research
Ridgefield, Connecticut, U.S.A.

INTRODUCTION

Numerous studies published in recent years have illustrated the potential value of wireline logs for source rock evaluation. The work done to date has generally focused on the qualitative identification of organic-rich formations or the quantification of organic matter. The advantages over a more traditional approach of analyzing cuttings include continuous sampling, greatly improved depth control, and enhanced vertical resolution. Although the quantification of organic matter is neither as accurate nor as precise as the analytical techniques used on core, it does eliminate the sampling bias and contamination that can be introduced by cuttings analyses. To date, there are only a few examples of wireline interpretation of organic matter maturity and none of type, but there are indications that the integration of existing techniques combined with some additional log data could lead to quantitative logs of total organic carbon as well as estimates of maturity and inferences of depositional environment.

The feasibility of interpreting organic matter from wireline measurements stems from its physical properties, which differ considerably from those of the mineral components of its host rock: lower density, slower sonic velocity or higher sonic transit time, frequently higher uranium content, higher resistivity, and higher hydrogen and carbon concentrations. Consequently, the logs used for source rock evaluation most commonly include density, sonic, gamma ray, and resistivity. The limitations inherent in the interpretation are primarily due to the fact that several of these properties are quite similar to those of pore fluids, and therefore solid organic matter in source rocks is sometimes difficult to differentiate from either water- or hydrocarbon-filled porosity. Some of these limitations can be eliminated by using pulsed neutron geochemical logs which provide measures of the carbon/oxygen ratio as well as concentrations of the major inorganic elements to estimate the total carbon, inorganic carbon, and organic carbon in the formation.

This chapter briefly reviews the "logging" properties of organic matter and presents a variety of wireline approaches that have been taken to evaluate potential source rocks. The advantages of the various techniques are presented along with some suggested directions for future applications.

RESPONSE OF WIRELINE LOGS TO ORGANIC MATTER

The qualitative response of wireline logs to organic matter (OM) present in the formation is presented in Table 1. A variety of examples that show log responses to volumes, types, and maturities of organic matter deposited in different environments are presented by Meyer and Nederlof (1984). A single example of a well in which most of the logs respond in an almost ideal way is taken from the organic-rich Kimmeridge shale in the North Sea and is presented here as Figure 1. The responses of the individual logs shown in Figure 1 will be explained in terms of organic matter content, and some of the techniques that have used these logs to evaluate source rocks will be reviewed.

Neutron and Density Logs and Organic Matter

The neutron and density logs both display significant deflections in the organic-rich Kimmeridge shale (Figure 1). The neutron log, which reads about 20 porosity units (pu) in the overlying Upper Cretaceous formation, increases to as much as 45 pu in the Kimmeridge. This increase is due to the high hydrogen concentration in the organic matter. For the purpose of

Table 1. Log responses to organic matter.

Log/Property	Response/Value for Organic Matter (OM)	Comments
Gamma ray (GR) or uranium (U)	High	High GR caused by U; can be linear with OM; U not always present.
Density (ρ)	~1 g/cm³	Similiar to pore fluids
Neutron	High	Due to H in OM
Sonic (Δt)	High transit time	Estimates vary from 150 to >200 μsec/ft
Resistivity (R)	High	Might not affect log response unless generated HCs occupy pores
Pulsed neutron	High C/O ratio	Most direct measurement of C; needs inorganic C correction

log interpretation, the neutron response is sometimes explained in terms of the hydrogen index (HI), which is defined as the ratio of the number of hydrogen atoms in a cubic centimeter of sample to the number of hydrogen atoms in a cubic centimeter of water; thus, water has an HI of 1. The range of hydrogen concentrations (~3–10 wt. %) reported for typical kerogens by Tissot and Welte (1984) would correspond to HI values of 0.3–0.9, with a typical type II kerogen having a value near 0.7.

Although the increase in the neutron log response has been observed in many organic-rich formations, it is generally not used as anything other than a qualitative indicator. The reason for this is the multiple sources of hydrogen in the formation. The other large contributors to hydrogen in the formation would be the clay minerals, which have HI values of 0.12 for illite, 0.36 for kaolinite, and 0.13 for montmorillonite with no interlayer water (Juhasz, 1981). For a montmorillonite with two layers of interlayer water, the HI would increase to about 0.6. Clearly, the similar HI values of organic matter and the coexisting clays complicate the interpretation of the neutron log for the purpose of source rock evaluation.

A significant decrease occurs in the bulk density of the formation in the Kimmeridge formation, with density dropping from about 2.45 to about 2.3 g/cm³. This decrease can be attributed to the presence of the organics, which have a low density, about 1 g/cm³, relative to that of their host minerals (~2.65–2.70 g/cm³). If one is willing to make the assumption that there is no water-filled porosity or that the porosity is constant throughout the formation, the density log can be used to estimate the organic content. Schmoker (1979) and Schmoker and Hester (1983) successfully implemented this concept for Devonian shales. They first compensated for the presence of a high-density mineral, pyrite, by assuming that it is proportional to the amount of organic carbon in the formation. They then solved for the volume of organic matter and converted to total organic carbon (TOC), with the result being the following equation:

$$\mathrm{TOC} = \frac{154.497}{\rho_b} - 57.261 \qquad (1)$$

where ρ_b is the bulk density of the formation. The technique worked fairly well (to within a few weight percent) in these formations where the TOC values average about 12 wt. %. At lower TOC values, below about 2–3 wt. %, the assumption that there is no porosity or that porosity is constant, combined with the implicit assumption that the only variations in matrix density are due to pyrite, would introduce such large relative errors that it would be impractical to apply such a technique.

Gamma Ray Logs and Organic Matter

The gamma ray log displayed in Figure 1 shows an increase from about 60 to >200 API units in the Kimmeridge formation. The elevated gamma ray is attributed predominantly to the presence of uranium in the organic-rich sediments. This would be seen directly in data from a spectral gamma ray tool which separates the total gamma ray into its thorium, uranium, and potassium components. The occurrence of high uranium in such sediments is due to the reduction of U^{+6} which is extracted from seawater and precipitated as U^{+4}. The reducing conditions responsible for the precipitation of uranium are also conducive to the preservation of organic matter, and hence correlations have frequently been noted between TOC and uranium. The practical application of using uranium measured from a natural gamma ray logging tool to predict TOC was proposed by Supernaw et al. (1978) in a U.S. patent. It should be noted, however, that the constant of proportionality between TOC and uranium can change considerably (see examples in Mendelson and Toksoz, 1985). Among the factors controlling this relationship are the rate of sedimentation, the concentrations of uranium and organic matter in the seawater, the type of organic matter, the uranium content of the surrounding minerals, and the nature of the sediment–water interface. Therefore, to use uranium or total gamma ray to predict TOC, the relationship must

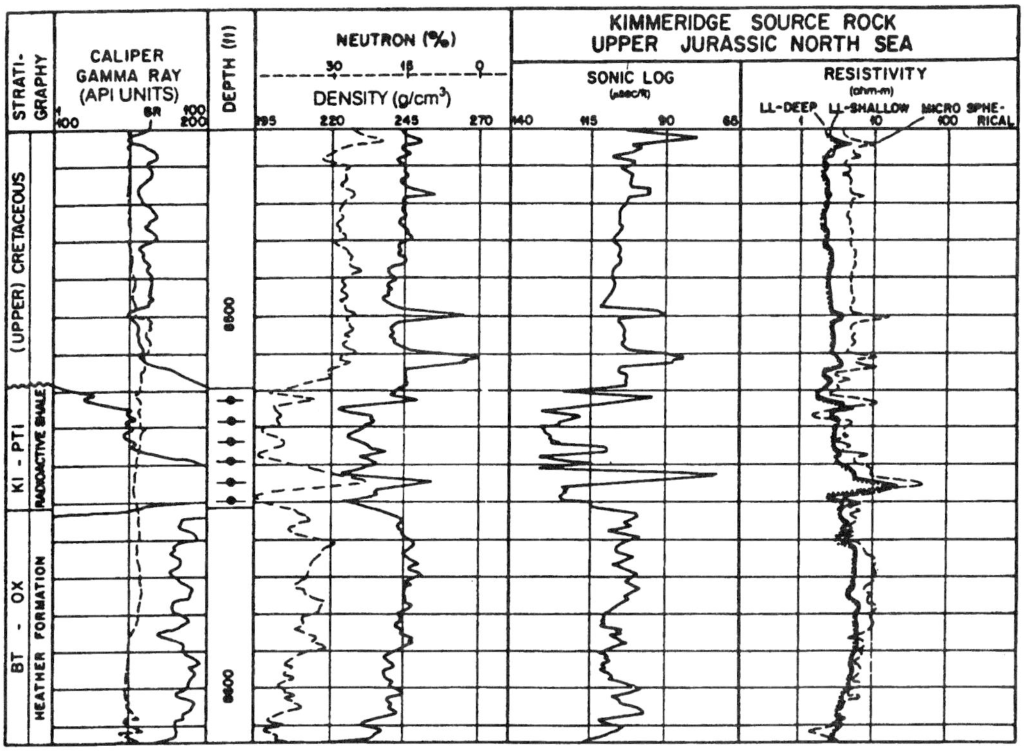

Figure 1. Gamma ray, neutron, density, sonic, and resistivity logs from the Kimmeridge shale, North Sea. The effects of the organic-rich shale, which is designated in the depth track, are indicated by the increased gamma ray, high neutron porosity, low bulk density, and high sonic transit time. The resistivity log shows only a marginal increase. (From Meyer and Nederlof, 1984.)

be locally calibrated and it must be restricted to a geologically uniform environment. It should also be noted that in many lacustrine environments, there is no appreciable uranium even in the organic-rich sediments (see examples provided by Meyer and Nederlof, 1984).

An example of using total gamma ray to predict the volume of organic matter was provided by Schmoker (1981) in the Devonian shale of the Appalachian basin. In that shale facies, it had been previously noted that there was indeed a proportionality between uranium and organic carbon. It was further noted that the concentrations of the other primary sources of natural radiation, potassium and thorium, remained relatively constant. Given these conditions, it was possible to construct an interpretation method in which gamma ray logs were calibrated to density logs and then converted into logs of organic content. The successful implementation of this technique required fairly detailed knowledge concerning the nature of the density/gamma ray relationship and the geographic variations in radioactivity. Schmoker showed that as long as the gamma ray/organic matter proportionality did not change, the estimation of organic content was generally within a few volume percent of the correct value, with values ranging from 0–20 vol. %. However, when the technique was applied to formations where the basic relationship was not valid, the estimates of organic matter were off by as much as 10 vol. %.

Sonic and Resistivity Logs and Organic Matter

Sonic transit times in organic matter have been estimated at values ranging from 150 to >200 μsec/ft (Mendelson and Toksoz, 1985; Autric and Dumesnil, 1985; Fertl and Chilingar, 1988; Dellenbach et al., 1983; Carpentier et al., 1989), whereas freshwater has a transit time of 189 μsec/ft, and the nonclay rock-forming minerals have values ranging from about 43 to 55 μsec/ft. In Figure 1, the sonic transit time log increases in the organic-rich section in a fashion quite similar to the neutron and density logs. In general, the sonic log has not been applied directly to the quantification of organic matter, but it does play a key role in a number of multivariate analyses, which will be presented later.

Organic matter has a high resistivity, but as long as it remains as a solid phase in the rock, it has little impact on the formation resistivity as measured by wireline devices. However, as the organic matter matures and begins to generate hydrocarbons, there can be a significant increase in resistivity. The Kimmeridge example shown in Figure 1 displays only a marginal increase in resistivity, but in other examples presented by Meyer and Nederlof (1984), the resistivity increases from tens to thousands of ohm meters in mature, oil-generating source beds. The resistivity log has been used in conjunction with other logs to indicate the presence of organic-rich deposits, but it is actually a better indicator of organic maturity than organic richness, and in some cases it has been used to predict organic maturity.

Pulsed Neutron or Geochemical Logs and Organic Matter

In addition to the wireline logs presented in Figure 1 and just described, there are a number of nuclear logging devices that can be grouped as geochemical logs. Included among these are the gamma ray and natural gamma ray spectroscopy logs that provide concentrations of thorium, uranium, and potassium. In addition, there are the pulsed neutron devices that can provide measurements of the carbon to oxygen (C/O) ratio in the formation, which can be used to indicate organic-rich deposits (Lawrence et al., 1984). Pulsed neutron logs can also be operated in conjunction with a natural gamma ray spectroscopy tool and an aluminum activation tool, as is done in the Geochemical Logging Tool (GLT*) service, to determine elemental concentrations of some of the inorganic elements (Al, Si, Ca, Fe, K, S, Ti, Gd, and sometimes Mg) in the formations (Hertzog et al., 1989). When these capabilities are combined, it is possible to use the C/O ratio to determine the total carbon in the formation and the other inorganic elements to predict the mineral carbon, thus making it possible to predict the TOC content quantitatively (Herron, 1986; Herron and Herron, 1990; Herron and Le Tendre, 1990). The techniques for doing this will be described later.

MULTILOG APPROACHES TO PREDICTING ORGANIC RICHNESS

Combinations of Gamma Ray, Resistivity, Sonic, and Density Logs

Using data from gamma ray, resistivity, sonic, and density logs from 15 wells in nine different countries, Meyer and Nederlof (1984) took a statistical approach to develop techniques to separate source rocks from nonsource rocks. For a cutoff between the two, they selected an organic content of 1.5 wt. % TOC. They used various combinations of logs (gamma ray/sonic/resistivity, gamma ray/density/resistivity, sonic/resistivity, and density/resistivity) to develop families of discriminant functions that could be applied to log data to identify source rocks in formations that are composed of either shales or other lithologies. To apply the technique, one converts the resistivity log data to 75°F (24°C), and then computes the appropriate discriminant scores, or D values, from the published equations. For example, the functions for shales are as follows:

$$
\begin{aligned}
D &= -8.094 + 0.739 \log_{10} GR + 3.121 \log_{10} \Delta t \\
&\quad + 0.399 \log_{10} R_{75^\circ F} \\
D &= 0.817 + 0.856 \log_{10} GR - 7.524 \log_{10} \rho_b \\
&\quad + 0.292 \log_{10} R_{75^\circ F} \\
D &= -6.906 + 3.186 \log_{10} \Delta t + 0.487 \log_{10} R_{75^\circ F} \\
D &= 2.278 - 7.324 \log_{10} \rho_b + 0.387 \log_{10} R_{75^\circ F}
\end{aligned}
\tag{2}
$$

A positive value of the discriminant score D would indicate a source rock, and a negative value of D would indicate a nonsource rock. The major advantage of such a technique is that it is a general function that does not require calibration on a well-by-well basis. It permits the user to examine a number of log combinations so that one log will not unduly influence the interpretation. The corresponding disadvantages are that it is not quantitative, and for a given location, it might not provide the optimum discriminators (Abrahão, 1989).

What is accomplished statistically by Meyer and Nederlof (1984) might also be performed manually using a series of crossplot techniques. An example is the work of Dellenbach et al. (1983) who proposed a series of three crossplots: resistivity and gamma ray, resistivity and sonic transit time, and sonic transit time and gamma ray, which are first used to establish the gamma ray and transit time signatures of a barren formation without any organic matter. Those values are then used to compute the parameter

$$I_x = (GR_{log} - GR_{barren\ shale})(\Delta t_{log} - \Delta t_{barren\ shale}) \tag{3}$$

and it is suggested that when performed on an individual basis, this parameter could be linearly calibrated to total organic matter. The similarity between the techniques of Dellenbach et al. (1983) and Meyer and Nederlof (1984) is illuminated in a study that applies the two techniques to a number of wells and concludes that, in general, the same organic-rich zones are identified (Autric and Dumesnil, 1985).

A slightly different approach to the interpretation of TOC from log data was also taken by Mendelson and Toksoz (1985), who approached the problem in two ways. The first approach was to treat the organic matter as a rock constituent with known properties and to predict TOC theoretically from log responses. Although they observed good qualitative agreement between log responses and organic richness, they implied that, in general, the physical properties of organic matter are too poorly characterized and the

*Trademark of Schlumberger.

physical properties of the rock matrix are too critical to solve the problem quantitatively. Their second approach was to perform multivariate regression analysis on five wells with mean TOC values ranging from approximately 3 to 10 wt. % using measured values of TOC along with sonic, neutron, density, and gamma ray logs. For each individual well, they achieved reasonably good predictions of TOC, but they found that none of the regressions could be applied successfully to the other wells. The advantage of such a treatment of the data is that it provides an optimized solution to the problem of quantifying TOC, but it does so at the expense of fairly detailed core (or cuttings) calibration, and the extent of its applicability remains unknown.

Quantitative Predictions of TOC from Sonic and Resistivity Logs

More recent interpretation schemes have returned to the petrophysical approaches of trying to interpret organic matter in terms of its theoretical log response (Carpentier et al., 1989; Stocks and Lawrence, 1990; Passey et al., 1989, 1990). Carpentier et al. have introduced a technique to interpret TOC from sonic and resistivity logs. The principle underlying their approach states that an increase in sonic transit time accompanied by a decrease in resistivity is attributed to an increase in clay content or water, whereas an increase in transit time coupled with an increase in resistivity indicates an increase in organic matter. Their technique uses the Wyllie time average and Archie resistivity relationships to set up the response equation for sonic and resistivity logs in a formation composed of organic matter, water, clay minerals, and nonclay rock matrix. It then uses core or cuttings TOC data from a key well (or formation) as a calibration for determining the Δt value for organic matter. Having optimized the solution, the authors apply the technique in the calibration zone to get a continuous TOC log that is in very good agreement with the core and cuttings data. They also provide examples of extending the solution to other wells in a given basin after having performed the calibration on one or two key wells, and again their predictions of TOC are reasonably good. In the analysis of these data, the authors reveal two of the shortcomings of relying on cuttings which can translate into two significant advantages of using wireline data. The first is the improved vertical resolution obtained when using wireline logs compared to cuttings. The second is the inaccuracy of cuttings data due to contamination from other zones, which they illustrate quite dramatically in a comparison of TOC values obtained from cuttings, sidewall cores, and cores.

Another interpretation of organic carbon content that relies on the same principles of sonic and resistivity logs has been presented by Passey et al. (1989, 1990). The interpretation technique consists of overlaying, or baselining, the sonic and resistivity logs in "clean" nonreservoir zones and using separations between the two curves to indicate the presence of organic matter. This is actually identical to a technique used in conventional log interpretation for qualitatively identifying hydrocarbon-bearing reservoir units. The major innovation of this work is the incorporation of organic maturity as an input to the interpretation, which makes it possible to interpret the separation between the two curves ($\Delta\log R$) in terms of TOC content and level of organic metamorphism (LOM). The key to the interpretation is the set of calibration curves originally derived from core data and duplicated in Figure 2. The equations required for this interpretation are

$$\Delta\log R = \log\left(\frac{R}{R_{\text{baseline}}}\right) + 0.02\left(\Delta t - \Delta t_{\text{baseline}}\right),$$
$$\text{TOC} = \left(\Delta\log R\right)10^{2.297-0.1688\text{LOM}} + 0.8. \tag{4}$$

Since the interpretation implicitly assumes no organic matter in the baselined portions of the log, a value of 0.8 wt. % TOC is taken as a background level of organic carbon and added to the computed result, as indicated in equation (4). An example of this interpretation implemented in the North Sea is presented in Figure 3, where TOC values of about 3 wt. % are predicted quite well by the log data.

The techniques of both Carpentier et al. (1989) and Passey et al. (1989, 1990) offer the advantages of continuous logs of organic richness with better resolution than can be obtained by cuttings. At best, these techniques offer a good quantitative value of TOC, and at worst, they provide excellent qualitative information on TOC. The approach of Carpentier et al. (1989) does requires calibration with TOC in a key geological unit, and it presumes that a geological similarity exists between the calibrated formation and the formations that are to be interpreted. However, the technique of Passey et al. has a predefined calibration among $\Delta\log R$, TOC, and LOM, and it requires LOM as an input.

A Geochemical Approach to Predicting Total Organic Carbon

A completely different approach to the quantification of organic matter is to use the C/O ratio measured by pulsed neutron logging devices to compute the carbon content of the formation directly, according to the relationship

$$C = \left(\frac{C}{O}\right)_{\text{formation}} \times O_{\text{formation}} \tag{5}$$

where the oxygen in the formation is an estimated

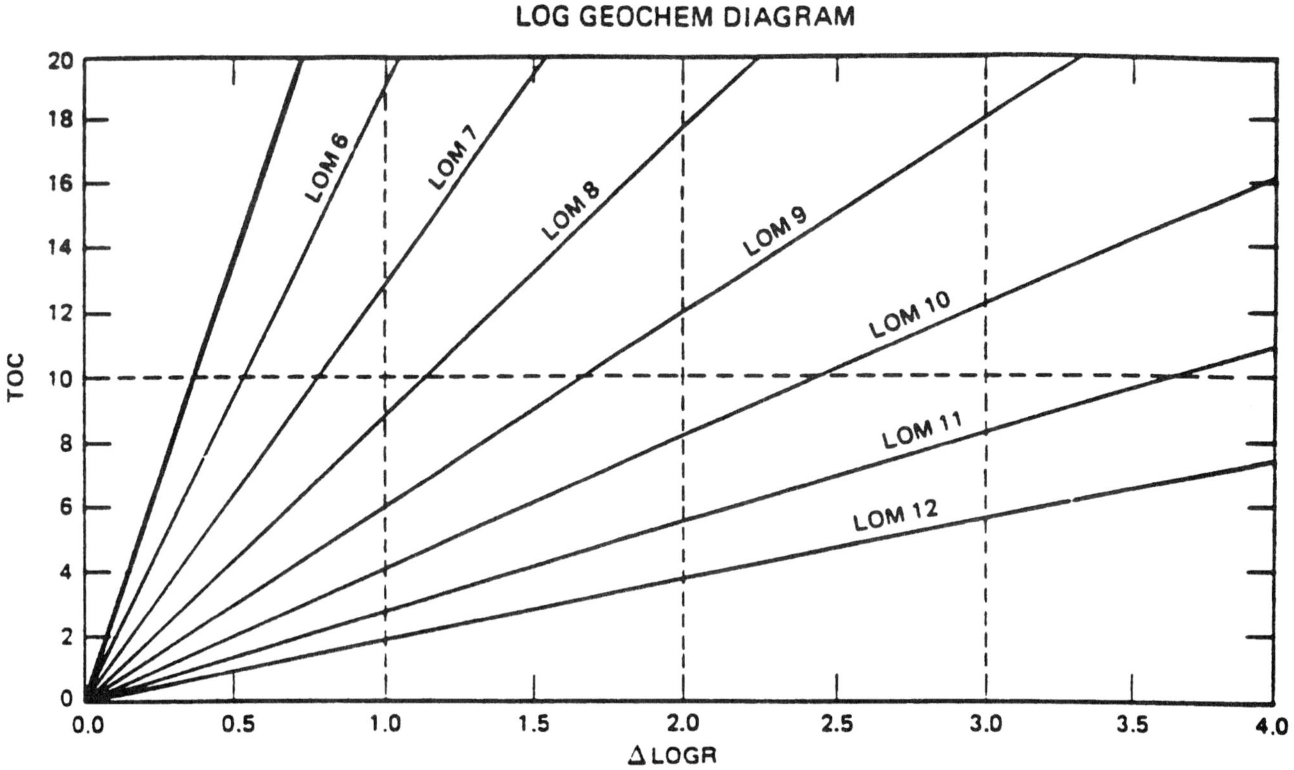

Figure 2. Diagram relating Dlog R to TOC and level of organic metamorphism (LOM). (From Passey et al., 1989, 1990.)

value that depends on the porosity of the formation (Herron, 1986, 1987; Herron and Herron, 1990; Herron and Le Tendre, 1990). The key to the interpretation is the estimation of formation oxygen which is accomplished in an iterative fashion. The formation is first partitioned using a density log into two components: a mineral matrix with 50 wt. % oxygen and a pore space filled with water that has 89 wt. % oxygen. The total carbon (TC) based on this solution is then treated as kerogen with an oxygen content of about 6 wt. %, and a new TC is computed. The calcium and magnesium logs, which can also be obtained from pulsed neutron measurements (Hertzog et al., 1989), are then used to determine the amounts of calcite and dolomite in the formation, and the corresponding value of inorganic carbon is subtracted from the TC to produce TOC. The major advantages of such an approach are that it is fairly sensitive to low amounts of organic carbon and it does not require calibration to core data. The precision on the measured C/O ratio is poor at conventional logging speeds, so higher quality logging data are acquired from stationary measurements or slow (or multiple) continuous passes. An example of this technique applied to a short interval of the Toarcian shale in the Paris basin is presented in Figure 4. The first two graphs show the TC and TOC determined from stationary log measurements and from core slabs taken over identical vertical intervals. The third graph displays the TOC log computed in the same manner but this time using log data taken from a single continuous pass at a logging speed of 600 ft/hr. The agreement between the core and log data is excellent for the stationary data, and it is still quite good for the continuous pass. While the organic contents are quite high in this example, the same degree of accuracy has been demonstrated using stationary C/O measurements for TOC values at the 1 wt. % level (Herron and Herron, 1990).

PREDICTING ORGANIC MATURITY FROM RESISTIVITY LOGS

Attempts to interpret organic maturity from logs have focused on deep resistivity measurements, and in general, researchers have credited Meissner (1978) for demonstrating the relationship between oil generation and shale resistivity. A good example of this relationship is provided by Smagala et al. (1984) in the Cretaceous Niobara Formation in the Denver basin. In that study, the authors first used vitrinite reflectance

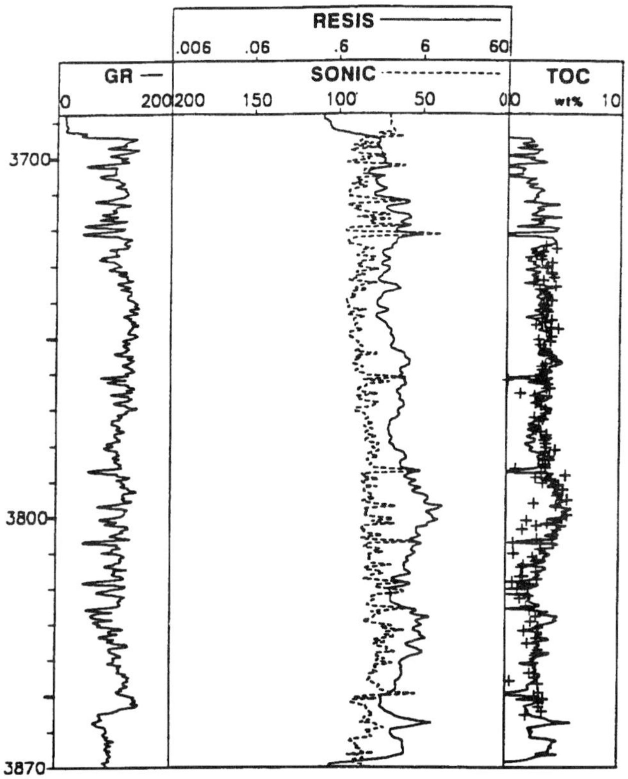

Figure 3. North Sea example from Passey et al. (1989, 1990) of the prediction of TC from sonic and resistivity logs for a formation with a maturity of LOM 9–10. The core data (+) are from closely spaced measurements made on a conventional core.

data to determine the log-linear relationship between vitrinite reflectance and depth and consequently to pick a representative vitrinite reflectance value for a given zone in the well. They then empirically correlated the vitrinite reflectance to the formation resistivity using data from 23 wells in the basin, as shown in Figure 5. The correlation was found to be good enough to apply to resistivity logs from other wells, and it enabled them to map maturity patterns throughout the Denver basin. Schmoker and Hester (1990) also used resistivity logs as an indicator of oil generation in the Bakken Shale where they found that a formation resistivity of 35 ohm m effectively discriminated between thermally mature and immature shales. They also offered the interesting observation that in the two organic rich members of the Bakken Shale, the 35 ohm m resistivity contour that represents oil generation corresponds to different values of thermal maturity, and they attributed this to different kerogen compositions in the two shales.

The success of these two studies can be attributed partially to the fact that the formations being interpreted have not undergone significant porosity reduction due to compaction, and consequently, the increase in resistivity is dominated by oil generation. Significant porosity variations would introduce a complicating factor to the interpretation.

FUTURE INTEGRATION OF WIRELINE MEASUREMENTS IN SOURCE ROCK EVALUATION

To date, wireline interpretation has focused primarily on the quantification of TOC and secondarily on the estimation of organic maturity. With the exception of the geochemical logging approach, the quantitative techniques rely on calibrations that either are made on a local basis (e.g., Mendelson and Toksoz, 1985; Carpentier et al., 1989) or are implicit in the model itself (Passey et al., 1989). The logs being used for most of these interpretations are fairly insensitive to the differences between kerogen, hydrocarbon, and water, and therefore the techniques are fairly insensitive to small volumes of organic matter. The geochemical logging approach offers a more direct and more accurate measurement of organic carbon content, particularly at low levels of organic carbon, but these types of logging data are not yet routinely obtained.

Organic maturity has been estimated from resistivity logs, but it has required calibration with substantial amounts of core data (Smagala et al., 1984; Schmoker and Hester, 1990). Passey et al. (1989, 1990) have established a broad-based empirical relationship between organic maturity, resistivity, and TOC. It seems clear that if a TOC log can be obtained independently of resistivity, the relationship described by Passey and coworkers could be exploited to solve for organic maturity.

Interpretations have not been developed for determining the type of organic matter. Passey et al. (1990) have suggested that in some instances it might be possible to infer the depositional environment of organic matter and then further infer the type of organic matter, and they provide an example of some thinly bedded coals and deltaic sediments. Along similar lines, Meyer and Nederlof (1984) and Abrahão (1989) both present examples of logs illustrating a cyclical depositional sequence commonly observed in lacustrine environments. High-resolution borehole imaging tools would also provide the ability to see small-scale depositional patterns.

Future enhancements to all aspects of wireline source rock evaluation can be envisioned by the inclusion of nuclear geochemical logs in the interpretation. Outputs from nuclear devices such as the GLT string include not only TOC but also a number of inorganic elemental concentrations that can be interpreted to characterize different depositional facies and

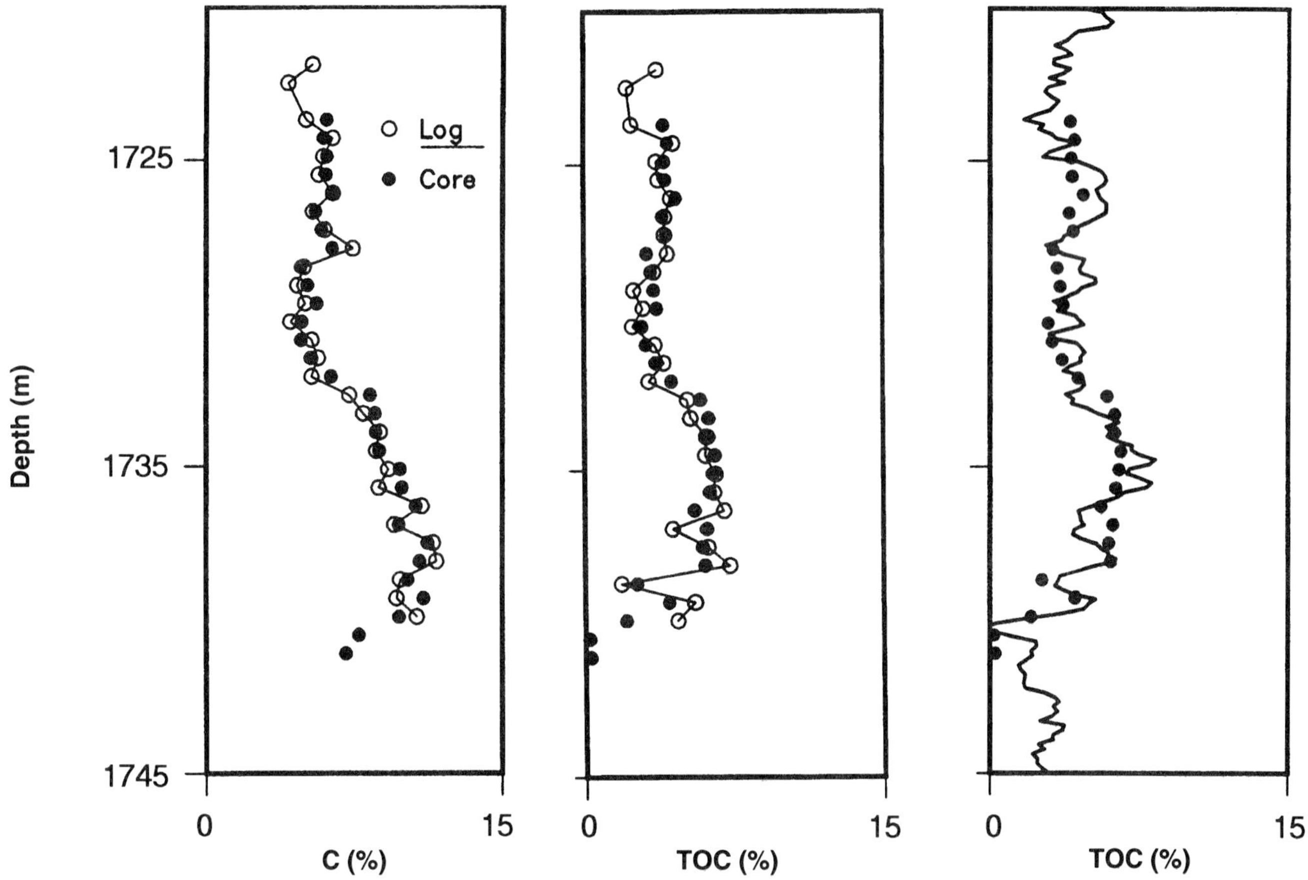

Figure 4. Paris basin example of TC estimated from stationary C/O measurements (circles connected by solid lines) and TOC estimated from stationary C/O measurements and a single pass at 600 ft/hr continuous C/O measurements. Core data (solid dots) are provided from comparison. Note the large difference between TC and TOC due to the relatively high concentration of carbonate minerals in the formation. (From Herron and Le Tendre, 1990.)

to begin interpreting the depositional environment (Primmer et al., 1990). For example, Berner and Raiswell (1984) demonstrated a technique for distinguishing freshwater from marine sedimentary rocks using the ratio of organic carbon to pyrite or total sulfur. In a later work, they further demonstrated how this ratio changes with maturation and might actually be used to predict thermal maturation (Raiswell and Berner, 1987). The elemental concentration logs also introduce the capability of determining formation mineralogy on a continuous *in situ* basis (Herron and Herron, 1990). This provides the opportunity to examine the relationship between clay mineralogy and depositonal environment as well as the relationship between clay diagenesis and organic maturity. The realization of these techniques will require additional research, but will undoubtedly enhance wireline interpretation of potential source rocks in the future.

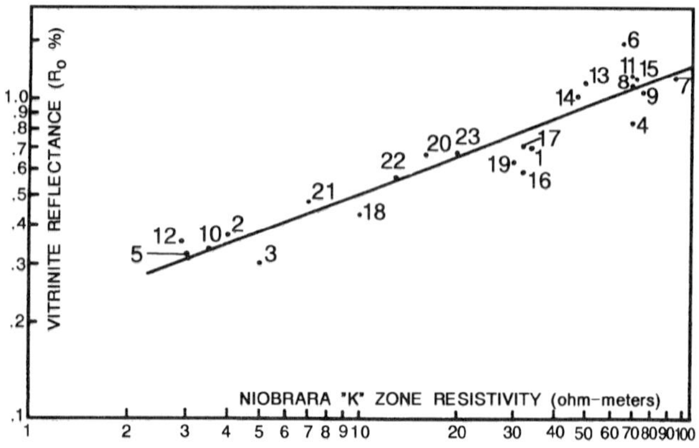

Figure 5. Relationship between vitrinite reflectance and resistivity determined from 23 wells in the Denver basin. (From Smagala et al., 1984.)

Chapter 14

Recent Advances in Near-Surface Organic Geochemical Prospecting for Oil and Gas

G. Saenz, K. H. Pannell, and N. E. Pingitore
The University of Texas at El Paso
El Paso, Texas, U.S.A.

INTRODUCTION

The field of near-surface organic geochemical prospecting for hydrocarbons is more than a half century old (Laubmeyer, 1933; Sokolov, 1933; Horvitz, 1939; Rosaire, 1940). During this time, much evidence has accumulated to validate its fundamental premise, that hydrocarbons move upward from petroleum accumulations at depth to the near-surface. Reviews of the technique are available in Sokolov (1970), Siegel (1974), Leythaeuser et al. (1980), Philp and Crisp (1982), Saenz (1984), and Horvitz (1985). Prospecting for oil and gas by geochemical means is nonetheless still viewed with skepticism in many quarters despite proprietary successes and published investigations that provide it with a sound scientific basis (e.g., Horvitz, 1954, 1972; Stroganov, 1969; Sokolov, 1970; Leythaeuser et al., 1980, 1982, 1983a; Saenz, 1984; Kross et al., 1986).

A large-scale project of geochemical exploration undertaken in Mexico during the last decade gave an excellent opportunity to establish the reliability of onshore surface prospecting. About 19,000 soil gas samples were taken in a marshy region over seismologically defined structures at depths from 2175 to 6500 m (7000 to 20,000 ft); sampling depths were 3, 15, and 30 m (10, 50, and 100 ft). This database, combined with the results of subsequent drilling of the structures, allowed us to address three important issues. First, is it possible to differentiate barren areas from producing areas reliably on the basis of the surface organic geochemical survey? Second, to what extent is such differentiation a function of sampling depth? Third, what are the respective roles of biogenic and thermogenic gases in this process?

In this chapter, we both review the previously published analysis of this data base and present new concepts and models for its interpretation. Studies (Saenz, 1980, 1984, 1987; Saenz and Pingitore, 1989) indicated that at a 3-m (10-ft) sample depth, producing and barren areas are difficult or even impossible to distinguish, whereas comparison of data from different sampling levels raises the success of distinction between barren and producing areas to approximately 80%. It is evident from these studies that the careful integration of geochemical prospecting into a conventional exploration program may provide important supplemental data that narrow the drilling area and reduce the dry hole risk.

EXPERIMENTAL METHODS

Sampling and Analytical Methods

Cores were obtained at depths of 3, 15, and 30 m (10, 50, and 100 ft), and cuttings at 0.5, 2, 14, and 29 m (1.5, 6, 45, and 95 ft). Sample holes to 30 m were drilled at surface intervals of 500 m (1640 ft); three additional holes to 3 m were drilled between the 30-m-spaced holes. Drills were chosen and modified according to the area requirements. Inorganic lubricant was employed to avoid hydrocarbon contamination. Core samples of 10 cm (4 in.) and an equivalent quantity of cuttings were canned, and the cans were filled two-thirds with water. All sample cans were sealed at the sampling site and were promptly sent to a laboratory close to the explored area.

In the laboratory, samples were energetically shaken and a 1-mL aliquot of the headspace gas was removed by means of a syringe. The composition (C_{1-7}) of this headspace gas was analyzed by gas chromatography using Perkin Elmer Sigma 2B and 3B chromatographs, as described by Saenz (1984, 1987). Methane isotopic analysis of carbon was performed at the Horvitz Reseach Laboratories and reported relative to the PDB standard.

Petroleos Mexicanos allowed this data base to be used for scientific purposes with the stipulation that the exact location of the sample holes and the seismic data be withheld.

Statistical Method

The statistical method used was discriminant analysis, a routine of the Statistical Package for the Social Sciences (SPSS), employing the Wilks stepwise method (Nie et al., 1975), which distinguishes between two or more groups of cases. The procedure is fully described elsewhere (Saenz, 1987; Saenz and Pingitore, 1989).

Concepts to Interpret Surface Organic Geochemical Data

The interpretation of surface organic geochemical data requires differentiation between biogenic and thermogenic gases. Biogenic activity produces gases (mostly methane) in continental swamps or marine sediments; such gases are associated with recent organic matter, not with oil (Schoell, 1983a). Gases of bacterial origin can be recognized by their high methane content ($C_1/C_n > 0.99$, where C_n is the sum of all hydrocarbons C_1 through C_7) and by their $\delta^{12}C$ enrichment of methane ($\delta^{13}C < -60‰$) (Stahl, 1973). At depth, natural gases originate by cracking or thermocatalysis of petroleum and/or organic matter in source rocks (Tissot and Welte, 1984). Hydrocarbon gases of oil and gas accumulations are rich in saturated hydrocarbons (alkanes or paraffins), but do not contain significant quantities of unsaturated hydrocarbons (alkenes or olefines) (Siegel, 1974). Biochemical processes of young sedimentary organic matter form minor amounts of paraffins (Horvitz, 1972). Paraffins do not exceed the amount of olefin generated from the sediments; therefore, the ethane/ethylene ratio may contribute to the differentiation between biogenic and thermogenic gases. Ethane/ethylene ratios exceeding 1 for more than one unit suggest the presence of thermogenic ethane.

Another concept used in the interpretation of the geochemical data is the fluctuation of hydrocarbon concentration with depth. Louis (1967) and Horvitz (1972) reported increasing quantities of hydrocarbons as one approaches the hydrocarbon pool. This has been confirmed in producing areas in Mexico (Saenz, 1984, 1987). However, it has been demonstrated that light hydrocarbons diffuse from near-surface strata into the atmosphere (Leythaeuser et al., 1979, 1980, 1982).

RESULTS AND DISCUSSION

Graphical Anomaly Analysis

Data from the marshy area of this study are complex, and only a careful and extensive examination of data from the various sample depths (3, 15, and 30 m) permits distinction of producing vs. barren structures. Hydrocarbon anomalies are present in most areas, overlying either producing or barren structures.

Concentration (ppm) graphs along sampling lines provide a way to visualize features that distinguish between producing and barren structures. Figure 1 represents a sample line crossing a productive area at all sample depths; Figure 2 is the same for a barren structure. Note how the hydrocarbon concentrations become increasingly different at depth. For instance, comparison of hydrocarbons from 3 m (10 ft) in the sample line in Figure 1 with hydrocarbons from 3 m in Figure 2 shows that virtually no differences are present. Differences are pronounced, however, when deeper samples are compared.

A closer analysis of the data reveals that, in general, gases found in producing areas at 15 m (50 ft) of depth are also exhibited at 30 m (100 ft). Those hydrocarbons often increase in concentration and lateral extent as sampling goes deeper. In Figure 1 it can be seen that at 3 m only three holes manifest gases other than methane, whereas at 15 m the number of holes bearing C_{2+} hydrocarbons has increased to six and at 30 m this number goes to nine. The average concentration of C_{2+} hydrocarbons at 3 m is 6 ppm, whereas at 30 m it is 18 ppm.

The opposite case is seen in barren areas; note in Figure 2 that hydrocarbons are erratic with depth. For example, hole 49 in Figure 2 contains only methane at 3 m. At 15 m, the same hole exhibits mainly C_4 and C_6 and minor amounts of C_3; at 30 m, hole 49 has retained only C_3. It is also evident that, in general, the concentrations of most hydrocarbons present decreased dramatically as depth increases.

It is of interest that even low hydrocarbon concentrations (1 ppm) may be meaningful if they are found at a 30-m depth of sampling. This finding constitutes an innovative concept for interpretation of data in surface prospecting.

From the analysis of gases present over producing structures (Figure 1), it can be observed that hydrocarbons are present as a continuous vertical track through depth (especially at 15 and 30 m). The continuous flow from the accumulation can apparently restore changes induced in the hydrocarbons by the near-surface environment. As a conclusion, it can be said that once equilibrium has been reached, hydrocarbons leaking from the reservoir accumulate in the soil faster than they can be laterally migrated, oxidized by microbes, or released to the atmosphere, especially at depths of 15 and 30 m. Consequently, lateral migration of hydrocarbons has no impact on pinpointing the accumulation at depth. This reinforces the importance of comparing concentration and depth.

Over barren areas (Figure 2), the distribution of hydrocarbons is erratic with depth. An additional important feature (not previously reported in the literature) present over nonproducing sites is the striking overabundance of C_5 and C_6.hydrocarbons. This could perhaps be explained by the lateral migration of hydrocarbons, by the presence of biogenic compounds, or by

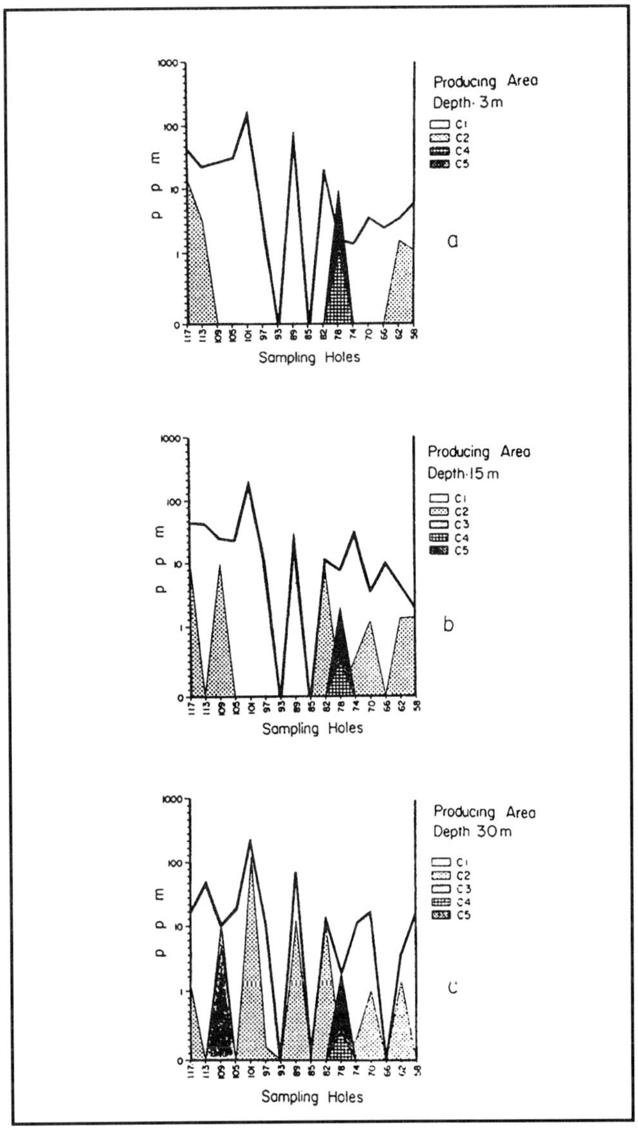

Figure 1. Concentration plots along sampling lines for producing area 1. The vertical axis represents concentration (in ppm). The horizontal axis represents the position of sampling holes in the field. Graphs (a), (b), and (c) correspond to 3, 15, and 30 m (10, 50, and 100 ft) sampling depths, respectively.

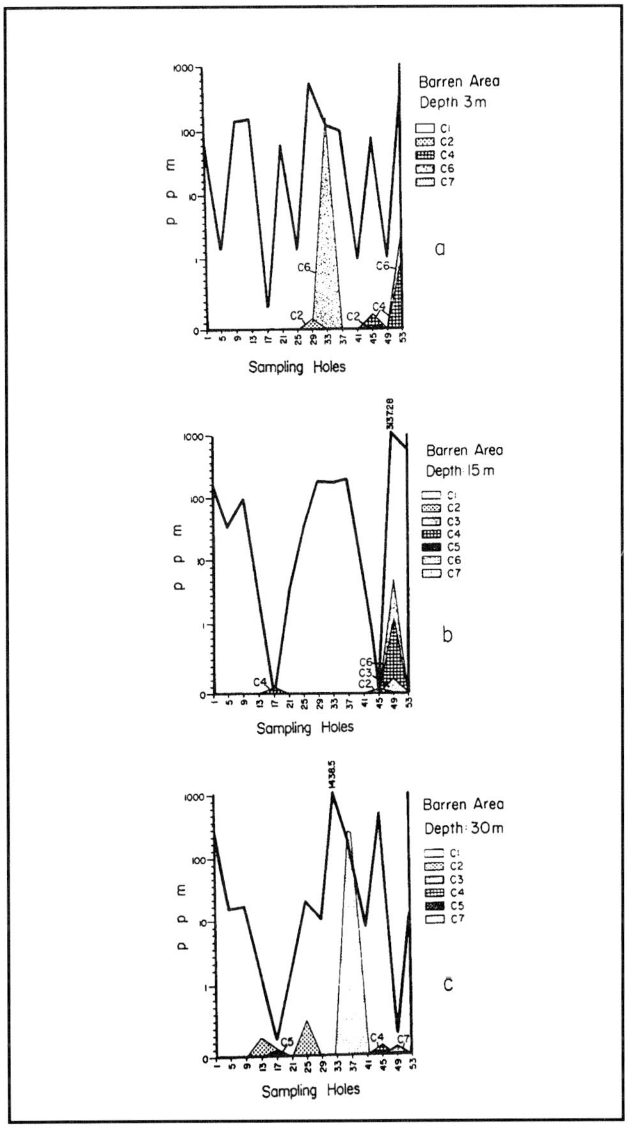

Figure 2. Concentration plots along sampling lines for barren area 1. The vertical axis represents concentration (in ppm). The horizontal axis represents the position of sampling holes in the field. Graphs (a), (b), and (c) correspond to 3, 15, and 30 m (10, 50, and 100 ft) sampling depths, respectively.

some other unique aspect of the study area.

From the analysis of these graphs, it was also found that a halo distribution pattern of gasometric manifestations is not evident in most cases, especially at 30 m of depth. As can be seen from Figure 1, there is no preferential concentration of gases at the edges at 30 m of depth on the sampling lines. This result agrees with Stroganov (1969), who found that the distribution of gases in rocks from depths of between tens and hundreds of meters rarely showed the development of halo anomalies and more often resulted in a bull's-eye target pattern.

Statistical Analysis

A second approach to the verification of the technique of near-surface organic geochemical exploration is multiple discriminant function analysis of the surficial gas data. This technique establishes the differences between two or more groups of samples and indicates the nature of those differences. The method also assigns additional "unknown" samples to the group(s) with which each shares the most similarities. Saenz (1987) and Saenz and Pingitore (1989) presented a retrospective multivariate statistical analysis of the utility of organic geochemical prospecting in distin-

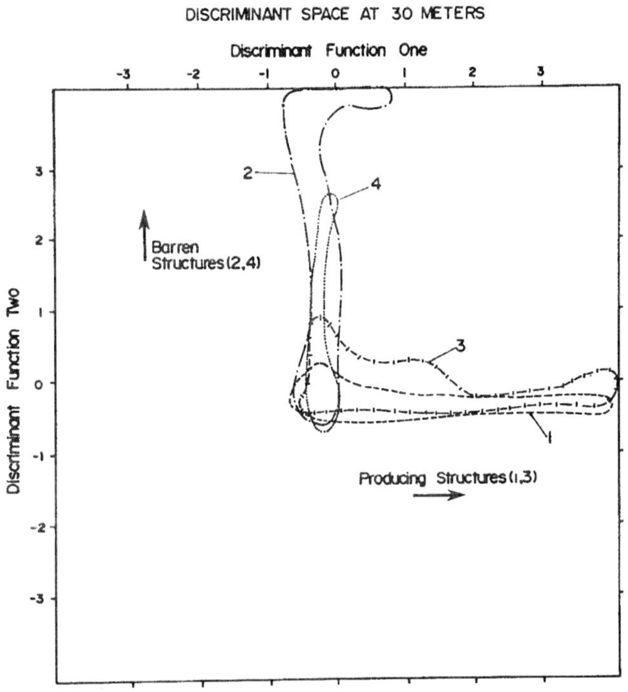

Figure 3. Samples taken at a 30-m (100-ft) depth over two producing and two barren structures plotted in Varimax-rotated discriminant space. Lines surround areas in which the individual samples plotted. The horizontal axis is discriminant function one, and the vertical axis is discriminant function two. Structures 1 and 3 are productive, while 2 and 4 are barren.

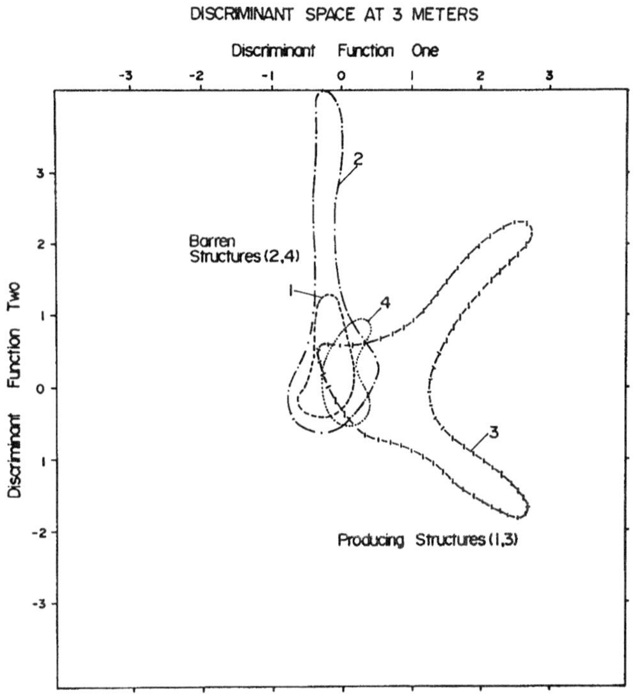

Figure 4. Samples taken at a 3-m (10-ft) depth over two producing and two barren structures plotted in Varimax-rotated discriminant space. Axes and labels are as in Figure 3.

guishing producing and barren structures. Two producing and two barren structures (chosen at random from the data base described earlier) were readily separated by discriminant function analysis of the C_{1-7} hydrocarbons collected at 30 m.

Figure 3 shows the Varimax-rotated discriminant scores (Saenz, 1987, and Saenz and Pingitore, 1989, used an unrotated plot) of each of the samples. Note how the samples from the producing structures (1 and 3) plot along the horizontal axis, whereas those from the barren structures (2 and 4) lie along the vertical axis of discriminant space. The discriminant axes represent the two directions in the seven-space of the C_{1-7} variables that best separate the producing from the nonproducing structures. Discriminant function one, the horizontal axis in Figure 3, had its highest loading from C_3, and discriminant function two, the vertical axis, had its highest loading from C_1, with C_6 showing a secondary influence. A discriminant function can be viewed as a direction created from a weighted combination of the original variables. A variable with a high loading on a discriminant function (e.g., C_3 on discriminant function one) contributes heavily to the direction and mathematical expression of that function and thus to the discrimination between the groups.

The importance of C_3 in the producing structures reflects the thermogenic gases that migrated vertically from the hydrocarbons trapped at depth. Carbon isotopic analyses of methane in the adsorbed form from near-surface samples taken above one oil-producing structure yielded ^{13}C values corresponding to a thermogenic origin of the gas (Saenz, 1984). The influence of C_1 on the barren structures likely represents biogenic gases, which form a background contribution over the entire region. The biogenic origin of C_1 is well known; in addition, Hunt et al. (1980) documented the generation of C_{4-7} alkanes and olefines from terpenoids by bacteria. Thus, the discriminant analysis is consistent with our understanding of the thermogenic and biogenic origin of near-surface hydrocarbons, as outlined in an earlier section of this report.

In contrast, multiple discriminant function analysis of the samples from 3 m permitted only a partial separation of producing from barren structures (Figure 4). Structures 1 and 2, a producer and nonproducer, respectively, both plot and overlap along the vertical axis. Structure 4 (barren) clusters about the origin and overlaps structures 1 and 2. Structure 3, a producer, is characterized by a trend along the horizontal axis. Thus, the statistical analysis at 3 m can distinguish some but not all of the producers. This finding is consistent with the graphical analysis of hydrocarbon concentration with depth presented previously.

In a final statistical test, Saenz (1987) and Saenz and Pingitore (1989) entered data from two additional structures, one producing and one barren, into the discriminant program as unknowns to be classified. The mathematical classification of these "unknown" structures was identical to their actual affinities, suggesting that the initial successes and failures of an exploratory drilling program can generate a valuable database for distinguishing producing and barren structures on the basis of the near-surface geochemical survey. As the exploration program progresses, each drilled prospect added to the data base should increase the reliability of the classification of future prospects. This bootstrap approach can be enhanced during the early stages by near-surface surveys of producing and dry holes proximal to the target area.

CONCLUSIONS

From the graphical anomaly analysis, it was found that continuity of hydrocarbons present through depth is a key to differentiating producing from barren structures. Generally, gases exhibited by producing areas at one depth are also present at deeper levels; this is especially valid for samples at depths of 15 and 30 m (50 and 100 ft). Those hydrocarbons very often increase in concentration and lateral extent as sampling goes deeper. Interestingly, low hydrocarbon values (<1 ppm) are meaningful if they are found at 30 m. Judging from the gases registered over producing structures, once equilibrium is reached, hydrocarbons leaking from the reservoir accumulate in soil faster than they can be laterally displaced, oxidized, or released to the atmosphere. This is most obvious at depths of 15 and 30 m and results in a continuous track of hydrocarbons in which lateral migration is inconsequential with respect to the location of the subsurface reservoir.

In barren areas, concentrations are erratic with depth. In nonproductive sites, there is an overabundance of C_5 and C_6 hydrocarbons. It was also found from the analysis of plots that halo distribution patterns of gasometric manifestations are not evident in most cases, especially at 30 m.

Multiple discriminant function analysis provides an unbiased statistical test of near-surface organic geochemical prospecting. Producing structures were easily distinguished from barren structures on the basis of C_{1-7} hydrocarbons collected at 30 m. Results from samples taken at 3 m were equivocal. C_3 gases were indicative of producing structures, and C_1 and C_6 hydrocarbons were associated with barren structures. Discriminant function analysis can be an important adjunct in a program of near-surface prospecting.

Chapter 15

Application and Theory of Soilgas Geochemistry in Petroleum Exploration

David M. Richers
Syracuse University
Syracuse, New York, U.S.A.

L. E. Maxwell
Marathon Oil Company
Houston, Texas, U.S.A.

INTRODUCTION

The successful application of surface and near-surface geochemical techniques in hydrocarbon exploration requires careful integration of many geological disciplines. Surface geochemistry can provide the explorationist with a means to screen large areas rapidly, economically, and qualitatively for overall petroleum source potential. Regions of favorable thermal environments for the formation of petroleum can be defined by identifying areas containing thermogenic hydrocarbons in soilgas. Soilgas techniques are also useful in determining areas of prospective accumulations prior to conducting additional and more costly exploration programs.

Historically, the presence of an active oil or gas seep was sufficient to deem an area as prospective. Seep detection today rarely involves visible evidence of hydrocarbons; rather, it deals with seeps of extremely small magnitude (microseeps) that are not directly visible to the unaided eye. Typically, microseeps are detectable through the use of sensitive analytical instrumentation. In addition, it may be possible to detect active or inactive microseepage from its secondary manifestations such as bleaching of rocks in the area of the seep, the presence of stressed vegetation, or the formation of authigenic magnetite. Examples of some of these secondary indicators can be found in studies conducted by Dalziel and Donovan (1980), Donovan (1974), and Furguson (1975). Interpretation of these secondary indicators is often difficult because other natural processes can produce the same physicochemical features. Furthermore, the quantity and composition of hydrocarbon gases in the soil may be affected by the local geology. Consequently, the true meaning and character of these microseeps can only be realized after very careful scrutiny of the geological environment in which they are found. Surface geochemistry is only one of many tools at the disposal of the explorationist, and it should always be applied in conjunction with other geological and geophysical techniques in the evaluation of a potential play. As such, geochemical "prospecting" is a misnomer; the term *geochemical screening* would be more appropriate to describe the true function of surface geochemical techniques. In other words, surface geochemistry is useful in delineating areas that may possess a higher probability of containing effective source rocks or hydrocarbon accumulations, but only so far as the explorationist has a thorough geological knowledge of the area.

It is foolhardy to attempt to predict the presence of economic hydrocarbon accumulations or to locate wells using surface geochemistry alone. Numerous near-surface hydrocarbon detection techniques, however, have been shown to be capable of distinguishing basins (or large portions of basins) that are productive from ones that are unproductive. Furthermore, many of these techniques can distinguish the type of production (oil, gas, or mixed oil and gas) in the basin. This ability has been recognized by Weisman (1980), Jones and Drozd (1983), Richers et al. (1982), Horvitz (1985), Klusman and Voorhees (1983), and others. In addition to highlighting prospective basins, near-surface soilgas data are useful in establishing favorable trends or fairways of potential production by delineating zones of active seepage of subsurface hydrocarbons. Such zones are commonly along faults and fractures. Surface geochemical techniques can help the explorationist select which of several frontier basins or which portions of a basin have the greatest chance of containing thermogenic hydrocarbons, as well as predicting their composition (oil, gas, or mixed). The premise that microseepage occurs and that useful information can be derived from its detection should now be a recognized fact.

Ideally, the explorationist will have one or more end-member controls (i.e., known fields or barren areas) to compare to the results of the geochemical survey. Otherwise, it is necessary to extrapolate results from similar but more distant areas, where an established data base exists, to the area in question. Comparison with poorly related provinces can prove frustrating because subtle differences between areas can often result in significantly different geochemical manifestations. Still, when end-member control is available, it is possible through careful comparison to determine prospective areas through the use of surface geochemistry. Advanced statistical studies utilizing the modern computer are useful in evaluating large volumes of multidimensional data. In the past when evaluating an area, it was expedient to use the data from only six or less specific hydrocarbon gases. Now it is possible through the use of modern surface geochemical techniques and multivariate statistics to include several dozen molecular species.

In addition to advanced statistics, there is a trend today toward integrating surface geochemical methods with other investigations. Published studies coupling microseep technology with remote sensing abound. The inclusion of micromagnetics, radiometrics, gravity, and magnetics is also more common. Thermal measurements such as heat flow are also being utilized. Berger (1985) showed that deep basement faults that may affect the geology of overlying strata can often be difficult to detect using standard seismic methods. Remote sensing data can often detect the surface expression of such faults. Landrum et al. (1989) and Richers et al. (1986) have demonstrated that microseeps are often present along the surface expression of such features. Patton and Manwaring (1984) identified a "hazy" anomaly on a Landsat image along the Syracuse fault over the Hugoton gas field in Kansas. They defined a hazy anomaly as an area of visible aberration on an image which was possibly due to the distortion of reflected light by vapors. In their study, they found a strong correlation between the concentration of soilgas and that feature. McCrossan et al. (1971) pointed out the strong association of high concentrations of hydrocarbons in the surface environment with photolineaments.

This chapter will not attempt to review the history of surface geochemistry thoroughly nor to describe all of the numerous technologies that have been applied (see review by Price, 1986). While the present discussion will cite references of many of the methods that are currently being used, the principle intent is to focus upon those concerning soilgas. After a discussion of the basic principles, various techniques, and data treatment, documented examples will be reviewed.

BASIC PRINCIPLES

The fundamental assumption in all near-surface hydrocarbon techniques is that thermogenic hydrocarbons generated and trapped at depth escape to the surface at varying rates. These hydrocarbons are themselves detectable, as are the products of their interaction with the near-surface environment. Numerous field studies both in the United States and abroad have shown this to be true. A further assumption is that the areal distribution, magnitude, and composition of the detected microseepage provides information concerning migration pathways and, in fact, can be useful in mapping faults and fracture systems.

The physical state of hydrocarbons during transport is not well known. Nevertheless, most of the models proposed for transport of these fluids from source to reservoir (aqueous transport in true solution, as micelles, as a discrete oil phase, as a gaseous phase, etc.) are applicable to the continued transport of hydrocarbons from these source beds and/or reservoirs to the near-surface environment. The controversial nature of theories of this migration towards the surface from entrapped hydrocarbons is well documented (Roberts and Cordell, 1980). The mechanisms that have been considered in soilgas geochemistry include diffusion, effusion, and separate phase transport via microbubbles.

Effusion is a term that has been applied to concepts of the migration mechanism by various authors (Donovan and Dalziel, 1977; Duchscherer, 1980). Effusion by definition refers to bulk flow in a continuous phase which is driven by a pressure differential. This concept actually describes macroseepage very well. However, microseepage cannot be explained by such a bulk flow pathway. Effusion can only grossly account for limited microseepage transport because of the chromatographic effects that have been observed in soilgases, .

Hydrocarbons that constitute microseepages are thought to exist in a separate phase from subsurface waters, as is the case for effusive transport. True effusion can occur over very short distances when sufficient hydrocarbons accumulate. However, it is more feasible that movement of microseep gases occurs as very small discrete bubbles (microbubbles) and not as a continuous hydrocarbon filament. The forces of buoyancy and capillary pressure drive the migration of hydrocarbon microbubbles. The buoyancy drive acts primarily in a vertical direction, and the rate of migration depends upon the permeability of the pathway. Faults, fractures, joints, and bedding planes present open pathways relative to pore space. Migration along these open pathways is relatively efficient and, therefore, should be manifested at the surface as high-magnitude microseepage.

Because of the preferred channeling along faults, fractures, and similar features (Landrum et al., 1989, Richers et al., 1986, Jones and Drozd, 1983), this mechanism can account for the discrete (localized) nature of many anomalies associated with micro- and macroseepage. Jones and Thune (1982) showed that gases escaping from subsurface coal retorts were detectable only hours after ignition. The gaseous products migrated along faults and updip along bedding planes. The localized, often linear nature of soilgas anomalies suggests that faults and fractures play an important part in the movement of these gases. Major linear features discernible on satellite remote sensing images, as well as on other remotely sensed mediums from Patrick Draw oil field, Wyoming, show such a relationship (Richers et al., 1982). The Lost River, West Virginia, Geostat study (Matthews et al., 1984) documents the relationship between anomalously high soilgas concentrations and linear features on imagery. Patton and Manwaring (1984) found a relationship between the presence of a hazy (spectral) anomaly on satellite imagery along the Syracuse Fault over the Hugoton gas field and over leaking hydrocarbons. There were also anomalously high soilgas concentrations reported along faults in the San Joaquin basin and in the Wyoming–Utah overthrust belt by investigators from Gulf Research Company (Jones and Drozd, 1983).

Diffusion is a relatively slow and widely dispersive process. Depending upon the particular lithology and geological conditions, the time required for diffusion to occur can sometimes be restrictive. Antonov et al. (1971) measured hydrocarbon diffusion coefficients for a variety of rock types from several hydrocarbon provinces in the Soviet Union. They discovered that the coefficients of diffusion vary over a wide range (from $n \times 10^{-3}$ to $n \times 10^{-8}$ cm^2/sec). Table 1 shows that diffusion of gas to the surface from a reservoir buried at 4000–5000 ft (1200–1500 m) would take in excess of 100 m.y., while diffusion of some of the light gas components would take 230–270 m.y. (Hunt, 1979; Smith et al., 1971). Clearly, if diffusion were the sole mechanism for vertical migration, seepage from deep source beds or reservoir rocks would not only exceed the age of the hydrocarbon accumulation but also could exceed the age of the host beds. If diffusion were the dominant process for vertical migration, the detection of soilgas anomalies in the near subsurface would be restricted to only very shallow accumulations. Data from many documented studies where source beds and/or reservoir rocks are present several thousands of feet below measured surface geochemical anomalies

Table 1. Losses of reservoir gas by vertical diffusion.[a]

Gas	Relative Molar Composition	Time To Reach Surface (m.y.)	Rate of Loss to Surface ($cm^3/m^2/yr$)
Methane	100	140	2
Ethane	7.5	170	0.16
Propane	4.3	230	0.06
Butane	2.1	270	0.02

[a]From Hunt (1979) and Smith et al. (1971).

suggest that diffusion is at best only a secondary process in hydrocarbon microseepage.

Diffusion may become important, however, once the gases have migrated to the near-surface environment. In some instances, diffusion may be responsible for the elevated "background" levels often observed in areas of accumulations. However, similar background levels could result from free hydrocarbons migrating outward and upward in areas of minute fractures, from dispersion of the gases by groundwater flow, or by passage of the gases through extreme thicknesses of unconsolidated alluvium. The widespread oily signature reported by Landrum et al. (1989) in Flathead Valley, Montana, might be an example of such a process.

SURFACE GEOCHEMICAL METHODS

Surface geochemical methods can be classified as direct or indirect. Direct methods involve the determination of the presence of dispersed petroleum compounds in the form of hydrocarbon gases or liquid residues in the soil, water, and rock. Indirect methods are based upon detection of chemical, physical, or microbiological changes in the soil, water, or rock associated with the oil and gas deposits.

The most common direct technique involves the collection and measurement of the light alkanes: methane through the butanes (C_1–C_4). Because of their relatively high vapor pressures, these light hydrocarbons are frequently found in sufficient concentrations for measurement within the free pore space of soil and rock. They also may be bound in the sediment matrix through absorption and adsorption. The mineral matrix functions as an accumulating sampler, and preparatory techniques for desorption of the gases such as disaggregation and acid extraction have been developed (Horvitz, 1985). Gas chromatography is the standard analytical technique for the quantification of hydrocarbons contained in free soilgas and those released from sediment samples. Newer sampling techniques that selectively adsorb heavier hydrocarbon compounds onto artificial adsorbents are also available (Klusman and Voorhees, 1983). These minimize the effects of variable adsorptive capacities of soils over different lithologies. By using artificial adsorbents, heavier components of the microseep are easily identifiable through analysis using mass spectroscopy as well as gas chromatography.

Stable isotope analysis is also a very important tool in surface geochemistry. Stable carbon isotope compositions of the light alkanes have been interpreted to indicate whether the gases are thermogenic or biogenic in origin (Fuex, 1977). Furthermore, stable hydrogen isotopic compositions enhance the ability to resolve the origins of hydrocarbon gases (Schoell, 1988, 1983). Unfortunately, unequivocal resolution cannot be achieved in all cases (Rigby and Smith, 1982; Smith et al., 1985). This uncertainty causes wide ranges in the stable isotope compositions of carbon and hydrogen in natural gases produced from coals.

Offshore methods have also evolved. The simplest direct offshore measurement is the subsea hydrocarbon "sniffer" (Sackett, 1977; Reitsema et al., 1978). Typically, a probe is towed near the sea floor from a ship. The water is pumped to a stripping chamber on the ship, where the hydrocarbons are exsolved either by passing an inert gas through the sample and/or by subjecting the sample to a slight vacuum. The gas is then directed to a gas chromatograph for analysis, and the water is discarded. This method is usually used to analyze for methane through the butanes, although pentane and heavier species are sometimes captured as well. Collection of sediment cores from the sea floor is also a common practice, either by drilling, gravity coring, or piston coring. Once a sample is collected, any of several standard analytical techniques can be employed, including light alkane gas chromatography, gas chromatography of extractable liquids, isotopic analysis, and scanning fluorescence (Brooks et al., 1986).

Indirect techniques are methods of detecting secondary effects of vertical hydrocarbon seepage, whether from an active source rock or from a reservoir. The analysis of spectral and visible imagery acquired by satellites or at high altitudes has proven useful for detecting tonal anomalies. Certain tonal anomalies have been shown to be associated with vegetation that has been stunted or stressed. Stressed vegetation may correlate to areas with high concentrations of hydrocarbons in the soil (Abrams et al., 1985). Geobotanical applications include the identification of plants that thrive in soils with high concentrations of hydrocarbon gases (Matthews, 1985). Other tonal anomalies may be associated with mineralogical alterations caused by hydrocarbon microseepage (Segal et al., 1985). It should be mentioned that some mineralogical alterations caused by microseepage may not be detectable by remote imagery. Bacterial oxidation of seeping hydrocarbons produces CO_2 that is enriched in ^{12}C. When sufficient CO_2 precipitates out as secondary carbonate cement in near-surface rock or soil, it can be detected by stable carbon isotopic analysis (Donovan, 1974).

Microbiology has also been utilized to identify areas indirectly that contain seeping hydrocarbons. These techniques were first reported by Mogilevskii (1938). Modern practices center on culturing thermogenic hydrocarbon specific bacteria (preferably obligates) from soil or groundwater samples. Reviews of this have been presented by Davis (1967) and Hitzman (1961).

Certain geophysical tools can also be applied in an indirect sense. Gamma radiometric surveys have shown some areal correlation between hydrocarbon accumulations and measured anomalies of gamma radiation flux at the surface (Weart and Heimberg, 1981). Most theories explaining this correlation are based upon the redox fixation of uranium above accumulations as escaping hydrocarbons are oxidized. Micromagnetic surveying and various applications of electromagnetics also attempt to detect metalliferous minerals or mineral suites that are precipitated via redox reactions coupled to the oxidation (abiotic or biotic) of seeping hydrocarbons. Certain micromagnetic anomalies have been attributed to authigenic magnetite where areal correlations with hydrocarbon accumulations have been demonstrated (Danziel and Donovan, 1980). Electromagnetic measurements such as induced polarization, resistivity, and magnetotellurics have been shown to respond to the presence of dispersed sulfide mineralization, which again may be the product of microseepage-induced redox fixation reactions (Oehler and Sternberg, 1984).

Table 2 describes some of the more common surface geochemical methods in use today. Although we have not tried to make this list inclusive, it is readily obvious that any of several methods can be useful to exploration.

ANALYSIS OF DATA

Regardless of the techniques used in a geochemical survey, it is necessary to scrutinize the resultant data for it to be useful. Simple comparisons of the signal strengths (i.e., concentrations of measured species in the soil) and/or compositions of the measured species can be made between an area of "known" potential and the survey area. If the geology is similar between the two areas and the purpose of the study is to grossly evaluate the presence or absence of migrated hydrocarbons, this may be adequate. This is particularly useful when an accumulation or source bed is known to be present in a basin and when the purpose of the survey is to identify prospective areas for the application of more costly exploration programs.

A less reliable evaluation results by comparing the survey results with data obtained from more distant study areas that may or may not have similar geological attributes. The source of this control data may be another survey conducted by the investigator using the

Table 2. Surface geochemical methods.

Method	C_1/C_4	C_{5+}	Heavy Hydrocarbons
Probe	Yes	Yes	No
Disaggregated soils	Yes	Yes	Maybe
Acid extracted soils	Yes	Yes	Maybe
Integrated adsorption	No	Yes	Yes
Microbiological	Indirect	Indirect	Indirect
Micromagnetics	Indirect	Indirect	Indirect
Remote sensing	Indirect	Indirect	Indirect
Fluorescence	No	Yes	Yes
Radiometrics	Indirect	Indirect	Indirect

same techniques and equipment, or it may be derived from published data for surveys of unknown design. The use of published empirical data can be misleading if one expects the geochemical response in the survey to match that of test area exactly.

Jones and Drozd (1983) published some empirical gas ratios based on thousands of measurements from several areas (Table 3). These ratios are useful in establishing the qualitative cutoffs between an oil-prone region and a mixed oil–gas or gas-prone region. However, there is a tendency for gas-prone areas to contain a higher proportion of methane in the soils than a mixed or oil-prone area. The exact values of the qualitative boundaries will vary depending on the local geology, the type of source beds in the area, and the maturity of the beds. Furthermore, hydrocarbons derived from multiple horizons will often obscure the interpretation, particularly if shallow gas reservoirs are present between the surface and deeper reservoir/source sections. Bacterial sources of methane can also pose problems, as can extensive permafrost conditions, playas, and water-logged terrain. Published data, therefore, should only be utilized as a rough "yardstick," while keeping in mind that variations in soilgas composition within the survey area may be more significant than any generalized empirical guidelines. In this way, "dry" and "wet" areas can be defined and that information alone can be used to answer the qualitative exploration questions.

Magnitude, or concentration, data can also be useful. Because microseepage is often highest in areas of fractures and faults, soilgas magnitudes can be interpreted as a means to map faults. Conversely, if two areas show similar fracture patterns and densities but only one shows high-magnitude microseepage, the relative qualities of hydrocarbon sources in the two areas can be inferred. Again, it is possible that a simple comparison of magnitudes between areas of known potential and the area under study can help determine zones of favorable potential provided that the geology over the two areas is constant. Typically, however, the geology varies significantly or is not understood well enough to assess the similarities.

The fact that faulting, fractures, and rock joints influence the flux of hydrocarbon microseepage has been documented by Richers et al. (1986) and Landrum et al. (1989). Figure 1 shows that, in general, the concentration of hydrocarbons in soilgas decreases as the distance of a sample from a fault increases. Understanding this relationship can be extremely important to the interpretation of a data set. Two generalized migration pathways exist: one is dominantly through faults or other open conduits and another occurs mainly through pore space. Mapping of magnitude data that contain samples representing both categories can produce artificial "anomalies" that only reflect the migration pathways. True, meaningful anomalies can be manifested through each migration route. A qualitative separation (i.e., in the mind of the interpreter) will often suffice. However, the data can be treated as subsets with separate statistical analyses.

More times than not, simple comparison between a known area and the survey area, or between the survey area and published data, is not possible without prior treatment of the data. For example, the effects of lithology or soil type may bias the results of a survey. If a thermally mature organic-rich shale exists in the near-surface, its contribution to the microseepage signal will likely mask that of deeper source beds. If this source is present locally, magnitude and composition data may appear anomalous as a result of the presence of such beds. Alternately, the presence of unfaulted evaporites might occlude a signal from a deep source, leading to erroneous conclusions. While not providing a perfect seal, evaporites can provide a migrational barrier that is effective enough so that the measurable signal at the surface is within the noise or background level (Jones et al., 1985; Patton and Manwaring, 1984). Therefore, soil type should be recorded for each sample site whenever possible, and the results of the soilgas analyses scrutinized for any correlations with soil composition. Samples should also be grouped and evaluated according to the lithologies over which they are positioned. At present, it is not possible to correct quantitatively for the effects of soil type. Semi-quantitative or qualitative corrections can be applied, however, and

Table 3. Empirical soilgas compositions product classifications.[a]

Product	Gas Wetness (%)	C_3/C_i x 1000	C_1/C_2
Oil	60–25	500–60	4–10
Mixed oil and gas	25–5	60–20	10–20
Dry gas	5	20–2	20–200

[a]Adapted from Jones and Drozd (1983).

should be accounted for in the interpretation. The use of artificial adsorbents effectively negates this consideration. Other errors may be introduced if sampling methods are not applied uniformly over the study area, or if the results of one technique used in the control area are compared with those in the study area which were acquired using different methods.

A constant concern with any large experimental data set is one of reliability or in effect, internal consistency. The simplest manner to evaluate this is to generate two variable crossplots of the various components. Typically, a fair linear correlation will exist between ethane, propane, and the butanes, reflecting a common thermogenic source. Methane may show a more discordant relationship because it may be a mixture of thermogenic and biogenic gases. When more than one source of gas is present, multiple linear trends can be observed. The generation of correlation matrices is useful as the intercorrelations among all variables are summarized. Subsampling individual trends and calculating correlation coefficients also yields useful information. Table 4 shows a typical correlation matrix for shallow soilgas data obtained from mechanically disaggregating soils from Flathead Valley and Townsend Valley, Montana. Based upon the rather high correlation coefficients among methane, ethane, propane, the butanes, and the C_{5+} fraction, a common thermogenic source for these gases would be indicated throughout the area. This is further supported by the generally linear relationship between ethane and propane shown in Figures 2 and 3.

Multivariate statistical analyses can be extremely useful for evaluating surface geochemical data (Saenz, Chapter 14, this volume). The increasingly large number of parameters that can be acquired from the chemical analysis of a single sample makes it cumbersome and very time consuming to interpret a data set "by hand." Some form of data reduction has become mandatory in almost all modern surveys. Factor analysis is an effective means of data reduction when performed and interpreted properly. Factor analysis can be applied to a single data set to determine which parameters are varying significantly across the survey area. However, a greater utility exists when it is applied to a local survey and a control survey. In such a case, the significant covariant factors within the combined data set can be determined. In either case,

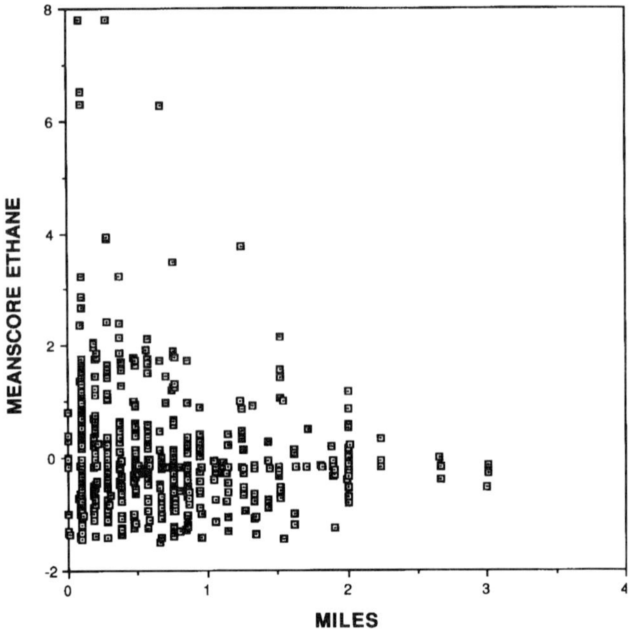

Figure 1. Magnitude of ethane versus distance from linear features. (From Richers et al., 1986.)

the interpreter is left with a lesser number of parameters to evaluate, hopefully excluding those with insignificant or random variation.

Discriminant and cluster analyses are also applicable to the large data sets involved. Both analyses mathematically plot the data in n-dimensional space, where n is the number of parameters considered for each sample. When a control survey is available over a known accumulation, for example, discriminant analysis will allow an evaluation of the degree of similarity between the individual samples of the test survey and a model that represents the soilgas composition associated with the known accumulation. The variances of all n variables can be treated simultaneously, eliminating the need for a large number of two-variable statistical evaluations. Cluster analysis is more applicable when a control survey is not available. Cluster analysis can facilitate the process of recognizing groups of samples within a survey which display significant similarity among themselves. This is often useful to discriminate local background from anomalies.

It is beyond the scope of this discussion to explain the intricacies of the theory and application of multivariate statistics. The blind use of any statistical results is dangerous, and a thorough understanding of these analyses is recommended before they are applied. Davis (1973) details the applications of discriminant, cluster, and factor analyses to geology. Additional discussions for choosing statistical thresholds, and data analysis are covered in the work of Koch and Link (1970).

Table 4. Correlation coefficients among light gases, Flathead Valley and Townsend Valley, Montana.

Flathead Valley	C_1	C_2	C_3	i-C_4	n-C_4	C_{5+}
C_1	1.000	0.890	0.878	0.891	0.903	0.800
n	200	200	200	200	200	197
C_2		1.000	0.996	0.995	0.994	0.810
n		200	200	200	200	197
C_3			1.000	0.998	0.997	0.808
n			200	200	200	197
i-C_4				200	200	0.809
n			200	200	200	197
n-C_5					1.000	0.816
n					200	197
C_{5+}						1.000
n						197

Townsend Valley	C_1	C_2	C_3	i-C_4	n-C_4	C_{5+}
C_1	1.000	0.118	-0.092	0.938	0.977	-0.031
n	231	55	22	4	6	195
C_2		1.000	0.983	0.976	0.902	0.203
n		55	22	4	6	44
C_3			1.000	0.990	0.945	0.417
n			22	4	6	20
i-C_4				1.000	0.905	0.361
n				4	4	4
n-C_5					1.000	0.752
n					6	6
C_{5+}						1.000
n						195

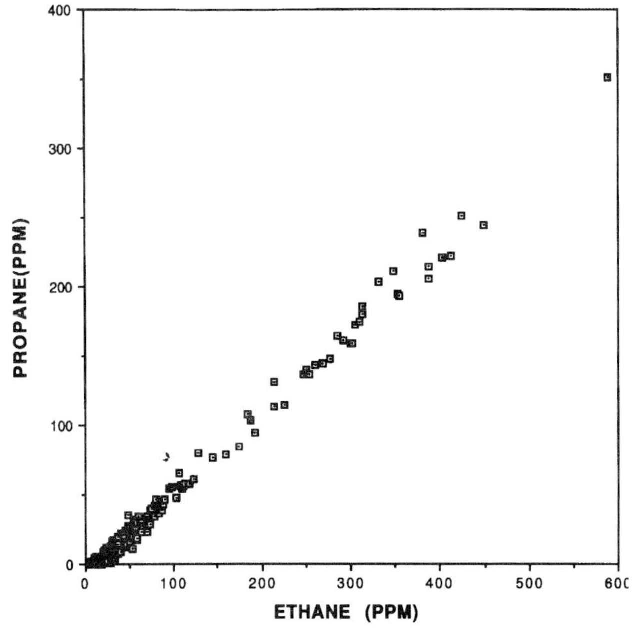

Figure 2. Crossplot of ethane versus propane in soilgas samples, Flathead Valley, Montana.

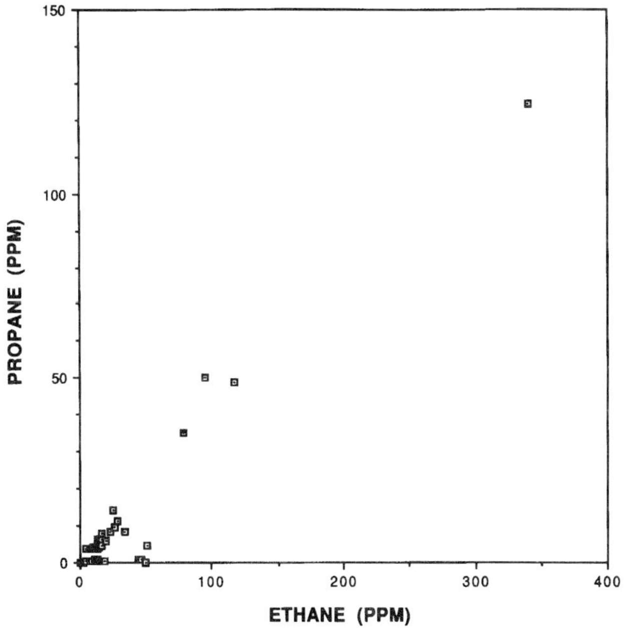

Figure 3. Crossplot of ethane versus propane in soilgas samples, Townsend Valley, Montana.

Table 5. Compositional data for the 1983 follow-up Patrick Draw soilgas study.

Sample Type	Gas Wetness (%)	C_3/C_1 x 1000	C_1/C_2
12-ft free soilgas	13.35	50.01	19.95
Number of samples	70	95	92
Classification	Mixed	Mixed	Mixed
4-ft free soilgas	4.33	9.07	91.99
Number of samples	219	221	210
Classification	Gas	Gas	Gas
4-ft acid extract soilgas	0.92	23.51	17.25
Number of samples	225	225	225
Classification	Mixed	Mixed	Mixed
4-ft disaggregation soilgas	0.91	26.05	26.52
Number of samples	246	246	246
Classification	Mixed	Mixed	Gas

CASE STUDIES

Patrick Draw, Wyoming

The purpose of this study was to determine whether gases leaking from a known subsurface reservoir preferentially followed fracture and fault pathways. In addition, the signature of the gases were tested to see whether it was possible to differentiate populations of data using discriminant analysis. Patrick Draw oil field is located about 45 miles east of the town of Rock Springs, Wyoming, along the eastern flank of the Rocks Springs uplift, a Laramide-age feature. Patrick Draw, along with the West Desert Springs and Desert Springs fields, are stratigraphic traps in the Cretaceous Almond Sandstone. Hydrocarbons are reservoired in barrier bar to back-barrier/lagoonal sandstones with lateral seals created by porosity–permeability changes in the Almond and vertical seals provided by the overlying Lewis Shale. The Table Rock field east of Patrick Draw is situated in the Almond as well, but it is a structural anticlinal trap.

Richers et al. (1986) found that the overall soilgas signature indicated a gas to gas/oil composition (Table 5). Furthermore, there is a preferential affinity toward higher soilgas concentrations as mappable linear features are approached (Figure 1). Ethane concentrations increase with decreasing distance from a lineament. The investigators concluded that there was a 70% probability that a sample collected over the field would correctly identify the field. When coupled with other exploration techniques, this fact could aid in the discovery of accumulations similar to Patrick Draw.

Patrick Draw also has the unique circumstance of being an area where a geobotanical anomaly might be associated with a soilgas seep. An area known as the "blighted sage zone" occurs over the western part of the field directly over the present-day gas cap. Sage collected in this area appears to be dwarfed, which can have a variety of causes. What is significant is that the amount of affected sage is quite large, over 4 sq mi. Furthermore, there are repeatable large-magnitude soilgas anomalies in the soil and soilair of the area. (Some researchers at the Jet Propulsion Laboratory showed this stress to predate the development of the field in excess of 90 years.) If this sage is being repeatedly stressed by hydrocarbons, then times of extreme climatic conditions might enhance the "stressed" appearance.

An independent study by researchers at the Colorado School of Mines was reported by Klusman and Voorhees (1983) over the Patrick Draw field. Small ferromagnetic wires coated with activated carbon were buried in a shallow hole for a 1- to 2-week period and allowed to adsorb hydrocarbons from the soil atmosphere. Collected species were made using a computerized Curie point pyrolysis mass spectrometer which allowed for mass identification up to mass number 240. Samples were obtained as close as possible to the locations of anomalous sites identified in the original geochemical survey conducted by Richers et al. (1982) (Figure 4). Klusman and Voorhees found significant fluxes of compounds with masses over 90 (Figure 5) as well as compounds with masses lower than 90 (Figure 6). The higher mass samples were found to be situated over the main Patrick Draw field, whereas the lower mass samples were present both over the oil field and the gas cap. The processing of such data can be quite cumbersome and involved, usually requiring sophisticated computer programs and multivariate statistics. However, since heavier molecules can be detected, such an effort is well warranted.

Lost River, West Virginia

Another example of geobotanical indicators of microseepage was reported by Matthews et al. (1984) for the Geosat Lost River Test site. The Lost River Field is a fractured Oriskany (Devonian) reservoir at the crest of the Whip Cove anticline. They also found an apparent rela-

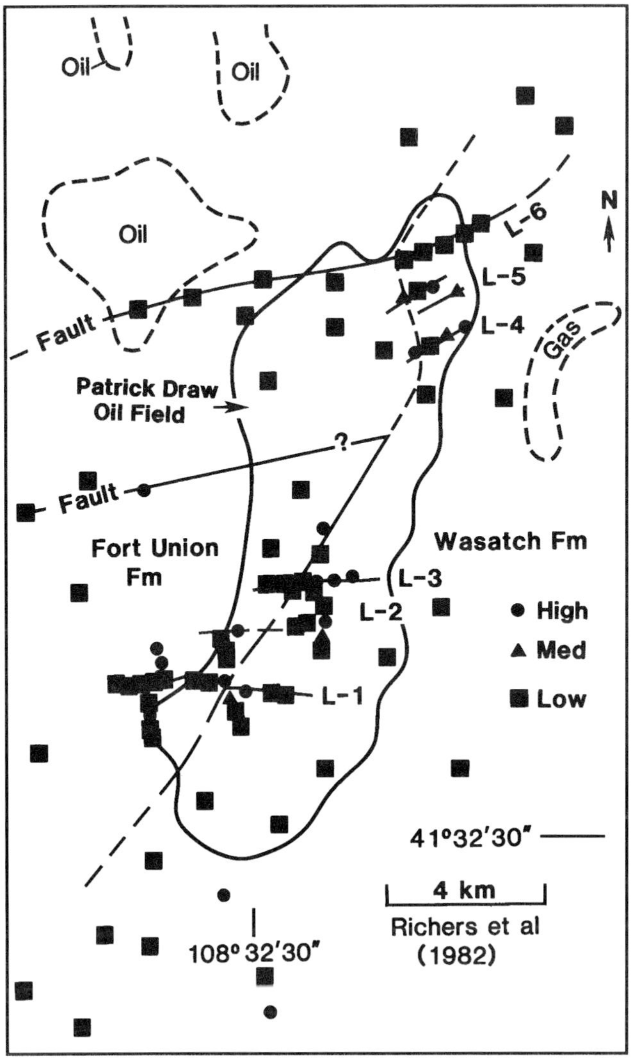

Figure 4. Combined significant anomalies for soilgas samples at Patrick Draw field, Wyoming. (From Richers et al., 1982.)

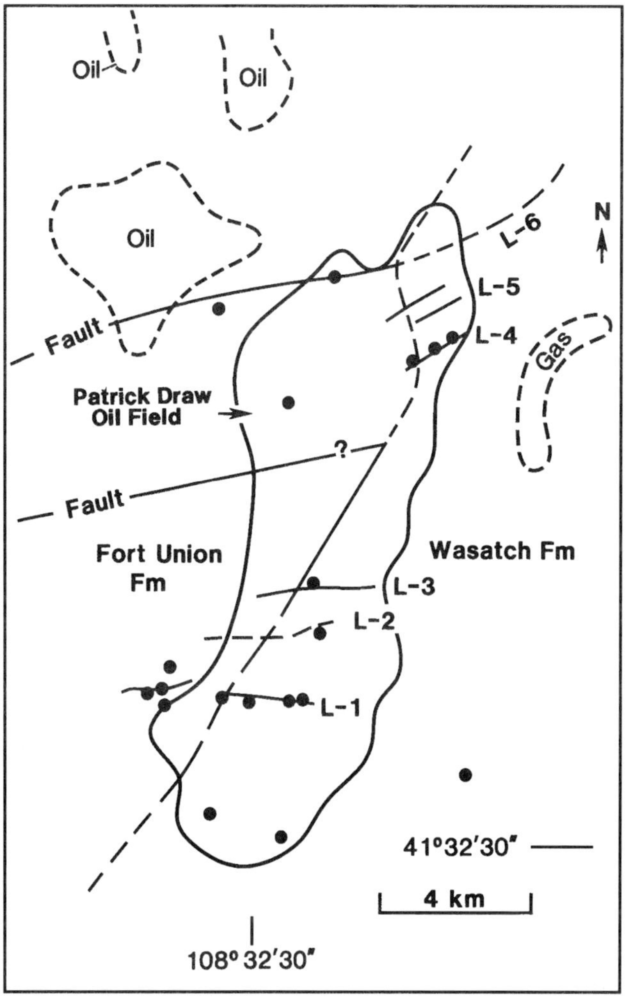

Figure 5. Sample sites with significant quantities of compounds with masses greater than 90, Patrick Draw field, Wyoming. (From Klusman and Voorhees, 1983.)

tionship between vegetation anomalies, soilgas anomalies, and hydrocarbon production. It was concluded that anomalous concentrations of maple trees in a climax hardwood oak–oak chestnut forest were directly correlative to high-magnitude soilgas signals along the crest and above the Oriskany reservoir (Figure 7). Thematic mapper simulator (TMS) imagery processed by Geosat participants at JPL helped delineate areas containing the maple trees. Lineament density analysis indicated that the greatest density of lineaments occurs along the crest of the Whip Cove anticline. Diecchio (1985) has shown that many of the Oriskany reservoirs situated in the Appalachian basin are either permeability stratigraphic pinchouts or are structure-related fracture porosity plays. Figure 8 is a simplified cross section of the Lost River field and its associated soilgas propane anomaly. East of the field where the reservoir strata and the Chattanooga Shale outcrop (the Marcellus, or Chattanooga, is the most likely source of the gas in the region), an additional high-magnitude propane anomaly exists. If one had no prior knowledge of the presence of this source rock facies in the near-surface, the "source rock" anomaly could be misinterpreted as a potential hydrocarbon accumulation in the subsurface.

Railroad Valley, Nevada

A combined remote sensing and surface geochemical survey of Railroad Valley, Nevada, was conducted by Jones et al., (1985). This study was designed to test the structural model proposed by Dolly (1979) and Foster (1979). The object of the study was to determine if the proposed structural control of the reservoirs in Railroad Valley can be detected by remote sensing and if surface geochemical data can be used to define those structures that most likely contain hydrocarbons. Railroad Valley

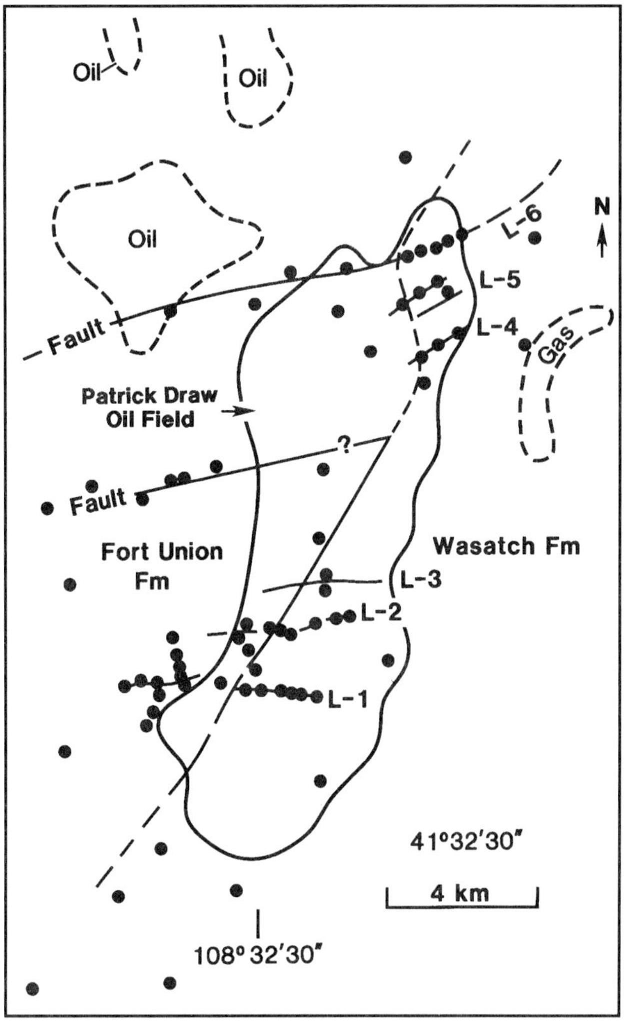

Figure 6. Sample sites with significant quantities of compounds with masses less than 90, Patrick Draw field, Wyoming. (From Klusman and Voorhees, 1983.)

currently has four producing fields: Eagle Springs (discovered in 1954), Trap Springs, Bacon Flat, and Grant Canyon. Total production is in excess of 10 million barrels of oil. Both Trap Springs and Eagle Springs fields are combination stratigraphic truncation/subcrop fault traps. Production is from matrix and fracture porosity in reservoirs of the lacustrine Sheep Pass Formation (Cretaceous and Eocene) and the Oligocene Garrett Ranch volcanic group (Foster, 1979; Dolly, 1979; Picard, 1960). They noted that production occurred only in the more elevated portions of lower subgrabens located near the edge of the graben forming the main valley. Higher adjacent blocks commonly showed extensive removal of reservoir facies by erosion. Foster and Dolly also noted that aerial photography and geomorphic and topographic studies accurately delineated these blocks. This ultimately lead to the discovery of the Trap Springs field. The Grant Canyon and Bacon Flat fields are also located in the structurally intermediate fault blocks, with production occurring from the Middle Devonian Simonson Dolomite (McCaslin, 1984)

The remote sensing study reported by Jones et al. (1985) used several formats, namely, synthetic aperture radar (SAR), Landsat-5, thematic mapper (TM), and multi spectral scanner (MSS). North–northeast striking lineaments are generally associated with faults bounding the main valley (Figure 9), and in fact, many springs occur along their strike accurately depicting the margins of the valley. A second major set of lineaments strikes east–west and probably represent basement fractures, such as the noted Pancake Range lineament. Two other sets of less extensive, but certainly pervasive, linear features strike northeast–southwest and northwest–southeast.

Samples of free soilgas were collected at depths of approximately 4 ft using probes. Mapping of the soilgas data (Figure 10) shows several high-magnitude anomalies updip from the fields. These anomalies are present over the major linear features that represent the block-bounding faults. Vertical migration of gases along faults, and subsequent leakage into unconsolidated sediments updip causes anomalies to appear not only along linear features (faults) but also adjacent to producing blocks.

Clarno Area, North-Central Oregon

Frequently, the primary concern of the explorationist in a frontier play is to discern the presence of source beds in an area, as well as to map the relative maturity of these beds. The Clarno area in north-central Oregon sits on the southern extreme of the Columbia River Basalt Plateau with inliers of sediments exposed through volcanics that cover everything to the north.

In a 1985 study, Sidle and Richers collected soilgas samples by drilling shallow holes (3 m or less) and placing a packer in the hole, then aspirating a sample as described by Jones and Drozd (1983). It was found that although source rock availability was limited to 13 Cretaceous inliers, soilgas signatures correlated with the quantity and quality of source materials (Figure 11). Pyrolysis data plotted on a modified Van Krevelen diagram (Figure 12) and solvent extractable data (C_{15+} hydrocarbons) plotted against total organic carbon (wt. %) (Figure 13) show that two discrete kerogen types were present in the area. Subsequent plotting of C_{2+} magnitude data from soilgas samples and C_3/C_1 x 1000 compositional data (Figures 14 and 15) show a high degree of correlation between the oil-prone inlier samples and the oil-prone soilgas samples and between the gas-prone kerogens and the gas-prone soilgas samples. Results from the central and northern portion of the study area suggest that potential source beds lie beneath the basalts to the north. There is a trend in both the outcrop geochemistry and the soilgas geochemistry indicating favorable potential in this direction.

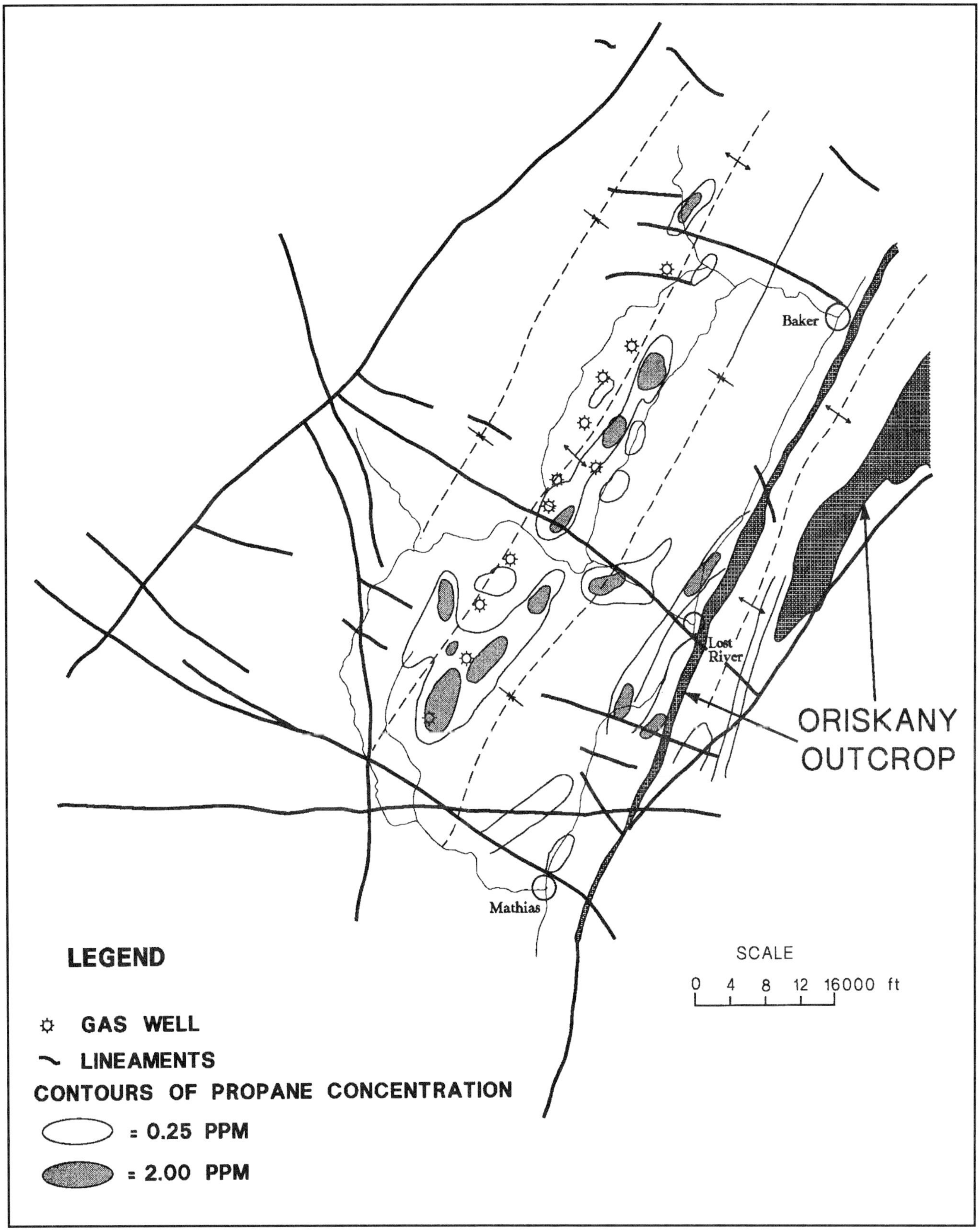

Figure 7. Distribution of propane in soilgas samples, Lost Rivers field, West Virginia. (From Matthews et al., 1984.)

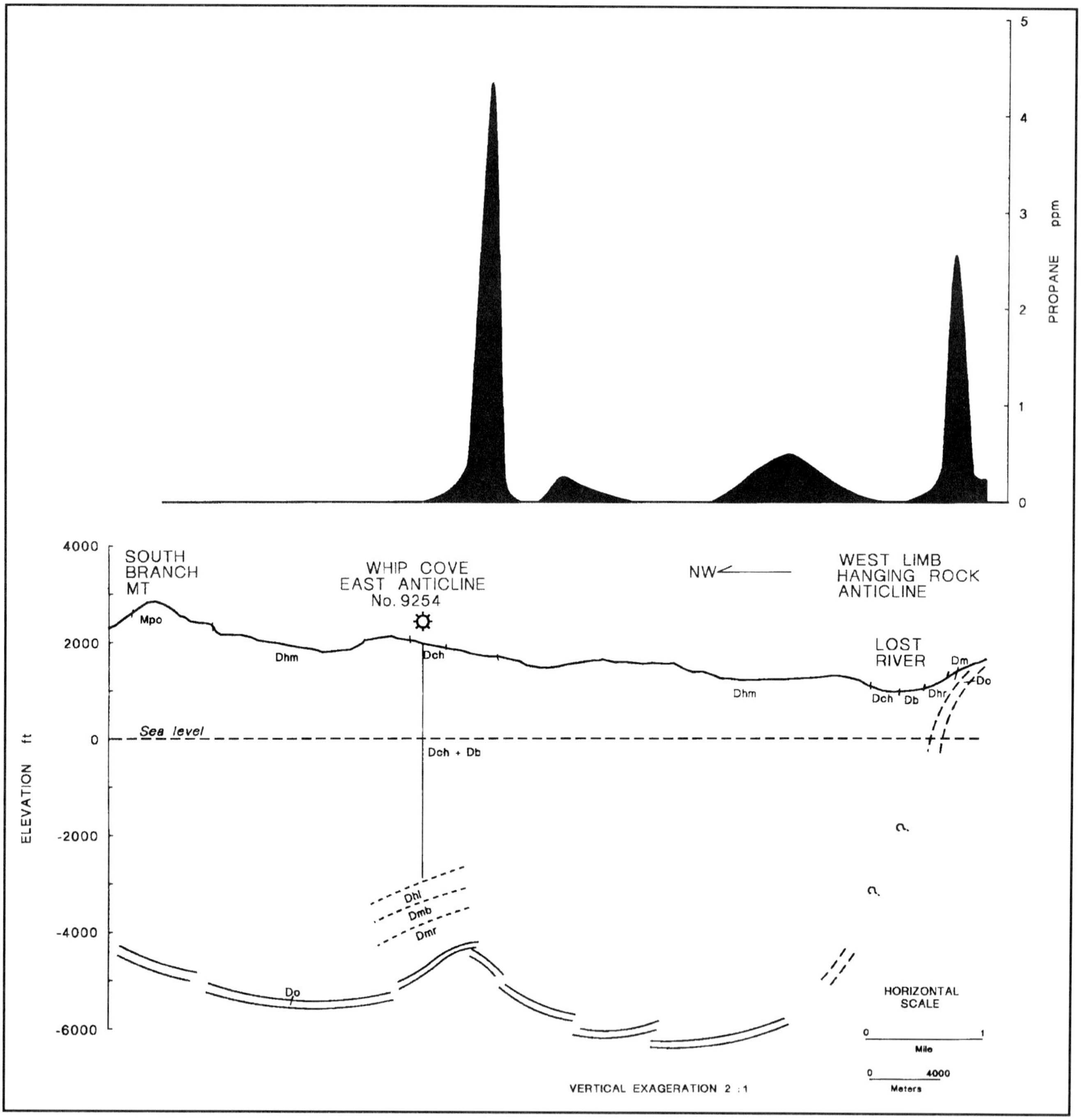

Figure 8. Cross section of soilgas propane, Lost Rivers, West Virginia. (From Matthews et al., 1984.)

SUMMARY

When applied judiciously, soilgas geochemistry is useful as an exploration tool to define prospective areas and areas with hydrocarbon source potential. Careful integration of soilgas data and other geological information results in maximum benefit. Simple statistics are often enough to determine the grade of areas with unknown potential, although the more complex an area is geologically, the more likely that advanced statistics and multiple soilgas techniques will be required. Undoubtedly as time continues, additional uses for near-surface geochemistry will be discovered that can aid in the evaluation and development of exploration plays.

Figure 9. ***(Facing page)*** **Detailed lineament analysis of a portion of Railroad Valley, Nevada. (From Jones et al., 1985.)**

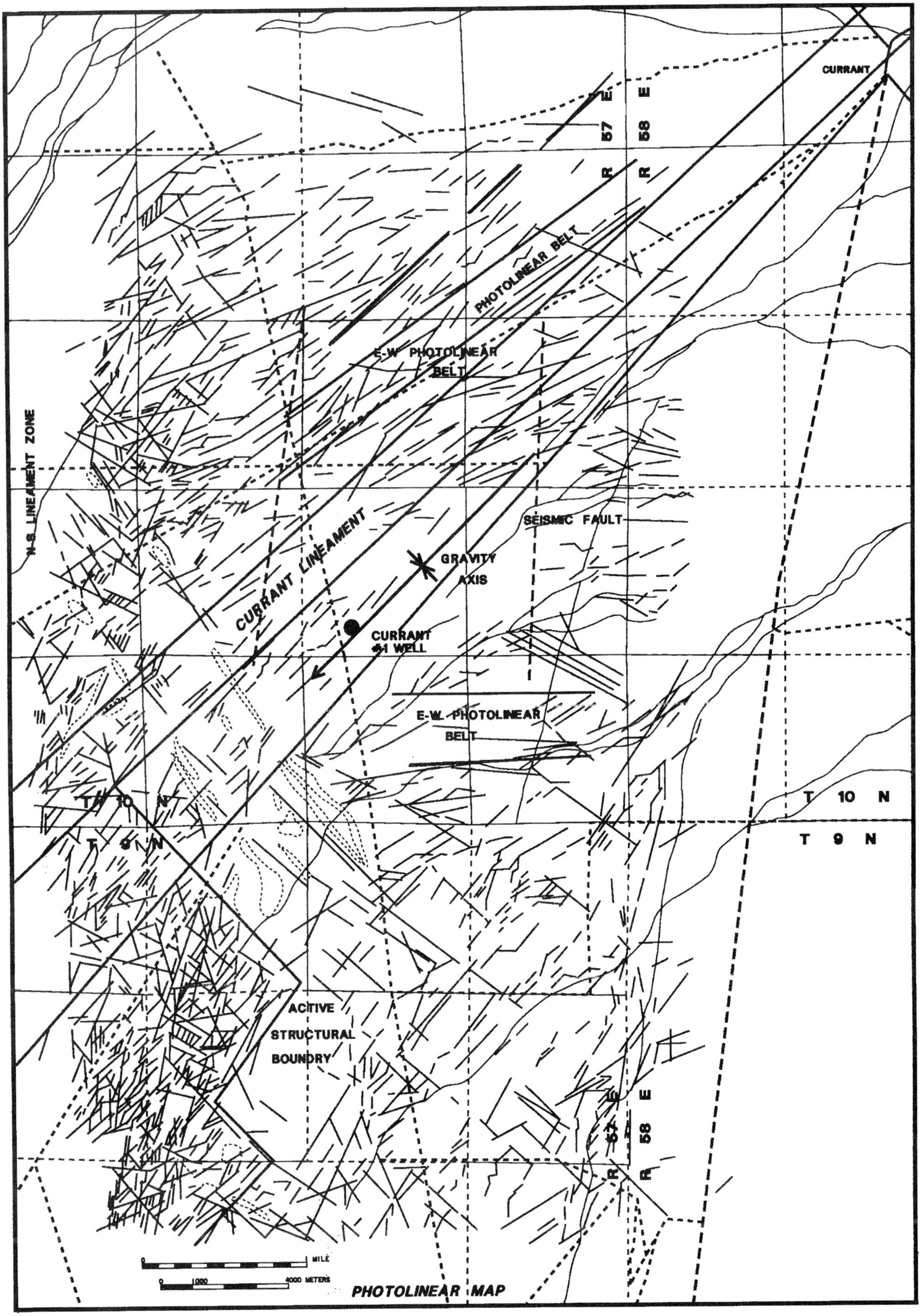
CURRANT
R 57 E
R 58 E
PHOTOLINEAR BELT
E-W PHOTOLINEAR BELT
N-S LINEAMENT ZONE
SEISMIC FAULT
CURRANT LINEAMENT
GRAVITY AXIS
CURRANT #1 WELL
E-W PHOTOLINEAR BELT
T 10 N
T 9 N
T 10 N
T 9 N
ACTIVE STRUCTURAL BOUNDRY
R 57 E
R 58 E
MILE
1000
4000 METERS
PHOTOLINEAR MAP

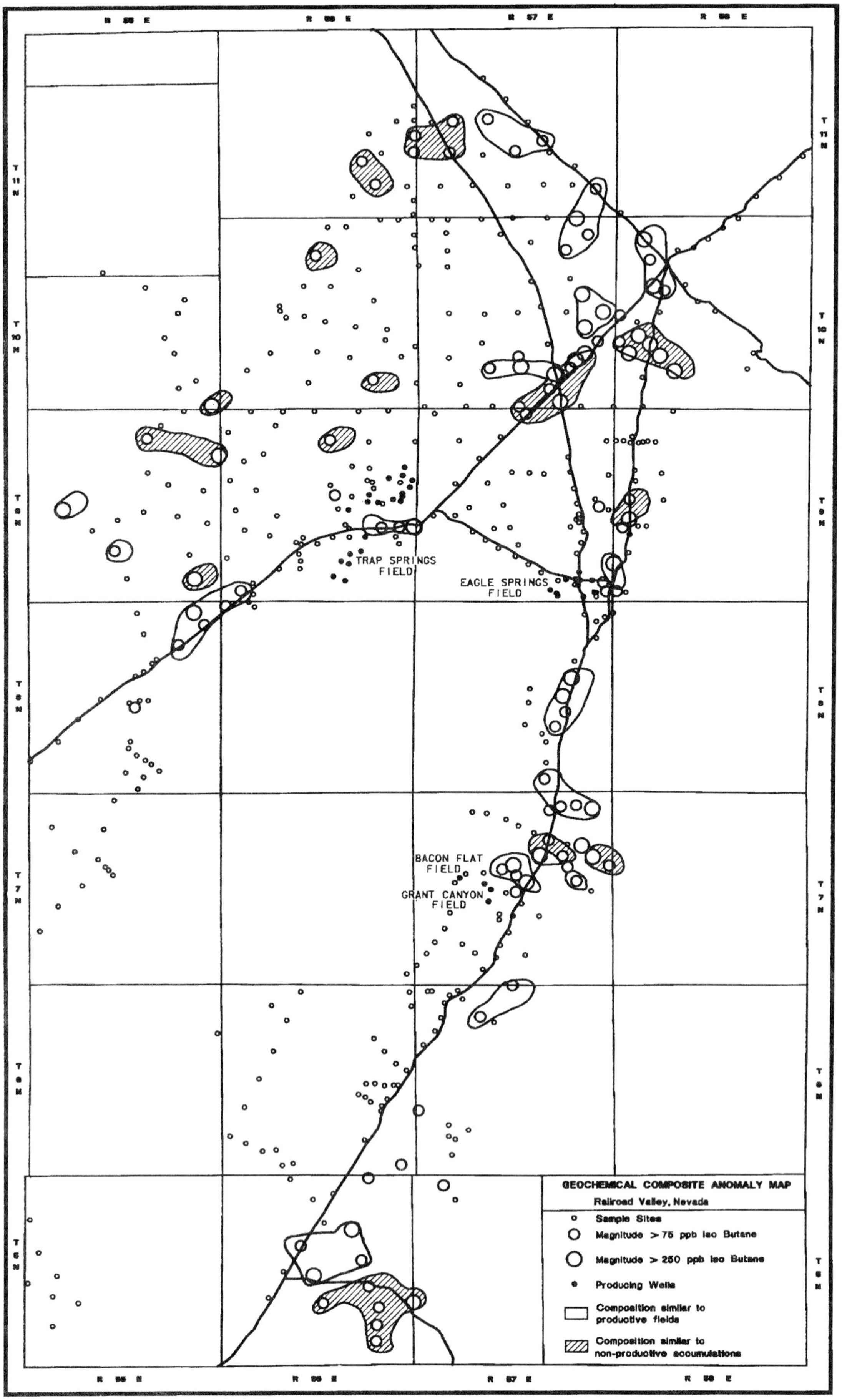
TRAP SPRINGS FIELD
EAGLE SPRINGS FIELD
BACON FLAT FIELD
GRANT CANYON FIELD
R 57 E
GEOCHEMICAL COMPOSITE ANOMALY MAP
Railroad Valley, Nevada
Sample Sites
Magnitude > 75 ppb Iso Butane
Magnitude > 250 ppb Iso Butane
Producing Wells
Composition similar to productive fields
Composition similar to non-productive accumulations

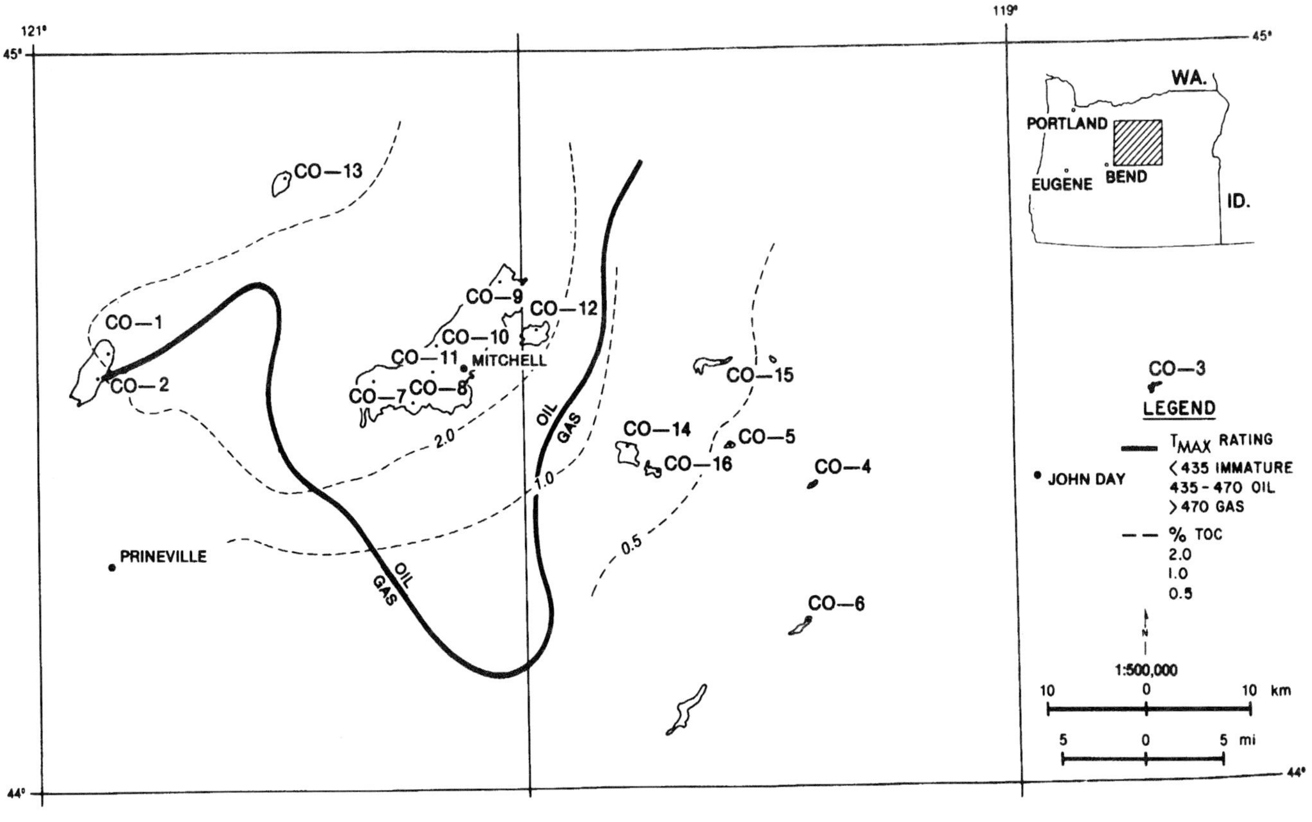

Figure 11. Map of Cretaceous inliers in the Clarno area, Oregon, plotted with TOC (wt. %) and T_{max} distributions. (From Sidle and Richers, 1985.)

Figure 10. *(Facing page)* Distribution of isobutane at Railroad Valley, Nevada. (From Jones et al., 1985.)

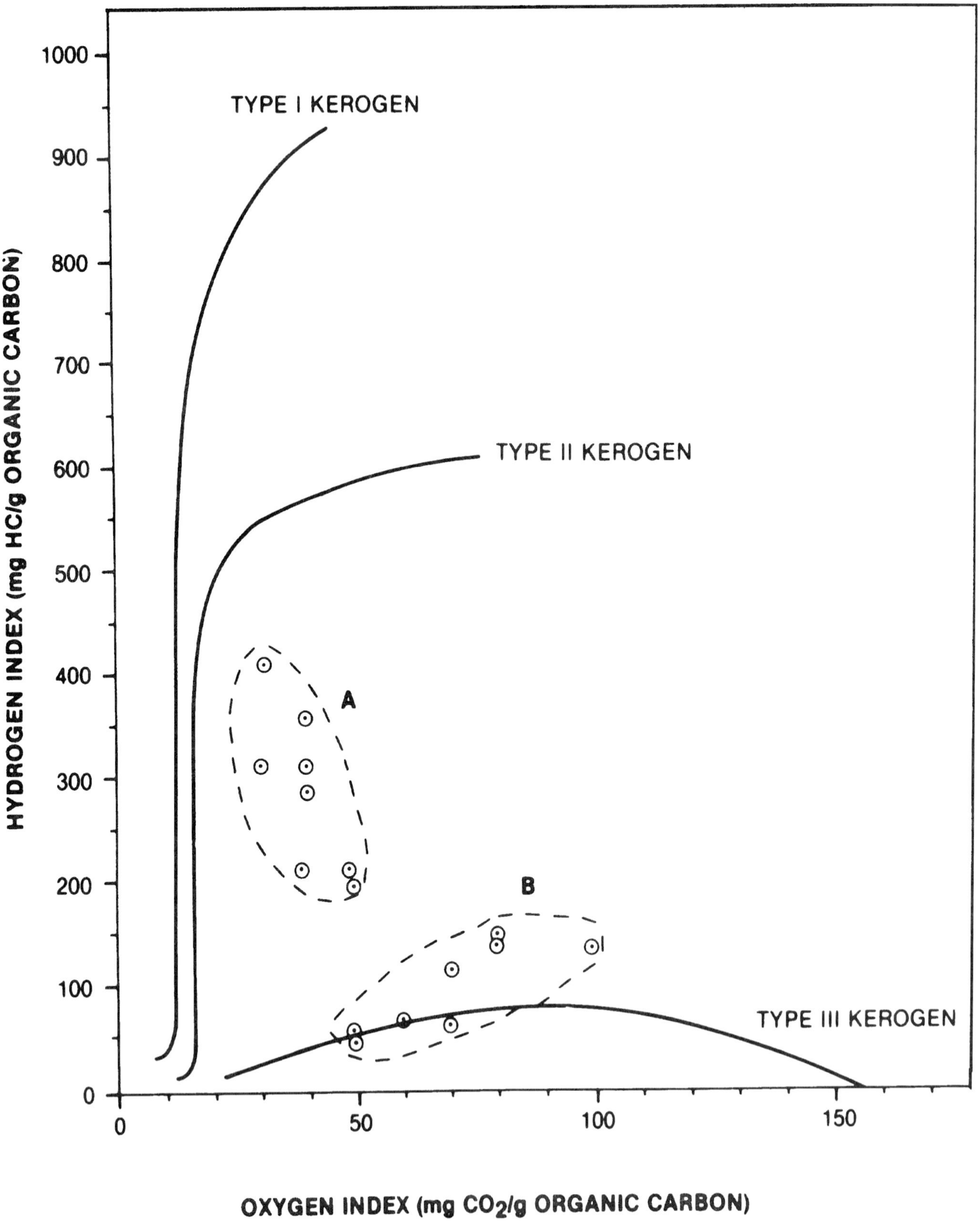

Figure 12. Modified Van Krevelan diagram of Cretaceous outcrop samples, Clarno area, Oregon. (From Sidle and Richers, 1985.)

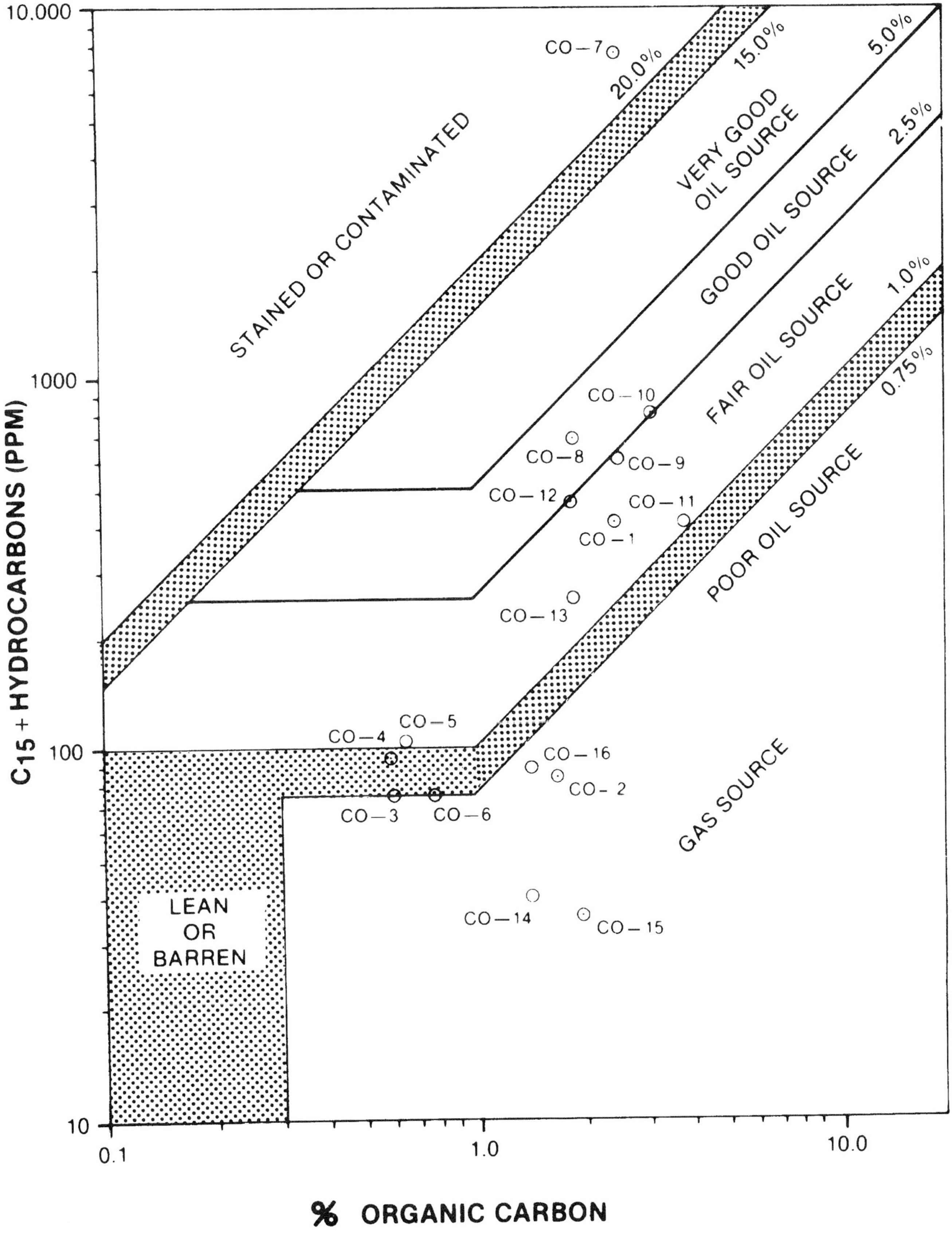

Figure 13. Source rock probabilities diagram, Cretaceous outcrop samples, Clarno area, Oregon. (From Sidle and Richers, 1985.)

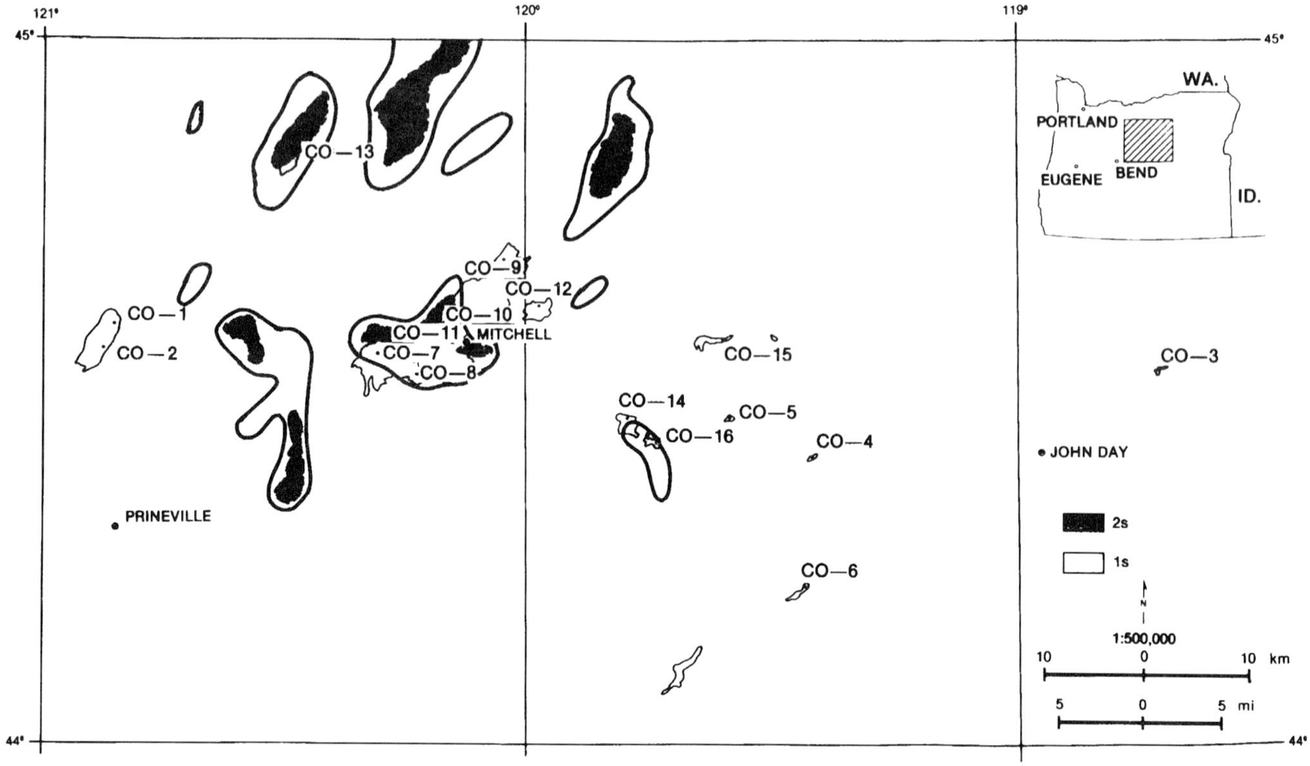

Figure 14. Distribution of C_{2+} fraction in soilgas, Clarno area, Oregon. (From Sidle and Richers, 1985.)

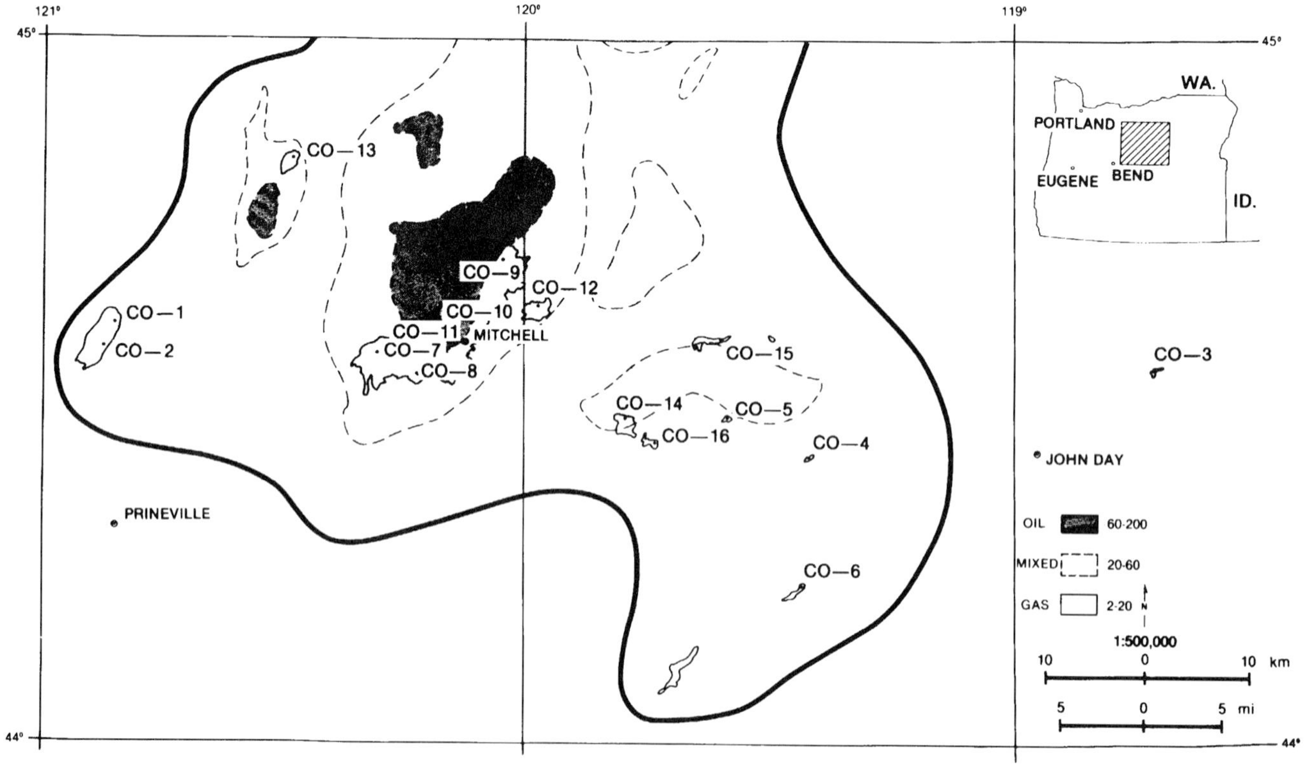

Figure 15. Distribution of C_3/C_1 x 1000 soilgas values, Clarno area, Oregon. (From Sidle and Richers, 1985.)

Glossary of Terms Applicable to Petroleum Geochemistry

Frederik W. Vlierboom
Occidental Petroleum Corporation
Bakersfield, California, U.S.A.

Note: Terms denoted by asterisks (*) are defined as entries elsewhere in the Glossary.

ABNORMAL PRESSURE: Any departure from hydrostatic pressure. Overpressures generally range above 12 kPa/m (0.53 psi/ft) and underpressures range below 9.8 kPa/m (0.43 psi/ft).

ABSORPTION: Penetration of a substance into the body of another (includes dissolution).

ACIDIFICATION: The process by which most of the inorganic mineral matrix of a rock is destroyed with acid to release the insoluble organic matter (kerogen) for further study. Hydrochloric acid (HCl) is used to destroy carbonates; hydrofluoric acid (HF) is used to destroy the silicates.

ACTIVATION ENERGY: The extra amount of energy a molecule must have before it can participate in a certain reaction (E in the Arrhenius equation*).

ACYCLIC: Having no rings.

ADSORPTION: The adhesion of a thin layer of molecules of gases, liquids, or dissolved substances to the surface of solids.

AEROBE: A bacterium that utilizes molecular oxygen for its metabolic processes.

AEROBIC: A term applied to bacteria or other microorganisms living or active only in the presence of molecular oxygen.

AIR SPACE: Synonym for headspace.*

ALBERTITE: An asphaltite* with specific gravity of 1.07–1.10 and 25–50% fixed carbon,* having brilliant luster, conchoidal fracture, hardness 1–2, and brown to black streak. It is practically insoluble in carbon disulfide but partly soluble in turpentine; decomposes prior to fusion.

ALCOHOLS: A class of oxygen-containing organic compounds having the structure R—OH, where R is a hydrocarbon radical.

ALGAE: General biological term for a large group of lower plants of single cell or cell aggregates, the majority living in water and containing chlorophyll* (autotroph). In geology, generally designating all phytoplankton, e.g., diatoms,* dinoflagellates, and seaweed.

ALGINITE: The coal maceral* of the exinite* group formed from algal remains. It is rare in humic coal* but is the principal constituent of the sapropelic* Boghead coal.*

ALICYCLIC: Referring to saturated cyclic hydrocarbons.

ALIPHATIC: Referring to all organic compounds characterized by open-chain structures.

ALKANES: Saturated* hydrocarbons with either straight or branched (by not cyclic) chains of carbon atoms. Alkanes have the general formula C_nH_{2n+2}, e.g., phytane,* $C_{20}H_{42}$.

n-ALKANES: Alkanes having a continuous, unbranched, noncyclic chain of carbon atoms. Also called straight-chain alkanes or normal alkanes.

ALKENE: An unsaturated hydrocarbon with at least one carbon–carbon double bond present.

ALKYL: The adjective form of alkane, made by dropping -ane and adding -yl. The same substitution can be made to convert the names for specific alkanes (such as propane) to names for attached groups (propyl).

ALLOCHTHONOUS: Originating elsewhere. Contrast with autochthonous.*

ALPHA: A designation meaning that the indicated group in a polycyclic compound is attached below the plan of the beta ring.

AMINO ACIDS: Amino acids are compounds having both amino (—NH) and carboxylic acid (—COOH) groups. They occur in nature in the free state and in proteins,* which are condensation products of amino acids. Most natural acids can be represented by

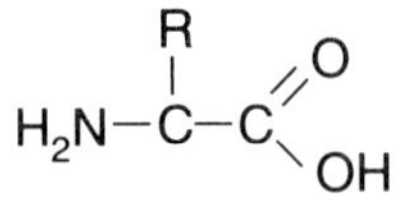

in which R represents an aliphatic, aromatic, or heterocyclic group.

AMORPHOUS ORGANIC MATTER: Kerogen particles exhibiting no distinctive morphology. Some amorphous organic material is apparently of algal origin; other examples represent highly degraded material of uncertain or perhaps eclectic origin.

AMU: See Atomic mass unit.*

ANAEROBE: A microorganism that functions under anaerobic conditions.

ANAEROBIC: A term applied to bacteria or other microorganisms that live and grow in the absence of molecular oxygen.

ANCHIMETAMORPHISM: Meaning high-age metamorphism and applied in those instances where reactions are interpreted as essentially governed by time. It is not recommended to use this term in connection with organic metamorphism.*

ANOXIC: Conditions where O_2 is absent or where the concentration of O_2 is very low (less than O.1 mL/L water).

ANTHRACITE: See Coal rank.*

ANTHRAXOLITE: A coal-like, lustrous, probably highly coalified asphaltite* rich in carbon (85–95%); hardness 3–4, specific gravity near 2; insoluble in organic solvents, practically infusible.

API GRAVITY: An arbitrary scale known as the API degree used for reporting the gravity or the density of a petroleum product. The degree API is related to the specific gravity scale (60°F/60°F) by the formula:

$$\text{Degree API} = \frac{141.5}{\text{Specific gravity at } 60^\circ\text{F}} - 131.5$$

AROMATICS: Hydrocarbons derived from or related to benzene,* C_6H_6. The benzene nucleus is the origin of a group of characteristic chemical properties known as the aromatic character of the molecule. Only one hydrogen atom is bound to each carbon atom of the ring; they may be replaced by various types of side chains. Typical aromatics are benzene, toluene, and xylene. The general structure is shown here.

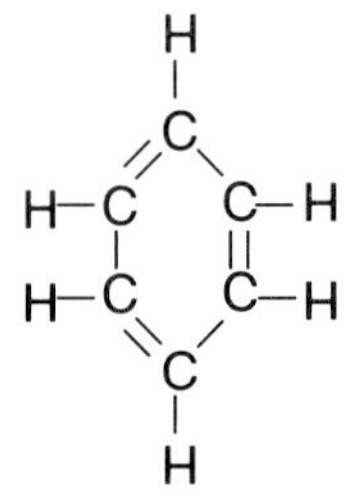

AROMATIZATION: The process of converting an alicyclic system to an aromatic one. Aromatization is an oxidative process that occurs during catagenesis and metagenesis.

ARRHENIUS EQUATION: $k = Ae^{-E/RT}$, where k is the rate constant, A is the frequency factor, E is the activation energy,* R is the universal gas constant, and T is the absolute temperature.

ASH: Inorganic residue obtained after combustion of fuels or caustobiolites.*

ASPHALT: A term applied to both native asphalt* and pyrogenous asphalts (i.e., man-made asphalt from thermal treatment of residual oils). They are generally hard, dark-colored, nonvolatile materials with low fusing points, low specific gravities (1.0–1.1), and low fixed-carbon values (4–20%). They are generally soluble in carbon disulfide, but are soluble only to the extend of 10–70% in petroleum naphtha.

ASPHALTENES: The portion of petroleum and of other bitumen that is soluble in solvents such as benzene, chloroform, and carbon disulfide (hence, soluble or extractable bitumen), but which is insoluble in low-boiling (C_{3-7}) alkanes. They generally contain more than 40 carbon atoms per molecule.

ASPHALTIC CRUDE OILS: Crude oils with a high content of asphaltenes, and often also of vanadium and sulfur, and with API gravities below about 35. They can be classified as asphaltic-paraffinic crudes or as asphaltic-naphthenic crudes.

ASPHALTIC ROCK: Predominantly sedimentary rocks containing in their porous space native asphalt.* An asphaltic sandstone is very often, but incorrectly, called a tar sand; asphaltic limestone is generally, but inadequately, called bituminous limestone.

ASPHALTITE: A group name embracing solid forms of native naphthabitumens,* which are harder and less fusible than true asphalt*. They are composed principally of hydrocarbons (substantially free from oxygenated bodies and crystallizable alkanes), either pure or associated with mineral matter. Asphaltites are derived either from naphthabitumen* or kerogen* and altered during or after migration.* They are usually found in veins and fissures. According to their solubility in carbon disulfide and benzene, they are divided into two groups: (1) largely soluble: gilsonite,* glance pitch,* and grahamite*; and (2) largely insoluble: elaterite,* wurtzilite,* albertite,* impsonite,* and anthraxolite.* It is possible that elaterite and wurtzilite represent immature kerogenous material; if so, they should be treated as sapropelites*. Albertite, impsonite, and anthraxolite may be asphaltites with a high maturity* acquired after migration.

ASSAY: See Fischer assay.*

ATOMIC MASS UNIT (AMU): A mass approximately equal to that of one neutron, used in describing the masses of atoms, molecules, or ions.

ATROMOTE: Fine (10 μm) detrital huminite particles of varying shape including cell wall debris or formless, dense, and almost homogeneous huminitic material. Forms matrix of brown coals. Bright red to red in transmitted light, very weak fluorescence.

AUTOCHTHONOUS: Originating in its present location (that is, deposited *in situ*). Contrast with allochthonous.*

BACTERICIDE: A chemical that kills bacteria.

BASE OF CRUDE OIL: The "base" of a crude oil is descriptive of the chemical nature of its main constituents. A paraffin-base oil contains predominantly paraffinic hydrocarbons, while an intermediate- or mixed-base crude contains roughly equivalent mixtures of paraffins and naphthenes (cycloparaffins). A naphthenic-base crude contains predominantly naphthene hydrocarbons. An asphalt-base crude is one containing a relatively high proportion of nonhydrocarbon constituents such as nitrogen, sulfur, and oxygen compounds. The term aromatic base is not used because there are no oils known to contain predominantly aromatic hydrocarbons.

BENZENE: See Aromatics.*

BICYCLANES: Saturated compounds having two condensed rings in their molecules.

BIOCHEMICAL PHASE: See Organic metamorphism.*

BIODEGRADATION: The alteration of organic matter, including oils, by microbes (bacteria); in the case of oils, usually producing poorer quality oils depleted in normal paraffins and with added sulfur.

BIOGENIC: Formed biologically by an organism or within an organism.

BIOGENIC GAS: (1) Natural gas, virtually all methane, produced by microbes (bacteria) in shallow rocks. Biogenic methane can be recognized by its relative abundance of the ^{12}C isotope; also called marsh gas. Hydrogen sulfide is also a biogenic gas. (2) Dry gas (virtually pure methane) formed by anaerobic microorganisms called methanogens.

BIOLOGICAL MARKERS: Organic compounds whose carbon structure, or skeleton, is formed by living organisms and is sufficiently stable to be recognized in crude oil or the organic matter of ancient sediments. Typical markers are the porphyrins, pristane, phytane, steranes, carotenes, and pentacyclic triterpanes.

BIOMARKERS: Chemical compounds derived from specific biological precursors. The trans-formation of precursor to biomarker can often be traced directly, permitting the utilization of biomarkers as environmental and maturity indicators. Important biomarkers include n-alkanes, isoprenoids, steranes, triterpanes, and porphyrins.

BIOPOLYMERS: Polymers created by enzymes. They have very regular, predictable structures. See also Geopolymers.*

BIS: Prefix meaning two.

BITUMEN: A term *not recommended* for specific purposes because of its wide range and greatly varying definitions. It comprises hydrocarbon mixtures of both artificial (pyrogenous) and natural origin. Natural bitumen includes all naturally occurring combustible materials rich in hydrocarbons, gaseous, liquid and solid, soluble or insoluble in carbon disulfide. The petroleum group is called naphthabitumen* and is derived from caustobiolites.*

BITUMINOUS COAL: A group of coal ranks* divided into high-, medium-, and low-volatile bituminous coal.

BITUMINOUS LIMESTONE: A nonspecific and *not recommended* term meaning either asphaltic* limestone (see Asphaltic rock*) or kerogenous* limestone (see Oil source rock*).

BOGHEAD COAL: A sapropelic coal* also called torbanite, chiefly consisting of the maceral* alginite.* Compact, brownish black to black, very tough and difficult to break; conchoidal to subconchoidal fracture.

BRANCHED CYCLIC FRACTION: Saturated hydrocarbons from a bitumen or crude oil from which the unbranched hydrocarbons (*n*-alkanes) have been removed.

BRITISH THERMAL UNIT (BTU): The quantity of heat required to increase the temperature of 1 lb of water by 1°F. See Calorific value.*

BROWN COAL: Low rank humic coal* intermediate between peat and bituminous coal, containing about 10–75% water. See also Coal rank.* It is differentiated into soft or unconsolidated brown coal and hard or consolidated. The soft one is made up of visible plant remains, while hard brown coal (lignite*) is compact, more homogeneous, and tough.

BURIAL HISTORY CURVE: An age–depth plot that traces the burial and tectonic history of a rock from the time of deposition to the present day, using the sea bottom as the datum. See also Geohistory diagram.*

C_{15+}: Essentially the same as bitumen, so named because it contains most of the compounds having 15 to 35 carbon atoms that were present in the original sample. Compounds having fewer than 15 carbon atoms are partially or completely lost during evaporation of the extraction solvent. See also Bitumen.*

CAI: See Conodont alteration index.*

CALORIE: The amount of heat required to raise the temperature of 1 g of water by 1°C.

CALORIFIC VALUE: The quantity of heat produced by a substances upon complete combustion at constant volume. Units are either calories* per gram (1 cal/g) or British thermal units* per pound (1.8 Btu/lb).

CANNEL COAL: A sapropelic coal* chiefly consisting of the maceral group* of exinite,* particularly sporinite.* It burns with a steady bright flame. Massive, tough, clean coal of fine compact grain, dull luster, usually with conchoidal fracture and good regular jointing.

CAPILLARY COLUMN: A very long, narrow-diameter tube often used in gas chromatography. Because of the column's narrow diameter, the stationary phase is coated directly on the walls of the column. See also Packed column* and Gas chromatography.*

CAPILLARY ENTRY PRESSURE: The pressure that opposes the deformation of a hydrocarbon globule and thus its movement into a constricting pore throat. It retards migration and thus can influence migration pathways and velocities. Capillary entry pressure is the force that ultimately leads to accumulation of hydrocarbons.

CARBOHYDRATES: A class of neutral compounds composed of carbon, hydrogen, and oxygen in units of five or six carbon atoms, usually connected by oxygen atoms, and having the general formula $CH_2O)_n$. Carbohydrates are formed by all green plants through photosynthesis. Includes sugars, starch, and cellulose, with the following general structure:

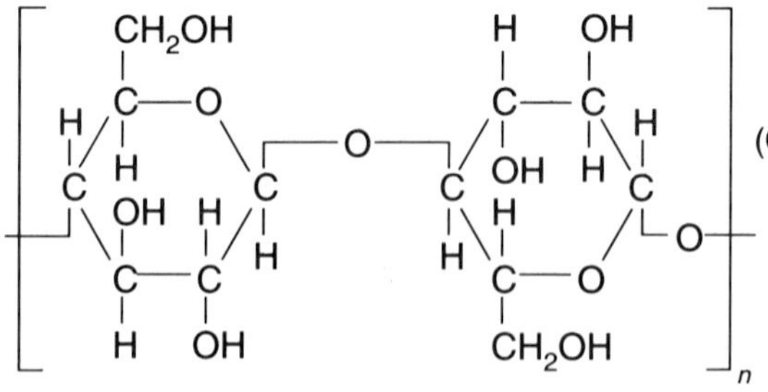

(Cellulose)

CARBONACEOUS: Pertaining to humic coal.*

CARBON CYCLE: The cycle through which carbon moves from its inorganic reservoirs in the atmosphere, in aqueous solution, and in carbonate minerals to the biosphere via photosynthesis, and finally back to the inorganic reservoirs via oxidative decomposition.

CARBONIZATION: Widely used instead of carbonification or coalification.* It is advised to use the term carbonization only for artificial processes.

CARBON ISOTOPES: See Isotopes.*

CARBON NUMBER: The number of carbon atoms in one molecule.

CARBON PREFERENCE INDEX (CPI): A carbon preference index value is defined as the mean of two ratios that are determined by dividing the sum of the concentration of odd-carbon-numbered *n*-alkanes by the sum of even-carbon-numbered *n*-alkanes over given concentration ranges. One ratio is obtained by dividing the sum of the percentages of the odd-carbon numbers 25–33 by the sum of the even-carbon numbers 26–34. The other ratio is obtained by changing the denominator to the sum of the percentages of 24–32. See also odd-predominance (OEP).*

CARBON RATIO: The ratio of fixed carbon* to the sum of fixed carbon and volatile matter in a coal. The so-called ratio is actually expressed as a percentage.

CARBON RATIO THEORY: The theory that the API gravity of oil increases (specific gravity decreases) as the carbon ratio of coals in the same area increases. As coals increase in rank, oils become lighter, eventually changing to gas. The carbon ratio is the ratio of fixed carbon to total carbon (fixed plus volatile) in a coal.

CARBOXYL: The functional group —COOH. The hydrogen in carboxyl groups is acidic.

CAROTENES: Hydrocarbon compounds belonging to the group of carotenoids.*

CAROTENOIDS: Naturally occurring pigments having 20 or 40 carbon atoms in an isoprenoid structure*; they have a widespread distribution in plants and animals and belong to the lipids.*

CASING HEAD GASOLINE: The liquid hydrocarbon recovered from casing head gas by adsorption, compression, or refrigeration. Casing head gas is the gas recovered at the surface from an oil well.

CATAGENESIS: The process by which organic material in sediments is thermally altered by increasing temperature. Catagenesis covers the temperature range between diagenesis and rock metamorphism, approximately 50–200°C (122–392°F).

CATALYSIS: The speeding-up of a chemical reaction by offering an alternative pathway having a lower activation energy.

CATALYST: A substance capable of accelerating a chemical reaction.

CATALYTIC CRACKING: The conversion of heavy hydrocarbons into lighter ones by the action of heat in the presence of an (acidic) catalyst (primarily by a carbonium ion process). See also Thermal cracking.*

CAUSTOBIOLITE: Literally, combustible biogenic rock. General name for sediments consisting of, or rich in, combustible native (organic) matter and its natural derivatives. Their classification is shown in Table 1.

Table 1. Caustobiolite classification

I. Sediments containing syngenetic, largely autochthonous organic constituents	
Oil/Gas Source Rocks = KEROGENOUS ROCKS	Gas Source Rocks = CARBONACEOUS ROCKS
Sapropelites	Humolites
Kerogenous shales, marls, and limestones	Humic coal
	Carbonaceous sediments
Sapropelic coal	
Oil shale	
II. Epigenetic, largely allochthonous, natural derivatives of kerogenous and carbonaceous rocks:	
NAPHTHABITUMENS	
From kerogenous rocks	From carbonaceous rocks
Natural gas	Natural gas
Petroleum	
Asphalt	
Mineral wax	
Asphaltite	

CENTIPOISE: A standard unit of dynamic viscosity, equal to 0.01 poise, the cgs unit of viscosity. Water at 20°C has a viscosity of 1.002 centipoise.

CHARCOAL: Dead carbon or carbonaceous residue. The nonvolatile residue with a very high fixed-carbon content obtained when a carbonaceous* material is partially burned or heated in the absence of air (pyrolyzed). It is insoluble in organic solvents.

CHEMICAL DIAGENESIS: Alteration of organic matter by attack with reactive chemicals; for example, sulfur or hydrogen sulfide can react with saturated hydrocarbons to form unsaturated ones or sulfur-containing bitumens.

CHITIN: A polysaccharide (carbohydrate polymer) with a structure similar to cellulose, except that two OH groups are replaced by CH_3ONH groups in each $(C{\cdot}H_2O)_6$ unit. Chitin forms the horny, hard, outer cover of insects, crustaceans, and parts of certain other invertebrates.

CHLORIN: A precursor molecule for a porphyrin.* The difference is that porphyrins have an aromatic structure, whereas chlorins have one double bond less than a completely conjugated aromatic structure. Chlorophyll* is a chlorin.

CHLOROPHYLL: The green pigment occurring in plants. Its presence is essential to photosynthesis. The chlorophyll molecule contains a phytol* remnant.

CHROMATOGRAPHY: A physical process of fractionation and separation involving the partitioning of a sample between stationary and moving phases, e.g., in a column. Table 2 shows the various chromatographic techniques.

Table 2. Chromatographic techniques

Chromatography Techniques	Moving Phase	Stationary Phase
LSC	Liquid	Solid
LLC	Liquid	Liquid
GLC	Gas	Liquid
GSC	Gas	Solid

CIS: A configuration in which two groups attached to a molecule are on the same side of the molecule. See also Trans.*

CLATHRATE: A chemical compound in which a loose molecule is trapped inside a crystalline network of surrounding molecules, usually of a different compound. See also Gas hydrates.*

COAL: Coal is divided into two groups: humic coal* and sapropelic coal*. (See also Caustobiolite.*) If no adjective is added, the term *coal* is used in the restricted sense of humic coal.* In the loose sense, coal is a combustible (organic) sedimentary rock, usually with less than 40% mineral (inorganic) matter (based on dry material), formed by the accumulation, decomposition and transformation of plant material. The nature of the original plant material determines the coal type,* while the degree of organic metamorphism (DOM*), maturity, or coalification determines the coal rank.*

COALIFICATION: The natural process of organic metamorphism* in humic coal* resulting in a relative increase in carbon content and a relative decrease in the content of hydrogen and other elements. See also Coal rank.*

COAL MACERAL: Generally homogeneous microscopic constituents of coal,* analogs to the minerals in inorganic rocks. The properties of most macerals change in the course of coalification.* All coal macerals have the suffix -inite. Their classification is shown in Table 3.

Table 3. Coal maceral classification

Maceral Group	Symbol	Maceral
Vitrinite	V	Collinite* Telinite* (Gelinite*) (Detrinite*)
Exinite* or liptinite*	E	Sporinite* Cutinite* Alginite* Resinite*
Inertinite	I	Sclerotinite* Semifusinte* Fusinite* Micrinite*

COAL MACERAL ASSOCIATION: See Microlithotype.*

COAL RANK: One of the various stages in natural coalification* indicating the degree of organic metamorphism (DOM*) of humic coal.* For low-rank coal, it is determined by the calorific value* of vitrite*; for high-rank coal, by the percentage of fixed carbon* of vitrite or its volatile matter.* Coal rank is not yet well defined in sapropelic coal. For humic coal, the two internationally used classifications are given in Table 4 (see Coal structure*).

COAL STRUCTURE: Humic coal,* not being a homogeneous material but an association of various coal macerals,* has been divided in coal petrography into various structures. The macroscopic fine structures or lithotypes* have the ending -ain, while the microscopic structures, called micolithotypes* or coal types,* always have the ending -ite, as shown in Table 4.

COAL TYPE: See Microlithotype.*

Table 4. Coal Classification.

Macroscopic Fine Structure Lithotype	Microscopic Fine Structure Microlithotype or Coal Type		Principal Groups of Constituent Macerals	
Vitrain	Vitrite	Homo-	Vitrinite	
Fusain	Fusite	geneous	Inertiite	Monomacerals
—	Liptite		Liptinite + exinite	
Clarain	Clarite		Vitrinite + exinite	
Durain	Durite	Hetero-	Inertinite + exinite	Bimacerals
—	Vitrinertite	geneous	Vitrinite + inertinite	
Transitions	Duroclarite		Vitrinite + exinite + inertinite	Trimacerals
	Clarodurite		Inertinite + exinite + vitrinite	

COLCORMET: A gadget developed by Shell Oil Co. for the shade comparison with palynomorphs* to estimate light translucency in coalification* studies.

COLLINITE: The apparently largely structureless coal maceral of the vitrinite* group appearing in reflected light as the most homogeneous part of humic coal.*

COLORIMETRY: See Pyrolysis–colorimetry.*

COLUMN CHROMATOGRAPHY: In petroleum geochemistry, chromatography carried out in a vertical column filled with a stationary phase, such as silica gel or alumina, and a mobile liquid phase, which initially is an aliphatic hydrocarbon-like *n*-hexane. The mixture to be separated is introduced at the top of the column, and the liquid phase is allowed to flow through the column by gravity into a collecting vessel at the bottom. The liquid phase is continually replenished at the top of the column. The saturated hydrocarbons* quickly pass through (elute from) the column. After a predetermined amount of *n*-hexane has flowed through, a more polar liquid phase, usually benzene,* replaces the *n*-hexane. The benzene rapidly elutes the aromatic hydrocarbons, which are collected in a second vessel. Successively more polar solvents (benzene methanol or chloroform) can elute many NSO compounds.* Asphaltenes* do not move easily through the column and are not recovered by column chromatography. This technique is the first step in most bitumen analyses because it separates the complex bitumen mixture into relatively homogeneous packages.

CONDENSATE: A liquid hydrocarbon mixture, consisting mainly of very light components, that is produced along with gas by pressure decrease (a phenomenon known as retrograde condensation) from reservoirs in which both gas and condensate exist as a homogeneous phase.

CONDUIT: A continuous migration pathway. Conduits can include coarse-grained rocks, fractured or jointed rocks, and possibly active fault planes.

CONJUGATED SYSTEM: A series of alternating single and double bonds between adjacent carbon atoms, as in the benzene molecule.

CONNATE WATER: Fossil reservoir water that has not been in contact with the atmosphere since deposition. It is high in chloride and calcium and frequently contains more than 100,000 ppm dissolved total solids.

CONODONT ALTERATION INDEX (CAI): A maturity scale with values ranging from 1 to 8 that is based on darkening of conodonts, which are microfossils found in many Paleozoic rocks. It is most valid and useful beyond peak oil generation.

COORONGITE: A substance resembling rubber, composed of algae, bits of plant tissue, and diatoms. It is often compared with elaterite* and is perhaps in asphaltite,* but is more probably a very immature sapropelite.*

CORRELATION: Comparison of gross or detailed chemical and physical properties of two or more samples of organic matter in an effort to compare their origins, diagenesis, catagenesis, migration, and alteration.

CORRELATION INDEX (CI): The correlation index was developed empirically by H. M. Smith of the U.S. Bureau of Mines. He set up an equation such that the addition of the reciprocal absolute boiling point and the specific gravity with appropriate constants was equal to 0 for the normal paraffins* and equal to 100 for the aromatic hydrocarbon benzene.* The equation is

$$CI = (48640/K) + 473.7G - 456.8$$

where CI is the correlation index, K is the average boiling point of the fraction in degrees Kelvin, and G is the specific gravity of the fraction at 60°F (16°C).

CRACKING: Breakdown of large organic molecules into smaller ones. It is a process usually associated with the formation of condensates, gasoline-range hydrocarbons, and gas from oil. It involves thermal decomposition of large organic molecules, such as kerogen and asphaltenes, to yield bitumen molecules of various sizes. In this usage it is similar to catagenesis.

C_R/C_T RATIO: The ratio of (residual) organic carbon* that is left after pyrolysis* under standard conditions (C_R) to the (total) organic carbon (C_T) before pyrolysis and after extraction.* The C_R/C_T ratio is comparable to, but not identical to, the carbon ratio* (fixed carbon*) of humic coals.* In contract to the carbon ratio, the C_R/C_T ratio can also be determined in rocks containing inorganic matter.

CRUDE OIL: A petroleum that is removed from the earth in liquid state or is capable of being so removed.

CRUDE OIL FRACTIONS: Fractions of crude oil based on viscosity and boiling point, which include gasoline,* kerosine,* gas oil,* lubricating oil,* and residuum* (see individual fractions).

CYCLANES: See Cycloalkanes.*

CYCLICS: Compounds having one or more rings in their structures. Most cyclic compounds in geological environments have six-membered rings, although five-membered rings also occur occasionally. Cyclic compounds can include alicyclics, aromatics, and naphthenoaromatics.

CYCLIZATION: Formation of cyclic structures by chemical reactions.

CYCLOALKANES: Also called cyclanes or cycloparaffins. Hydrocarbons of the general formula C_nH_{2n}, containing one or more saturated* rings (cyclic chains) of carbon atoms, but no aromatic* rings. See also Naphthenes.*

CUTINITE: A coal maceral* of the exinite* group formed from cuticles (the waxy layer formed on outer walls of epidermal plant cells).

CUTTINGS GAS: The natural gas that remains in well cuttings upon their arrival at the shale shaker. The amounts found in cuttings are either run at the well site or canned at the well site and sent to the laboratory for analysis. After removal of the drilling mud, the cuttings are crushed in a blender and the released gas is analyzed. It is used in the estimation of the relative source potential of the rock units in the subsurface.

DEASPHALTENING: The process by which asphaltenes* are separated from rock extracts in the laboratory (by diluting with nonpolar solvents) or from oils in the reservoir (by dilution with increasing amounts of methane).

DEAD CARBON: See Charcoal.*

DECARBOXYLATION: Elimination of carbon dioxide from the carboxyl group (—COOH).

DELOCALIZATION: Freedom of movement of electrons through a conjugated system. Delocalization greatly increases the stability of a molecule compared to a nondelocalized analog. Delocalization also increases the stability of an ion formed from a molecule containing delocalized electrons or of excited-energy states of such a molecule or ion by spreading the excess charge or energy (which represents a stress on the system) over a greater number of atoms.

DENSINITE: Fine (<10 μm) detrital huminite particles of varying shape, including cell wall debris or formless, dense, and almost homogeneous huminitic material. Forms matrix of brown coals. Bright red to red in transmitted light, very weak fluorescence.

DESTRUCTIVE DISTILLATION: See Pyrolysis.*

DETRINITE: A coal maceral* of the vitrinite* group observed in brown coal* only. Possibly the precursor of collinite.*

DIAGENESIS: The process involving biological, physical, and chemical alteration of the organic debris in sediments without a pronounced effect from rising temperature.

DIATOMS: Microscopic plants having silica skeletons. See also Algae.*

DIESEL FUEL: A refined petroleum distillate (a gas oil) containing mainly saturated hydrocarbons of 13 to

17 carbon atoms. It is often used as an additive in drilling muds. Also called diesel oil or dieseline.

DIFFUSION: The slow movement of material from an area of high concentration or pressure to areas of low concentration or pressure. It may be an ancillary mechanism for expulsion of hydrocarbons from a source rock. Diffusion plays a role in hydrocarbon movement in the subsurface, but as a dispersive force, it cannot account for accumulation. It will in fact help destroy accumulations because much leakage is diffusional.

DISPARITY COEFFICIENT: A mathematical quantity, based on the composition of selected C_6 and C_7 hydrocarbons, which indicates the degree of correlation between crude oil and source rock hydrocarbons or between hydrocarbons from different groups of crude oils. The lower the value of this quantity, the better the correlation. See also Similarity coefficient* and Rank correlation.*

DISPROPORTIONATION: Conversion of a single starting material into two products. One of the products will be oxidized relative to the starting material; the other will be reduced. The process thus represents an internal oxidation–reduction (redox) system. It is important in kerogen catagenesis and cracking. The light hydrocarbon products are reduced and rich in hydrogen, whereas the residue (asphaltics, dead carbon, etc.) is oxidized and hydrogen poor.

DISTILLATE: The hydrocarbon fluid produced from natural gas processing. It is denser than condensate and is generally run into the tanks with the crude oil. Gravities range from 50° API upward.

DITERPANES: Hydrocarbons formed from two terpane (four isoprene) units. Many have three six-member rings. They often are derived from resinite.

DOM: Degree of organic metamorphism. A scale based on the coal rank* of humic coal*. In the high ranks, the units are equal to fixed carbon* of vitrites*, while for the lower ranks, they are adapted to the calorific value.* The true layer DOM (TL DOM) of a sediment is the DOM that a thick layer of humic would have obtained when subjected to the same burial history. See also Coal rank* and LOM.*

DOUBLE BOND: A bond between two atoms in which four electrons are shared instead of the normal two. The most common double bonds encountered in petroleum geochemistry are carbon–carbon and carbon–oxygen. Double bonds are stronger and of shorter length than single bonds.

DRILL STEM TEST (DST): A test of the productive capacity of a well while it is still full of drilling mud. The testing tool is lowered into the hole attached to the drill pipe and placed opposite the formation to be tested. Packers are set to shut off the weight of the drilling mud, and the tool is opened to permit the flow of any formation fluid into the drill pipe, where the flow is measured.

DRY GAS: Natural gas consisting principally of methane and devoid of readily condensible constituents such as gasoline. Dry gas contains less than 0.1 gal natural gas liquid vapors per 1000 ft^3 (1.3 L/100 m^3).

DSRA: Abbreviation for detailed source rock* analysis.

DYSAEROBIC: A process occurring in a dysoxic environment.

DYSOXIC: An environment depleted in oxygen but not quite anoxic.

EFFECTIVE SOURCE ROCK: See Source rock.*

ELATERITE: A rare, massive, amorphous, rubbery, dark brown hydrocarbon or asphaltite* ranging from soft and elastic when fresh to hard and brittle when exposed, and having a conchoidal fracture. Gives a brown steak and melts in candle flame without decrepitation.

ELECTRON SPIN RESONANCE: A technique for estimating maturity of kerogen by measuring the number of unpaired electrons (free radicals) in the kerogen. At higher maturity levels the increased aromaticity stabilizes unpaired electrons.

ELEMENTAL ANALYSIS: Quantitative analysis of the various elements present in a sample. The elements most commonly analyzed in petroleum geochemistry are carbon and hydrogen; oxygen, sulfur, and nitrogen are next.

ELUATE: A solution of a fraction eluted (desorbed) from a chromatographic* column by means of a specific solvent, e.g., pentane eluate and toluene eluate containing, respectively, saturated* and aromatic* hydrocarbons.*

ENTRY PRESSURE: See Capillary entry pressure.*

EOM: See Bitumen.*

EOMETAMORHISM: Meaning early meta-morphism. Synonym for the geochemical phase of organic metamorphism.* Not recommended for use.

EPIGENETIC (SECONDARY): As applied to the organic matter in a sedimentary basin, epigenetic includes derivatives (e.g., petroleum and gilsonite) from the primary (syngenetic) sedimentary organic matter.

EPIMER: Isomers* that differ only in configuration at one carbon atom.

EPIMERIZATION: The conversion of one epimer into the other.

EQUILIBRIUM ISOTOPE EFFECT: The effect derived from the fact that the heavier isotopes of an element prefer to exist in more oxidized compounds, the lighter isotopes in the more reduced compounds. Thus, for any reversible reaction that has come to equilibrium, the isotopic distribution among the components is dependent on the relative stabilities of the compounds rather than on their rate of formation. In organic geochemistry, the equilibrium isotope effect is important primarily in the carbon dioxide–carbonate equilibrium. See also Kinetic isotope effect.*

ESR: See Electron spin resonance.*

ESTERS: Compounds formed by the combination of an acid (generally organic) with an alcohol. Thus, acetic acid and ethyl alcohol give an ester called ethyl acetate. Esters of long-chain fatty acids and alcohols are called waxes (to be distinguished from petroleum waxes*).

ETCHING: As applied to rock samples, etching refers to the successive treatment of a flat rock surface with hydrochloric acid, water, and hydrofluoric acid to expose the organic material and its original texture.

ETHERS: Organic compounds in which two hydrocarbon radicals are bridged by an oxygen atom. The best known ether is $C_2H_5—O—C_2H_5$ (diethyl ether).

EUXINIC: From the Latin *Pontus Euxinus*, or Black Sea, famous for its sapropelic* conditions with stagnant, chemically layered, partly sulfurated waters. Euxinic environments favor sapropelic* facies.

EXINITE: Synonym for liptinite, but applied only to hard coal macerals.* The coal maceral* group including sporinite, cutinite, alginite, and resinite. These are not necessarily all exines, but they appear to have similar properties.

EXPULSION: Transfer of oil and/or gas from fine-grained source rock into other formations or into fractures. The often seen term *primary migration* should be discouraged.

EXTRACT: The material removed from a mixture by the selective action of a solvent.

EXTRACTABLE ORGANIC MATTER: Organic matter that can be extracted from sedimentary rocks by means of organic solvents such as benzene methanol, ether, and chloroform. The percentage extract represents mainly the high-boiling (above ~325°C) compounds; the low-boiling compounds are lost during the removal of the solvent by evaporation. See also Extract.*

EXTRACTION: Removal of bitumen from a rock matrix. Extraction can be accomplished by a solvent or by heat.

FATS: Esters of saturated fatty acids* and glycerol. See also Lipids.*

FATTY ACID: An organic acid containing carbon, hydrogen, and oxygen and having the following general structure:

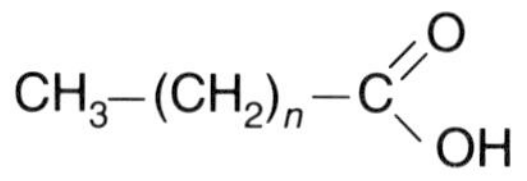

FID: See Flame ionization detector.*

FIRST-ORDER REACTION: A reaction the rate of which is proportional to the first power of the concentration of only one of the reactants.

FISCHER ASSAY: Actually, modified Fischer assay. A method used to determine the yield of shale oil from an oil shale* by the low-temperature pyrolysis* or destructive distillation* of a rock sample.

FIXED CARBON: The percentage of coal's organic carbon that remains after pyrolysis* of that coal under prescribed conditions (ASTM method D121-30). Fixed carbon is numerically equal to 100 minus the percent volatile matter on a water-free and ash-free (waf) basis

FLAME IONIZATION DETECTOR (FID): A detector used in gas chromatographs in which molecules emerging from the column are burned in a hydrogen flame and the ions produced are counted as a measure of the flux of material off the column.

FLUORESCENCE: The property of emitting radiation as the result of absorption of radiation from some other source. The emitted radiation persists only as long as the exposure to radiation. The fluorescent radiation always has a wavelength longer than that of the absorbed radiation.

FRAGMENT ION: An ion formed within a mass spectrometer* by decomposition of the molecular ion.

FRAGMENTOGRAM: A trace obtained from a GCMS instrument in which all compounds that yield a fragment ion of a specified m/z ratio are recorded, and all compounds that do not give a fragment with that particular m/z value are ignored. The detector response is directly proportional to the number of fragment ions with the correct m/z ratio. The number of fragment ions, in turn, depends upon the concentration of the precursor molecule an the frequency with which the precursor breaks down to yield to particular fragment ion.

FREE RADICAL: An unpaired electron in a molecule.

FREQUENCY FACTOR: The frequency factor as used in the Arrhenius equation* is related to the number of properly oriented molecular collisions per second per unit volume.

FULVIC ACID: Humic substances of relatively low molecular weight that are soluble in both aqueous acid and base. See also Humic acids,* Humin,* and Kerogen.*

FUNCTIONAL GROUP: Any group of atoms containing a functionality.*

FUNCTIONALITY: Any portion of an organic molecule that is not a carbon–carbon or carbon–hydrogen single bond. Examples include double bonds and any heteroatom.*

FUSINITE: A coal maceral* of the inertinite* group chiefly found in fossil wood charcoal, showing well-defined cellular structure.

GAMMA FACTOR: Temperature factor used in calculating TTI* values by Lopatin's method.*

GAS CHROMATOGRAM: The output data in analog form from a gas chromatograph.*

GAS CHROMATOGRAPH (GC): An instrument for performing separation of hydrocarbon mixtures. The mixture is introduced via syringe into the heated injection port, where it is vaporized immediately. An inert gas, usually nitrogen or helium, flows through the injection port and sweeps the vapor onto and through the column, which is a narrow-diameter metal or glass tube. More volatile, less polar components pass through most rapidly. The separated components are detected, usually by a flame ionization (FID)* or thermal conductivity detector (TCD),* and are recorded on a gas chromatogram.*

GAS CHROMATOGRAPH–MASS SPECTROMETER (GCMS): A sequential combination of these two instruments. The gas chromatograph (GC)* separates components, which then pass individually into the mass spectrometer (MS).* Coupling the two instruments permits separation by GC and identification of compounds by MS without human intervention, thus greatly increasing sensitivity and accuracy.

GAS DEADLINE: See Hydrocarbon deadline.*

GAS HYDRATES: Crystalline compounds in which the ice lattice of H_2O expands to form cages that contain the gas molecules. Methane hydrates can hold 8 methane molecules in each of 46 H_2O molecules, a formula of CH_4 5.75H_2O.

GAS OIL: A general name for a variety of refined petroleum distillates* boiling mainly in the range of about 230°–300°C (446°–572°F) and containing hydrocarbons with about 13 to 17 carbon atoms per molecule. The term is sometimes applied to fractions boiling even higher.

GAS/OIL RATIO (GOR): The amount of free and dissolved gas a reservoir contains relative to the amount of oil; usually expresses as standard cubic feet of gas per barrel of oil.

GASOLINE: The portion of petroleum boiling in the temperature range of 30°–200°C (60°–392°F) and having 4 to 10 carbon atoms.

GAS SOURCE ROCK: A caustobiolite* capable of generating and expelling natural gas in the past, the present, or the future. Carbonaceous rocks are capable of releasing (dry) gas but incapable of releasing oil. Kerogenous rocks can release oil *and* gas but are usually simply called oil source rocks.* In the postmature stage, however, they can only generate and release gas. See also Natural gas.*

GC: See Gas chromatograph.*

GCMS: See Gas chromatograph–mass spectro-meter.*

GELINITE: A coal maceral* of the vitrinite* group found only in a low-rank humic coal and sapropelic coal. Probably a gelified collinite.*

GEOCHEMICAL PHASE: See Organic metamorphism.*

GEOHISTORY DIAGRAM: An age–depth plot tracing the burial and tectonic histories of a rock from the time of deposition to the present day. It differs from a burial history curve* in that the datum is sea level rather than the sea floor, thus permitting water depth to be shown at all times.

GEOPOLYMER: A polymer* formed in the geosphere as the result of chemical combination of small molecules. Their structure is random, almost completely lacking the strict ordering of biopolymers.* Because of their irregular structures, they are not susceptible to microbial attack. Geopolymers include fulvic acids,* humic acids,* asphaltenes,* and kerogens.*

GEOTHERMAL DIAGENESIS: The alteration of organic material in rocks in response to the increasing temperatures at depth. See also Eometamorphism.*

GEOTHERMAL GRADIENT: The change in temperature in the earth's crust with depth, usually expressed in °C/100 m or °F/100 ft. It is proportional to the heat flow* and inversely proportional to the thermal conductivity* of the rocks involved.

GILSONITE: A soluble asphaltite* with specific gravity of 1.03–1.10 and 10–20% fixed carbon,* having a black color, brilliant luster, brown streak, and conchoidal fracture. Known chiefly in Utah, U.S.A., in vertical veins varying in width from a few millimeters to several meters. (Obsolete synonyms are uinthahite and uintanite.)

GLANCE PITCH: A soluble asphaltite* with a specific gravity of 1.1–1.15 and 20–35% fixed carbon,* having a black streak and brilliant conchoidal fracture. When sufficiently free of mineral matter, it is also called Manjak.

GRAHAMITE: A soluble asphaltite* with a specific gravity of 1.15–1.20 and 35–50% fixed carbon,* having jet-black luster, conchoidal fracture, and black streak. It is fusible and brittle and occurs in veins; often mixed with up to 50% mineral matter

GRAPHITIZATION: The process leading to graphite, which is the ultimate polycyclic* aromatic material. Graphitization is what happens to kerogen* during metagenesis* when aromatization* is very advanced.

HALOCLINE: See Pycnocline.*

HC: See Hydrocarbons.*

HEADSPACE GAS: Refers to gas expelled from canned cuttings samples. Cuttings collected at the well site are canned with an added bactericide. In the laboratory, the expelled gas is taken from the can with a syringe and analyzed. Also called airspace gas.*

HEAT FLOW: The quantity of heat (cal) that passes a unit area (cm) of a body in a unit time (sec). Together with the thermal conductivity* of rocks, it defines the geothermal gradient.*

HEAVY HYDROCARBONS (HHC): A term applied to the total of saturated* and aromatic* hydrocarbons boiling above 325°C (617°F) separated by liquid–solid chromatography (see Chromatography*) from the extractable organic matter.*

HEAVY OIL: Oil or tar that has a low API gravity* (less than about 15°–20° API, depending on various factors, particularly economic ones). Heavy oil can be formed in two distinct ways: by biodegradation* (aided sometimes by water washing or evaporation) of normal crude oils and as oils generated early from some sulfur-rich kerogens. See also Tar.*

HEMPEL DISTILLATION: A method of distilling an oil into 15 fractions and the residuum.* The first fraction consists of everything distilling up to 50°C (122°F). The next 9 fractions are taken at intervals of 25°C (45°F) up to 275°C (527°F) under atmospheric pressure. The last 5 fractions are distilled under vacuum at 40 mm Hg pressure. The eleventh fraction consists of all materials boiling up to 200°C (392°F), and the succeeding 4 fractions are taken at intervals of 25°C (45°C) up to 275°C (527°F) at 40 mm Hg. The base of the crude and the quantities of gasoline and other constituents in the crude can be calculated from the distillation data.

HERBACEOUS: Organic material of land plant origin, especially that which is rich in lipid components. Herbaceous organic matter is normally considered to have good oil source potential.

HETEROATOMS: Any atoms other than carbon and hydrogen found in kerogen,* petroleum,* bitumen,* and natural gas.* The most common heteroatoms are nitrogen, sulfur, and oxygen.

HETEROCOMPOUNDS: Organic compounds containing, besides carbon and hydrogen, other elements such as nitrogen, sulfur and/or oxygen. The term is used colloquially in source rock* investigations to indicate extractable organic matter* from which the saturated* and aromatic* hydrocarbons* have been removed.

HETEROCYCLIC COMPOUNDS: Ring compounds in which one or more of the carbon atoms in the ring are replaced by an atom of another element such as nitrogen, sulfur, or oxygen.

HIGH-PERFORMANCE LIQUID CHROMATOGRAPHY (HPLC): A type of column chromatography performed under high pressure to achieve a more efficient separation of compounds within a particular class. It is most commonly used with porphyrins* and sometimes with hydrocarbon biomarkers.*

HHC: See Heavy hydrocarbons.*

HILT'S LAW: The observation that the volatile carbon in coal decreases proportionally with depth of the coal in normal stratigraphic sequences. The nonvolatile or fixed carbon or coals increases with increasing depth and temperature.

HOMOLOGOUS SERIES: A series of compounds with the same type of structure in which each compound is distinguished from its preceding one by having one more CH group. The members of the series bear close chemical relation to each other.

HOMOLOGS: Compounds that are members of the same class but which differ in the number of carbon atoms they contain. For example, *n*-pentane and *n*-hexane are homologs, as are pristane and phytane.

HOPANES: Pentacyclic triterpanes* whose precursors occur in some terrestrial plants and in many microorganisms. They are useful indicators of depositional environment and thermal maturity.

HPLC: See High-performance liquid chromatography.*

HUMIC ACIDS: Organic compounds, generally dark brown, which occur in soils and sediments and dissolve in or are peptized by aqueous alkaline solutions; they generally precipitate from these solutions upon addition of acid.

HUMIC COAL: Coal, in the strict sense. The fossil remnants of vegetable debris, chiefly woody material, often having maintained their morphological organization with little decay. Major element is carbon (60–100%), with hydrocarbon between 1 and 6%. Humic coals are subdivided according to coal type* and coal rank.*

HUMIC ORGANIC MATTER: Organic matter derived primarily from the decomposition of woody plant material and having a low hydrogen content (<6%) even in the uncoalified portion of the subsurface.

HUMIC SUBSTANCES: Organic matter with a relatively low hydrogen/carbon ratio, composed largely of woody and cellulosic material of terrestrial origin. Humic substances generate a little oil and some gas, but do not have high-generative capacities for hydrocarbons.

HUMIN: Humic material of high molecular weight that is insoluble in both aqueous acid and base. See also Fulvic acids,* Humic acids,* and Kerogen.*

HUMOLITE: A sediment consisting of, or rich in organic matter of, predominantly humic coal* type carbonaceous rock. See also Caustobiolite.*

HYDRATES: See Gas hydrates.*

HYDROCARBON DEADLINE: Maximum depth or temperature at which oil or gas is present in economic quantities in a particular area.

HYDROCARBONS (HC): Compounds composed only of carbon and hydrogen. In the petroleum industry, a general term that includes a variety of accumulations of hydrogen-rich organic matter having a predominantly hydrocarbon character, for example, crude oil,* natural gas*, natural asphalts,* and asphaltites*.

HYDROGENATION: The addition of hydrogen to a chemical compound, usually by reaction with the electrons in a double bond. For example, ethene (C_2H_4) + $H_2 \rightarrow$ ethane (C_2H_6).

HYDROGEN INDEX (HI): An indication of the remaining hydrocarbon-generative capacity of a kerogen, as measured by Rock-Eval pyrolysis.* Hydrogen index is expressed in mg HC/g TOC. See also Oxygen index.*

HYDROPHILIC: Literally, water loving. Hydrophilic compounds or functional groups are polar and thus miscible with water.

HYDROPHOBIC: Literally, water fearing. Hydrophobic compounds are nonpolar and thus immiscible with water.

HYDROSTATIC GRADIENT: The pressure increase with depth of a liquid in contact with the surface. The gradient for freshwater is 9.8 kPa/m (0.433 psi/ft).

HYDROXYL: The functional group —OH.

HYPERSALINE: Any water with a dissolved salt content higher than that of normal seawater.

IMMATURE: See Mature.*

IMPSONITE: A black variety of asphaltite* with a specific gravity of 1.10–1.25 and 50–85% fixed carbon; slightly soluble in carbon disulfide, almost insoluble in turpentine; black streak, infusible.

INDIGENOUS: As applied to hydrocarbons in the subsurface, indigenous refers to hydrocarbons* formed in the rocks in which they occur.

INERTINITE: The group of coal macerals* comprising micrinite,* semifusinite,* fusinite,* and sclerotinite,* based on certain similarities in the technological properties. The term does not imply that they are more inert than the other macerals, particularly with respect to coalification.*

INERTODETRINITE: All particles in coal that display a higher reflectivity than the corresponding vitrinite and cannot be classified as micrinite,* macrinite,* or sclerotinite.* It shows no cellular structure, but displays a well-defined multisided and angular outline. Its size varies from <20 to 30 μm. Light gray white color in reflected light, predominantly opaque in transmitted light, only showing dark red to brown border in very thin sections.

IR: Abbreviation for infrared.

ISOALKANES: Alkanes* with branched chains of carbon atoms, for example, pristane,* which has this structure:

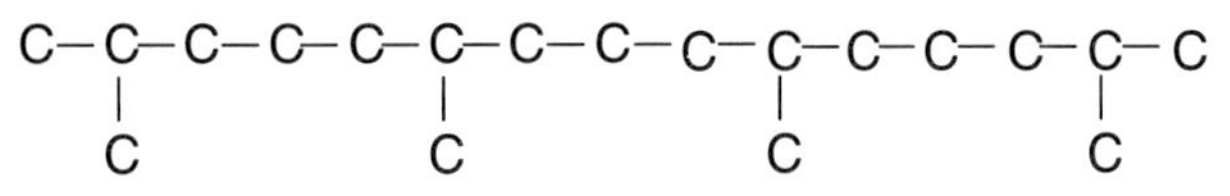

ISODOM: On geological basin maps, the lines connecting points of equal DOM.* If volatile matter* is mapped, the contours are then called isovol. (Obsolete synonym is isocarb.)

ISOMERIZATION: The conversion of a compound into another compound containing the same atoms in the molecule but arranged in a different way. For example, *n*-butane may be converted to isobutane by isomerization:

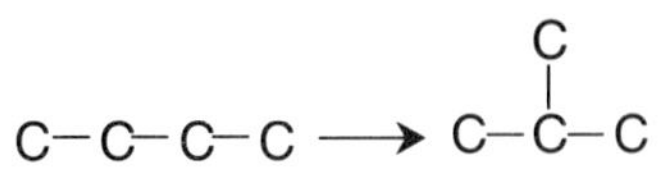

ISOMERS: Molecules that have the same number and kinds of atoms but are different substances. Structural isomers differ in the way atoms are linked together; for example, *n*-butane and isobutane. Stereoisomers differ in the spatial arrangements of groups; for example, cis and trans ("boat" and "chair") isomers of 1,2-dimethylcyclopropane. Optical isomers are nonidentical mirror images comparable to right- and left-handed gloves; for example, D- and L-lactic acid.

ISOPARAFFINS: See Isoalkanes.*

ISOPRENE An unsaturated hydrocarbon (C_5H_8) with the following carbon skeleton:

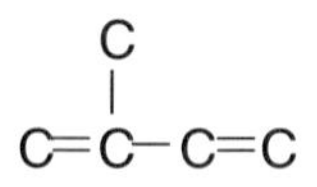

ISOPRENOID INDEX: A maturity index based on the following concentration ratio of different isoprenoids* following isoprenoid isoalkanes*:

$$\frac{C_{19} + C_{20}}{C_{15} + C_{16}}$$

The ratio decreases strongly with increasing maturity* of the sediment.

ISOPRENOID ISOALKANES: Alkanes* with methyl groups (CH_3 groups) attached to every fourth carbon of the otherwise continuous chain of carbon atoms:

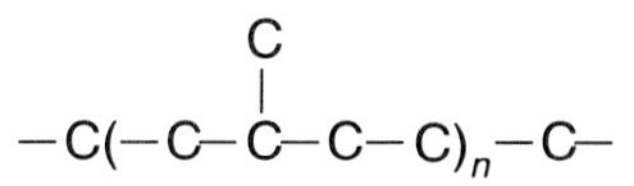

ISOPRENOIDS: Molecules related structurally or genetically to the building block molecule isoprene*:

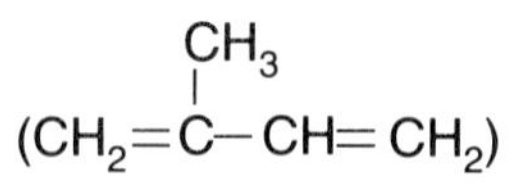

and existing in petroleum in the form of isoparaffins,* naphthenes* (including poly-naphthenes), and aromatics* (including polyaromatics).

ISOTOPE PROFILE: A plot of isotope values measured for several fractions of a bitumen or an oil or for members of a homologous series (*n*-alkanes, for example). The measured values are generally presented in agraphical (profile) format that shows the trend among the fractions measured. Profiles are useful in correlations.

ISOTOPE RATIO: The ratio of two isotopes* in a compound. In geochemistry the following ratios are of special importance: the ratio $^{12}C/^{13}C$ which may be indicative of the type of organic matter, and the ratio of the oxygen isotopes $^{16}O/^{18}O$, which may be used as an indicator of the paleotemperature.

ISOTOPES: Elements having an identical number of protons in their nuclei, but differing in the number of their neutrons. Isotopes have the same atomic number and the same chemical properties, but different atomic weights.

ISOVOL: See Isodom.*

KEROGEN: (1) The solid, insoluble organic matter of sapropelites* (kerogenous rocks), which yields oil on destructive distillation.* It is the product of the biochemical alteration of lipid-rich organic matter. Its insolubility in organic solvents such as benzene, chloroform, and carbon disulfide is related to its extremely high molecular weight. See also Caustobiolite.* (2) After Hunt, the disseminated organic matter of sedimentary rocks that is insoluble in nonoxidizing acids, bases, and organic solvents. The organic matter initially deposited with unconsolidated sediments is not kerogen but a precursor that is converted to kerogen during diagenesis.* Sapropelic* kerogens yield oil and gas on heating, while humic kerogens yield mainly gas. Kerogen includes both marine- and land-derived organic matter, the latter being the same as the components of coal.

KEROGENOUS: Pertaining to kerogen*; of sapropelic* origin.

KEROSENE (KEROSINE): A petroleum distillate with a boiling range of about 180°–230°C (392°–500°F) and with about 11 to 13 carbon atoms per molecule.

KINETIC ISOTOPE EFFECT: The effect derived from the fact that the lighter isotopes* of an element react more rapidly than the heavier ones. Thus, any irreversible reaction in which 100% of the reactant has not yet been consumed will show an enrichment of the light isotope in the products. See also Equilibrium isotope effect.*

KINETICS: The study of the rates at which chemical reactions proceed and the dependence of these rates on various factors, such as time, temperature, pressure, and concentrations of the reactants.

LECO CARBON ANALYZER: An instrument in common use for measuring TOC* values by combustion of the organic carbon and subsequent measurement of the carbon dioxide produced.

LIGNIN: Highly aromatic organic polymer* that forms the structural material of land plants. It has a very low potential to source oil and only a slight potential for gas.

LIGNINS: Chemical constituents of trees and woody plants. They are substances of high molecular weight built up of structural elements containing benzene nuclei and methoxyl groups ($—O—CH_3$).

LIGNITE: Loosely synonymous with brown coal.* In the strict sense, lignite is hard or consolidated brown coal or the individual pieces of wood enclosed in brown coal.

LIPIDS: A general name for plant and animal products typified by esters* of higher fatty acids* but including certain other oil-soluble, water-insoluble substances. The term *fat** (or vegetable or animal oil, if liquid) is usually confined to esters of fatty acids with glycerol, and the term *wax** to esters with other alcohols*. The term *lipids* includes fatty acids, alcohols, steroids,* terpenes,* and carotenoids.* They are generally characterized by solubility in solvents such as ether, benzene, and chloroform and have slight or no solubility in water.

LIPTINITE: Synonym of exinite, but applied to brown coal macerals. A coal maceral group that is the dominant organic constituent of boghead coals.* Liptinite macerals include sporinite, cutinite, resinite, and alginite, which are derived from spores and pollen, cuticles, resins, and algae, respectively. Bituminite is an amorphous liptinite maceral. Liptinite is widely disseminated in sediments and is an important source of crude oil.* The kerogen* of oil shales* is mostly of liptinitic origin.

LIPTOBIOLITE: See Sapropelite.*

LIPTODETRINITE: Collective term for constituents of the liptinite* group, which, because of their finely detrital condition and/or small particle size, can no longer be assigned with certainty to one or other macerals of the liptinite group. Have varying forms, sharp-edged splinters, thread-like structures, rounded particles, often only 2–3 μm. In transmitted light, are white, yellow, reddish yellow; in reflected light, black, dark gray, or brown; show strong fluorescence.

LIQUID CHROMATOGRAPHY: An analytical method in which components of a mixture are separated and measured by dissolving them in a liquid and then passing them through a column of solid sorbing material that has a different affinity for the different components. The liquid emerging from the column is collected in a number of fractions in which the separated components are identified or measured by chemical or physical methods.

LITHOSTATIC GRADIENT: The total pressure increase with depth caused by rock grains and water. It averages about 24.4 kPa/m (1.08 psi/ft).

LITHOTYPE: The macroscopic fine structure of humic coal*; also called ingredient, rock type, banded ingredient, or banded constituent. See also Coal structure.*

LOM: Level or organic metamorphism. An arbitrary scale used by Shell Oil Co. in the United States to indicate the progress of the thermal alteration of organic matter in the subsurface. LOM is defined as being linear to the maximum temperature of a sediment, with an LOM value of zero assigned to organic matter subjected to a temperature of 70°F (21°C) and an LOM value of 10 assigned to organic matter subjected to a sufficient temperature history to produce an opacity or light absorption of 62% for the spore type Sco 31, using the EPR coalification equipment. See also DOM* and Coal rank.*

LOPATIN'S METHOD: A method developed in 1971 by N. V. Lopatin for predicting the thermal maturity of rocks from a detailed knowledge of their burial and thermal histories.

LOW-GRAY: Term applied to the vitrinite population having the lowest reflectance value. Except where caving has occurred, the low-gray population should represent the indigenous, first-cycle material.

LUBE OIL: A general name for a variety of refined, nonwaxy, nonasphaltic petroleum fractions boiling above about 300°C (572°F).

LUBRICATING OIL: The fraction of crude petroleum having a viscosity above 50 sec S.U. at 38°C (100°F) and a boiling range between about 332° and 421°C (630° and 790°F).

M+: The molecular ion formed in a mass spectrometer.* See also Molecular ion.*

M/Z: Ratio of mass to charge for an ion in a mass spectrometer.* The charge is virtually always +1, so the m/z value is actually the mass of the ion in amu.

MACERALS: Elementary organic units of coal that can be distinguished under the microscope. See also Coal maceral.*

MACRINITE: Shows no cell structure, occurs isolated or as ground mass in isolated form, rounded outlines. Size mostly above 10 μm generally opaque or brownish red.

MANJAK: See Glance pitch.*

MASS SPECTROMETER (MS): An instrument used for identifying chemical compounds. The compound is vaporized in the inlet system and then bombarded with large amounts of energy. The energy knocks one electron out of some of the molecules, forming ions with essentially the same mass as the original compounds. The ions are then accelerated into a magnetic field whose intensity allows only certain masses to pass through (the remainder being deflected to one of the magnets). The intensity of the magnetic field is varied rapidly so that ions of a wide range of mass/charge ratios (m/z) can be detected. The pattern of ions passing through the magnetic field is called the mass spectrum. Mass spectrometers are also used to determine isotope ratios. Because molecules containing heavy isotopes are heavier than normal, so are the ions formed in the mass spectrometer. Mass spectrometry is used in fingerprinting, for maturity determinations, and for environmental determinations. See also Gas chromatograph–mass spectrometer.*

MASS SPECTRUM: The recording of the ions present and their relative abundances in a sample that has been analyzed in a mass spectrometer.*

MATURE: As applied to the organic matter of fine-grained rocks, the organic matter is mature when its indigenous,* extractable hydrocarbons are petroleum-like in composition. Specifically, the *n*-alkane* carbon number distribution and the naphthene* ring number distributions are used as indices of maturity. Organic matter that has not yet reached the mature stage is called immature. Mature organic matter that has been subjected to a sufficient temperature for most of the heavier hydrocarbons (^{14}C) to have been converted to light hydrocarbons and dead carbon is referred to as being postmature.

MATURITY: Similarity of extractable rock hydrocarbons to petroleum in chemical composition.

MATURITY INDEX: Chemical parameter indicating maturity* of an oil source rock, e.g., the isoprenoid index,* the naphthene index,* and the *n*-paraffin index.*

MERCAPTANS: Compounds containing the sulfhydryl (—SH) group. They are sulfur analogs of the alcohols in that the oxygen in the alcohol (—OH) group is replaced by sulfur.

METAANTHRACITE: The highest rank of humic coal.* See also Coal rank.*

METAGENESIS: The late stages of thermal maturity when gas generation and cracking* predominate.

METAMORPHISM: Modification of original composition in response to temperature, pressure, radiation, catalysts, etc. See also Organic metamorphism.* The transformation of preexisting rocks into new types by the action of heat, pressure, stress, and chemically active migrating fluids. Metamorphism usually begins at temperatures above 200°C (392°F). At such temperatures, the organic matter is already reduced to a low-hydrogen carbon residue capable of yielding only small amounts of gas.

METERORIC WATER: Fresh surface water entering a subsurface sedimentary section through permeable outcrops. It is high in sodium, bicarbonate, and sulfate and usually contains less than 10,000 ppm dissolved total solids.

METHANE: CH_4, the first member of the homologous series of paraffins:

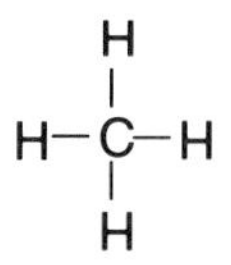

METHANE HYDRATES: See Gas hydrates.*

METHANOL: Methyl alcohol, the simplest of the class of organic compounds known as alcohols*: CH_3OH.

METHYLPHENANTHRENE INDEX (MPI): Ratio of several methylphenanthrenes, tricyclic aromatic* molecules found in the aromatic fractions of oils and bitumens. The index is related to vitrinite reflectance* by the empirical equation $R_o = 0.60(MPI) + 0.37$.

MICELLE: An aggregation of a few molecules in which the polar portions of the molecules align themselves close together. The nonpolar parts of the molecules can thus also lie near each other. In a polar medium, the polar parts of the micelle are pointed outward, interacting favorably with the medium, while the nonpolar parts are together in the middle, protected from the medium. In a nonpolar medium the roles would be reversed. Oil could form micelles since most rocks are wetted with polar water molecules, and oils do contain some polar components in addition to nonpolar hydrocarbons. Micelle formation would increase the compatibility of water and oil.

MICRINITE: An opaque coal maceral* of the inertinite* group showing no cellular structure. Often divided into fine-grained micrinite and massive micrinite. The origin of this maceral has not yet been traced.

MICROBES: Microscopic organisms such as bacteria and methanogens, which play very important roles in diagenesis.*

MICROBIAL: Referring to living microorganisms, including bacteria.

MICROFRACTURES: Tiny fractures that open up temporarily as a result of overpressuring in source rocks, at least partially in response to hydrocarbon generation. Microfractures may be important pathways for expulsion. Because they apparently heal rapidly and without leaving a trace in most cases, microfractures are not well understood.

MICROLITHOTYPE: Synonym for coal type* and coal maceral association*; the microscopic fine structure of humic coal*.

MIGRATION: Movement of oil and/or gas in the subsurface after expulsion*. Terms such as primary and secondary migration are generally discouraged. See also Remigration.*

MINERAL WAX: A species of a bitumen having a characteristic luster and unctuous feel, composed principally of saturated hydrocarbons* and containing considerable amounts of crystallizable paraffins,* the nonmineral constituents being soluble in carbon disulfide. An example is ozocerite. See also Paraffin wax.*

MOLE: The mass numerically equal to the molecular weight. It is most frequently expressed as the gram molecular weight, i.e., as the weight of 1 mol expressed in grams. The molecular weight of methyl alcohol (CH_3OH) is 32 (based on atomic weight of C = 12, H = 1, and 0 = 16); 1 g mol of methyl alcohol is 32 g.

MOLECULAR ION: The ion formed in a mass spectrometer* by the initial loss of a single electron from a molecule. Its mass is essentially the same as that of the parent molecule. The molecular ion can subsequently decompose to give fragment ions.

MOLECULAR SIEVES: Zeolites used in the laboratory

separation of *n*-alkanes from branched and cyclic saturated hydrocarbons. See also Urea adduction.*

MORETANES: Pentacyclic triterpanes* that are not very stable thermally. Their presence is a good indicator of thermal immaturity, although some oils sourced from rocks rich in high-sulfur kerogens* may still contain moretanes.

MPI: See Methylphenanthrene index.*

MS: See Mass spectrometer.*

MUDLOGGING: A process in which the amount and composition of gases in the mud returning up the annulus of the drill string from the bit are determined. These values are recorded on a mudlog which reveals relative amounts of gas in fine-grained sediments (shale gas which reflects source character) as well as in reservoir rocks (indications of gas bed ring zones). The mudlogger also is used as a warning device to prevent the mud from becoming too heavily "cut" with gas and causing impending blowouts.

NAPHTHA: The portion of petroleum boiling in the temperature range of 100°–200°C (212°–392°F) (as defined by the U.S. Bureau of Mines).

NAPHTHABITUMEN: The natural, gaseous, liquid and solid mixtures of hydrocarbons and hydrocarbon-like compounds derived from caustobiolites.*

NAPHTHENE: Synonym for cycloalkane* and cycloparaffin. A hydrocarbon ring with the molecular formula C_nH_{2n}. Cyclopentane (C_5) and cyclohexane (C_6) ring structures are the most common in petroleum. Condensed or polycyclic naphthenes contain rings in which two or more carbon atoms are shared. Tetracyclic and pentacyclic naphthenes contained four and five rings, respectively, fused together.

NAPHTHENE INDEX (NI): A maturity index; the sum of the percentages of one- and two-ring naphthenes in the 420°–470°C (788°–878°F) isoparaffin–naphthene concentrate of petroleum or of rock extracts.

NAPHTHENES: A term often used by petroleum chemists to indicate the cycloalkanes* occurring in petroleum.

NAPHTHENIC ACIDS: The organic acids that are found in abundance in naphthenic crudes and in associated waters.

NAPHTHENIC CRUDE OILS: Nonwaxy crudes with high naphthene* content and low content of gasoline; API gravities* are usually less than 35°.

NATURAL GAS: A naturally occurring gas in most cases consisting mainly of the lighter hydrocarbons, especially methane.* It is believed to originate from two sources: (a) humic coal* (predominantly methane) and (b) kerogenous* rocks along with or after the formation of petroleum. Natural gas consisting principally of methane and ethane and devoid of the heavier hydrocarbons is called dry gas.* See also Caustobiolite.*

NATIVE ASPHALTS: A class of naturally occurring bitumens* with low fusing points having specific gravities of about 1.0–1.1 and fixed carbon values of 4–20%.

NONCARBONATE CARBON (NCC): See Organic carbon.*

NORMAL ALKANES: See *n*-Alkanes.*

NORMAL PARAFFINS: See *n*-Alkanes.*

NSO COMPOUNDS: See Heterocompounds.*

NSOs: Nonhydrocarbon organic components of crude oils and extracts with a nitrogen, sulfur and oxygen atom in their structures in a position normally occupied by a carbon atom.

NUCLEAR MAGNETIC RESONANCE (NMR): A type of spectroscopic analysis used in organic geochemistry to distinguish aliphatic carbon atoms from aromaticones. Application has been rather rare but may increase in the future in kerogen studies.

ODD-PREDOMINANCE (OEP): Refers to the phenomenon that *n*-alkanes* with an odd number of carbon atoms predominate over those with an even number of carbon atoms. Also called Carbon preference index (CPI).* In general, *n*-alkanes in the molecular weight* range of C_{25} to C_{35} are taken into consideration. Sediments showing odd-predominance are considered to be immature.* Follows this equation:

$$OEP = \frac{C_i + 6C_{i+2} + C_{i+4}}{4C_{i+1} + 4C_{i+3}}^{(-1)^{i+1}}$$

OIL DEADLINE: See hydrocarbon deadline.

OIL FLOOR: The maturity VR* of 1.30 % R_o* of reservoir or associated sediments above which commercial oil occurrences are extinguished owing to thermal cracking* of oil.

OIL GENERATION WINDOW: The range of depths, temperatures, maturities, or less commonly, times during which oil generation was or is occurring in a particular area.

OIL PRESERVATION WINDOW: The range of depths, temperatures, or maturities at which oil is thermally stable in a particular area.

OIL SHALE: A sapropelite* containing a variable amount of kerogen* (from 10 to 67%) which on destructive distillation* yields shale oil.* The term is a well-known misnomer because the rock is not really a shale nor does it contain oil. The shaly appearance comes from the foliated arrangement of the organic matter within the mineral matrix. In principal, oil shale can be considered a rich oil source rock.* See also Caustobiolite.* An oil shale containing more than 67% kerogen (or less than 33% ash*) is classified as sapropelic coal.*

OIL SHOW ANALYZER: An instrument that performs both TOC* measurement and Rock-Eval pyrolysis* in a single operation.

OIL SOURCE ROCK: A kerogenous* rock capable of generating and expelling petroleum* in the past, the present, and/or the future. Depending on the maturity, one can distinguish immature*, mature,* and postmature* oil source rocks. (Shell Oil Co. prefers the term *potential* for immature oil source rock.) An oil source rock is also always a gas source rock.* See also Caustobiolite.*

OIL WINDOW: See Oil generation window* and Oil preservation window.*

OLEFINS: A class of unsaturated hydrocarbons of the general formula C_nH_{2n}. These are not normally found in more than trace quantities in crude oils. See also Alkanes.*

OM: Abbreviation for organic matter.

OML See Oxygen-minimum layer.*

ORGANIC: (1) A chemical compound containing one or more carbon atoms. However, carbonates and metal carbides are not considered organic. (2) Originating in or derived from an organism.

ORGANIC CARBON: Synonym for noncarbonate carbon.* The carbon that is left in a sediment after removal of soluble organic matter (by extraction) (see also Extractable organic matter*) and of carbonates (by treatment with HCl).

ORGANIC METAMORPHISM: A colloquial expression for metamorphism of organic matter. It comprises all physical and chemical alteration processes after the death of an organism during sedimentation and in the subsurface. Two phases are distinguished: (a) a biochemical phase, which is conversion of organic matter mainly influenced by microbial processes in the shallow subsurface, and (b) a geochemical phase, which is conversion primarily influenced by thermal processes leading to products of increased thermal stability and ultimately to methane and graphitic carbon. Organic metamorphism is in general limited to subsurface temperatures of less than about 250°C (~500°F).

ORGANIC NETWORK: Term used for the continuous network of organic material as it occurs in source rocks.

OVERCOOKED MATURE: Synonym for postmature. See Mature.*

OVERMATURE: Hydrocarbon generation has already occurred. The term usually refers to oil generation.

OXIDATION: Loss of electrons by an atom, ion, or molecule during a chemical reaction.

OXIDIZING AGENT: A molecule, atom, or ion that promotes oxidation in some other material. The oxidizing agent is itself reduced during the reaction. See also Reducing agent.*

OXYGEN INDEX: The amount of CO_2 produced from kerogen during Rock-Eval* pyrolysis.* Oxygen index is measured as mg CO_2/g TOC and is supposed to be related to the oxygen content of the kerogen. See also Hydrogen index.*

OXYGEN-MINIMUM LAYER: Layer of water within which the dissolved oxygen content is lower than in the over- and underlying water layers. It is caused by excessive demand for oxygen by decaying organic matter falling from the photic zone above.

OZOKERITE: A native mineral wax, yellow to dark-brown in color, composed of the higher members (C_{22} to C_{29}) of the C_nH_{2n+2} (paraffin*) and C_nH_{2n} (naphthene*) series. Ozokerite is believed to be a shallow alteration product of a paraffinic oil. See Paraffin wax.*

PACKED COLUMN: A relatively large-diameter gas chromatographic* column that is packed with a solid material upon which the stationary phase is coated. See also Capillary column.*

PAH: Abbreviation for polycyclic aromatic hydrocarbon.

PALYNOMORPHS: A term embracing pollen* and spores* (sporomorphs*) and also including dinoflagellates and acritarchs. Their light translucency is used in palynology as a rough indicator of coalification*.

PARAFFIN DIRT: A yellow-brown, gummy, soil organic matter composed of nitrogenous-humic-cellulosic remains of plant material heavily impregnated with fungi, yeasts, actinomyces, and bacteria. The word *paraffin* in this term is a misnomer since this substance contains less than 3% lipid material. The waxy appearance is caused by living and dead microbial cells. Paraffin dirt tends to accumulate in moist soils hear hydrocarbon seeps.

PARAFFINIC CRUDE OILS: Waxy crudes with low naphthene content.

n-PARAFFIN INDEX (R_{29}): A maturity index; the ratio of the concentration of *n*-C_{29} to an average of the concentrations of *n*-C_{28} and *n*-C_{30} in a rock extract or in an oil.

PARAFFIN–NAPHTHENE (P-N) HYDRO-CARBONS: The mixture of saturated hydrocarbons including both paraffin* (nonring) and naphthene* (ringed or cyclic) structures; these often are separated as one fraction in liquid chromatography.*

PARAFFINS: See Alkanes.*

n-PARAFFINS: See *n*-Alkanes.*

PARAFFIN WAX: Solid, waxy material made up mainly of *n*-paraffins with 17 or more carbon atoms per molecule. Includes (a) petroleum wax (derived from petroleum); (b) mineral wax (naturally occurring naphthabitumens,* e.g., ozokerite,* hatchetite, and scheerite) (see also Caustobiolite*); and (c) montan wax (consisting mainly of long-chain alcohols, acids, and esters extracted from peat,* lignite, etc.).

PDB: Belemnites from the Cretaceous PeeDee Formation of South Carolina. PDB is the standard for comparing carbon isotope ratios.*

PEAT: The lowest rank of humic coal* (some researchers exclude peat from coal rank* and call it a precursor of coal).

PER MIL: Parts per thousand, written as ‰. Used in expressing the difference in isotope contents between a sample and the standard.

PETROGENIC GAS: A gas associated with petroleum or derived from crude oil by diagenetic processes as distinguished from biogenic processes as distinguished from biogenic gas produced by the action of bacteria on organic matter in relatively shallow sediments.

PETROLEUM: Liquid naphthabitumen* composed mainly of hydrocarbons,* along with varying amounts of other soluble compounds containing not only carbon and hydrogen but also sulfur, oxygen, and nitrogen atoms. Includes crude oil and condensate. See also Caustobiolite.*

PETROLEUM WAX: See Paraffin wax.*

PETROLIFEROUS: Referring to rocks containing or yielding petroleum.

PHENOL: A group of organic compounds containing a hydroxyl (OH) group attached to an aromatic ring. The simplest member of the phenol family.

PHOTIC ZONE: The depth range within which photosynthesis occurs in marine or lake waters. The thickness of the photic zone is seldom greater than 200 m (660 ft), and is limited by light penetration, which in turn is limited by turbidity.

PHYTANE (PH): An isoprenoid* isoalkane* with 20 carbon atoms, having the following structure:

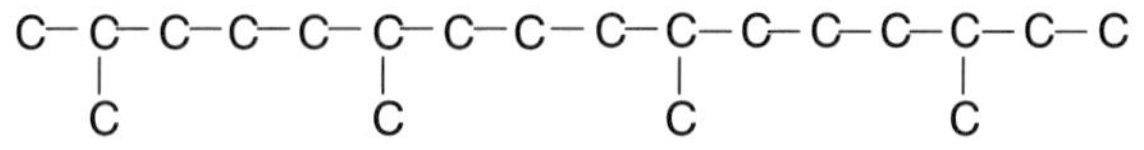

PHYTOL: An isoprenoid* unsaturated* alcohol* with 20 carbon atoms, having the following structure:

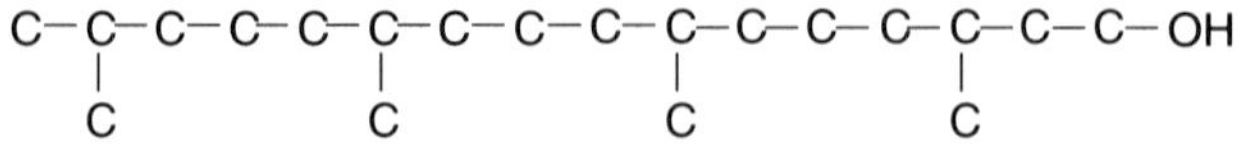

PHYTOPLANKTON: Unicellular photosynthetic organisms that mainly live in marine or lacustrine waters. They are responsible for approximately half of the annual photosynthetic productivity on the earth. They are also called algae and include coccolithophorids, diatoms, and dinoflagellates.

PITCH: Undistilled residue from the thermal treatment of organic matter (destructive distillation,* pyrolysis,* etc.); generally of viscous to solid consistency; comparatively nonvolatile; solubility in CS_2 depends considerably on severity of thermal treatment (may be approaching insoluble charcoal, coke, or dead carbon in nature).

PLANKTON: All aquatic organisms freely floating or drifting in the water, as opposed to necton (which propels itself) and benthos (which is attached to or crawls on the basin floor). Although the term includes organisms such as seaweed, most planktonic organisms are microscopic. Phytoplankton is plant plankton and zooplankton is animal plankton.

PNA ANALYSIS: The determination of relative amounts of paraffins, naphthenes, and aromatics in an oil or in fractions of a crude oil. This method is applied mainly to gasoline fractions as a means of classifying crude oils.

POLLEN: The individual grains or microorganisms produced in the male apparatus in the flowers of higher (seed) plants; usually a fine yellow dust. See also Sporomorphs.*

POLYCYCLIC: Having many rings in its chemical structure.

POLYMERIZATION: A chemical reaction that produces very large molecules by a process of repetitive addition of small molecules. Polymerization processes always lead to a broad mixture of macromolecules with different molecular weights.

POLYMERS: Large molecules consisting of many small subunits. Biopolymers* have regular structures, whereas geopolymers* have irregular unique structures formed from a wide variety of subunits.

PORPHYRINS: A group of organic compounds, generally red. The two most important types are the DPEP (or phyllo) type and the MEP (or etio) type. The first type is derived from chlorophyll,* whereas the second type can form from DPEP by further thermal metamorphism or from hemin (blood pigment). In crude oils they generally are bound to vanadium and nickel (porphyrin/vanadium or porphyrin/nickel complexes). In sediments they may occur metal free (free porphyrins) or bound to iron.

POSSIBLE SOURCE ROCK: See Source rock.*

POSTMATURE: See Mature.*

POTENTIAL SOURCE ROCK: See Source rock.*

POUR POINT: The temperature at which crude oil will not flow when a tube containing it is first heated in a bath to dissolve all wax and then cooled slowly. At cooling intervals of 3°C (5°F), the tube is held in a horizontal position until there is no flow for 5 sec.

PRIMARY MIGRATION: The movement of oil and gas out of fine-grained source rocks into permeable reservoir rocks.

PRISTANE (PR): An isoprenoid* isoalkane* with 19 carbon atoms and of the following structure:

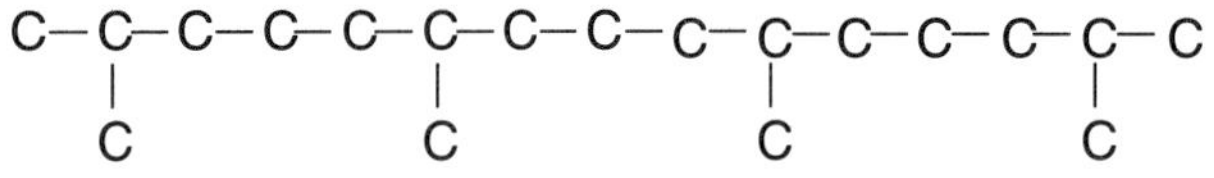

PRODUCTION INDEX: See Transformation ratio.*

PROGRAMMED TEMPERATURE CHROMATOGRAPHY (PTC): Gas chromatography carried out beginning at a low temperature and progressing to higher temperatures according to a predetermined temperature program. See also Gas chromatograph.*

PROTEINS: Naturally occurring complex condensation products of amino acids* with the following basic arrangement:

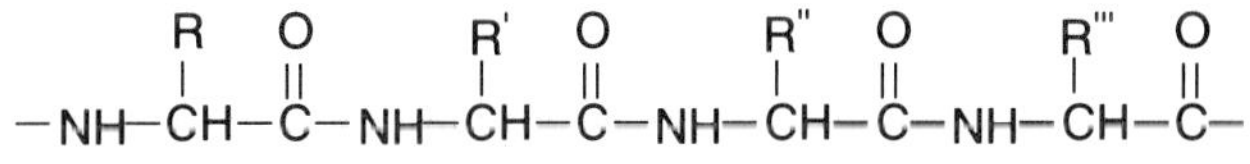

Proteins are essential constituents of all living cells. They are readily broken down by degradation processes.

PROXIMATE ANALYSIS: The determination by standard methods of moisture, volatile matter,* fixed carbon* (by difference), and ash* in coal. For all low-rank coals, the determination of the calorific value* should also be added. See also Ultimate analysis.*

PSEUDO ACTIVATION ENERGY: A parameter used to describe the average activation energy of a system consisting of numerous different chemical reactions. See also Activation energy.*

PTC: See Programmed temperature chromato-graphy.*

PYCNOCLINE: A sharp density discontinuity between water layers caused either by thermal stratification (thermocline) or salinity differences (halocline).

PYROBITUMEN: A term not recommended for use because it is a confusing nongenetic name covering some insoluble asphaltites* (asphaltic pyrobitumen), humic coal* (nonasphaltic pyrobitumen), and occasionally sapropelites as well.*

PYROLYSIS: (1) Decomposition of organic matter by heating in the absence of oxygen. This process is the basis for several methods of source rock analysis (C_R/C_T, pyrolysis-GSC,* Rock-Eval* pyrolysis, etc.). (2) Destructive distillation,* nearly identical with definition (1), although generally not carried out in a completely oxygen-free atmosphere. Applied to coals and oil shales (e.g., Fischer assay*), but also to sedimentary rocks (e.g., pyrolysis-fluorescence,* pyrolysis-sniffing,* test tube pyrolysis, and pyrolysis-colorimetry*).

PYROLYSIS-COLORIMETRY: The analytical technique of test tube pyrolysis of a rock followed by the determination of the color of the dissolved tarry compounds produced by the pyrolysis. Like pyrolysis-fluorescence,* this method provide a raid identification of pyrolyzable organic matter of a rock. It is developed for field use.

PYROLYSIS-FID: The analytical technique of programmed temperature pyrolysis* followed by flame ionization detection (FID)* of the hydrocarbon products. The pyrolysis temperature at which maximum generation of hydrocarbons occurs provides an indication of maturity.

PYROLYSIS-FLUORESCENCE: The analytical technique of test tube pyrolysis of rock followed by the fluorometric determination of the fluorescing compounds produced by the pyrolysis. This method provides a rapid indication of the thermally available oil of a rock.

PYROLYSIS-GSC: The analytical technique of pyrolysis followed by gas-solid chromatography (GSC) as a means of determining the pyrolysis temperature at which maximum ethane generation occurs. This provides an indication of maturity for organic matter having reached a maturity value of VR/E* 1.32 or more.

PYROLYSIS-MS: The analytical technique of programmed temperature pyrolysis followed by mass spectrometric* determination of some of the products. This provides an indication of both type and maturity of organic matter.

PYROLYSIS-SNIFFING (PF): The analytical technique of rapid, high-temperature pyrolysis of rock sample followed by the semiquantitative detection of the total hydrocarbon gases produced. The detection is accomplished by means of a flame ionization detector.* A laboratory method for the rapid screening of rocks for the presence of pyrolyzable organic matter.

PYROLYTIC GAS RATIOS: The gas ratios of ethane, ethylene, propane, propylene, isobutane, and *n*-butane relative to methane obtained by pyrolysis of organic matter. It allows the determination of the type of organic matter.

R: Chemical designation for rectus (right), which indicates the relative positions of the four groups bonded to a carbon atom. There is not necessarily a correspondence between the *R,S* system and the *X,B* system.

R_a: Vitrinite reflectance measured in air. This technique is used mainly by Soviet workers. Charts exist that convert R_a values to R_o.*

R_o: Vitrinite reflectance* measured in oil immersion.

RANK: See Coal rank.*

RANK CORRELATION: A mathematical method in use for correlating oils and extracts or different oils. It is based upon the ranking of the hydrocarbons in the fraction boiling between 30° and 120°C (86° and 248°F) according to their quantities. It results in the rank correlation coefficient, which expresses the degree of correlation. The value of this coefficient is essentially limited to the range –1 to +1; the closer to +1, the better the correlation.

REDUCING AGENT: An atom, molecule, or ion that promotes reduction of another substance. The reducing agent is itself oxidized in the process.

REDUCTION: The gain of electrons by an atom, molecule, or ion during a chemical reaction. See also Oxidation.*

REFLECTANCE: See Vitrinite reflectance.*

REMIGRATION: The movement of hydrocarbons from an existing trap into a new one or to the surface due to changes in physical conditions and/or tectonic mobilization.

RESIDUUM: The residue obtained from the distillation of crude oil* after all fractions, including lubricating oils,* have been taken off.

RESIN: Petroleum resins are the fraction of residuum* that is insoluble in liquid propane but soluble in normal pentane. Plant resins are terpenoids* ranging in molecular size from sesquiterpenes (C_{15}) to tetraterpenes (C_{40}). They contain the olefinic double bonds of the isoprene* building block that causes the liquids to polymerize and oxidize to hard resins on exposure to air. Balsam and mastic are plant resins.

RESINITE: A coal maceral* of the exinite* group formed from resinous substance (plant resins).

RETENTION TIME: The length of time a compound takes to come out of a chromatographic column. Retention time will vary greatly as chromatographic conditions (flow rate, temperature, nature of stationary and mobile phases, etc.) are changed.

ROCK-EVAL: An instrument for carrying out pyrolysis of rocks and sediments. The Rock-Eval is in common use throughout the oil industry. Information on both kerogen type and maturity can be obtained with it.

RSRA: Abbreviation for rapid source rock analysis. See Pyrolysis-fluorescence,* Pyrolysis-sniffing,* and Pyrolysis-colorimetry.*

S: Chemical designation for sinister (left), which indicates the relative positions of the four groups bonded to a carbon atom. There is not necessarily a correspondence between the *R*,*S* system and the *X*,*B* system.

S_0: Thermally extracted hydrocarbons obtained during Rock-Eval analysis mainly consisting of gas molecules. This system is only used on the Rock-Eval oil analyzer.

S_1: Thermally extracted hydrocarbons obtained during Rock-Eval analysis prior to the onset of true pyrolysis. These hydrocarbons are roughly similar to bitumen or EOM. See also S_2.*

S_2: Hydrocarbons generated by kerogen decomposition during Rock- Eval pyrolysis. See also S_1.*

S_3: Carbon dioxide released by kerogen decomposition during Rock-Eval pyrolysis.

S_4: Residual carbon obtained by combustion of the remaining carbon after pyrolysis (Rock-Eval* with TOC* module).

SAPROPEL: From the Greek *sapros*, meaning rotten. Unconsolidated aquatic sedimentary deposit rich in lipid organic matter derived chiefly from phytoplankton and putrefied under reducing conditions. It can be compared with the peat stage of humic coals. When consolidated it forms kerogenous rocks or sapropelite.*

SAPROPELIC: Derived from or pertaining to sapropel.*

SAPROPELIC COAL: A caustobiolite* rich in lipid organic matter, with less than 33% mineral matter. There are two main types: boghead coal* (torbanite*) and cannel coal.* These are not coal in the strict sense but sapropelites* with a higher content of hydrogen than humic coal* (6–12%). On destructive distillation* they give a high yield of oil. Sapropelic coals are distinguished from oil shale* by having less than 33% mineral matter.

SAPROPELIC ORGANIC MATTER: The decomposition and polymerization products of high-lipid organic materials, such as spores and planktonic algae deposited in subaquatic muds (marine or lacustrine) under predominantly anaerobic conditions.

SAPROPELITE: Synonym for liptobiolite. A kerogenous* sediment. See also Caustobiolite.*

SATURATED: A term applied to any compounds in which all carbon–carbon bonds are single bonds.

SATURATES: Saturated hydrocarbons. A general term including normal and branched alkanes* and cycloalkanes (paraffins* and naphthenes*). Saturates are the nonaromatic hydrocarbon fraction of an oil.

SCLEROTINITE: A coal maceral* of the inertinite* group formed by fungal tissue, chiefly sclerotia.

SCI: Spore color index. Color scale from 1 to 10 for measuring changes in spore color for maturity measurements (CITCO system).

SECONDARY MIGRATION: The movement of fluids within the permeable rocks that eventually leads to the segregation of oil and gas into accumulations in certain parts of these rocks.

SEMIANTHRACITE: See Coal rank.*

SEMIFUSINITE: A coal maceral* of the inertinite* group, a constituent intermediate between vitrinite* and fusinite*, showing well-defined woody structure. The term *merofusinite* has been proposed as an improvement.

SHALE OIL: Oil obtained by the destructive distillation* of oil shale*.

SILL: The point in a restricted basin through which the deepest waters enter into the basin.

SIMILARITY COEFFICIENT: A mathematical quantity, based on the composition of selected C_6 and C_7 hydrocarbons, to indicate the degree of correlation between crude oil and source rock hydrocarbons or between hydrocarbons from different groups of crude oils. Its value is essentially positive but lower than 1. The closer to 1, the better the correlation.

SINGLE BOND: A chemical bond in which two electrons are shared. Most bonds are single bonds.

SOM: Structureless organic matter. Bacterially degraded organic matter and/or bacterial remains forming a cloudlike mass of organic matter.

SORPTION CAPACITY: The capacity to hold hydrocarbons, both liquid and/or gaseous.

SOURCE ROCK: A rock that has generated and given up hydrocarbons during diagenesis and compaction, thus satisfying one of the four requirements for petroleum accumulation (i.e., source, reservoir, trap, and preservation). A source rock has realized the potential of a "potential source rock" that contains chemical precursors of hydrocarbons but must experience adequate thermal exposure to generate these hydrocarbons. If used without a prefix, oil source rock is generally meant. See Oil source rock* and Gas source rock.*

SOUR CRUDE: A crude oil containing objectionable amounts of hydrogen sulfide and mercaptans.*

SOUR GAS: Natural gas that contains objectionable amounts of hydrogen sulfide and mercaptans.*

SOXHLET: A device used for extracting bitumen from rock samples. Its design allows clean solvent to reflux continually through the powdered rock sample while the extracted bitumen is accumulating in the reserve plot.

SOXHLET EXTRACTION: A method which involves a piece of glassware that repetitively soaks a sample with portions of freshly distilled solvent and collects the extracts by periodically syphoning the sample compartment to a heated reservoir, from which the solvent is distilled back into the sample compartment.

SPECTROFLUORESCENCE: A method of involving an analytical instrument in which certain organic compounds (particularly aromatic hydrocarbons and NSOs*) are determined by their response upon exposure to monochromatic (single color or wavelength) light; both the intensity and color of the resulting fluorescence, as well as the wavelength of the light that causes fluorescence, are important in this determination.

SPECTROSCOPY: Various techniques for analyzing samples by looking at a range of energies (wavelengths) given off or absorbed, or by observing a range of ions produced, as in mass spectrometry.*

SPORE: The reproductive cell of the lower plants including fungi, mosses, and ferns, with a wall extremely resistant to acidic destruction.

SPORE DARKENING: See TAI* and SCI.*

SPORINITE: A coal maceral* of the exinite* group formed from spore exines flattened parallel to stratification.

SPOROMORPHS: Group term for recent or fossilized spores* and pollen,* of particular importance for coalification studies.

STERANES: Tetracyclic saturated hydrocarbons* derived from the steroids* present in all organisms. They are useful in determining maturity and the type of organic matter and can be of value in oil typing and oil correlation.

STEREOCHEMISTRY: The spatial arrangement of atoms in a molecule.

STEROIDS: Condensed cyclic products of plant and animal origin with one ring of five carbon atoms and three rings of six carbon atoms, thus showing the following carbon skeleton:

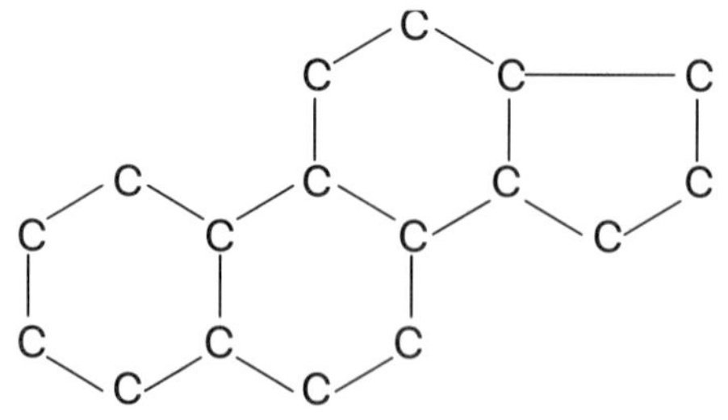

STEROLS: Steroids* containing an alcohol* group (—OH).

STRATIFICATION: Development of nonmixing water layers of different densities as a result of temperature or salinity differences.

STRIPPING: The removal, generally by a current of gas, of certain constituents from rocks and chromatographic columns.

SUBBITUMINOUS COAL: Glossy black coal intermediate between brown coal* and bituminous coal* with a transition into the latter. See also Coal rank.*

SURFACE PROSPECTING: An exploration survey method in which surface or near-surface phenomena are measured. These phenomena should be either directly or indirectly attributable to oil or gas seeping from below. In the most common method, surface or near-surface soil or sediment samples are analyzed for gaseous or other light hydrocarbons or other presumed secondary products (iodine, radioactivity, etc.) of petroleum seepage. Alternately, hydrocarbon detectors or scintillation (radioactivity) detectors are moved along roads or flown on a grid over an area looking for "anomalies."

SWEET CRUDE: A crude oil containing negligible amounts of hydrogen sulfide and mercaptans.*

SWEET GAS: A natural gas containing negligible amounts of hydrogen sulfide and mercaptans.*

SYNGENETIC: As applied to the organic matter in a sedimentary basin, syngenetic has been used to include mainly the insoluble organic matter of fine-grained sediments (e.g., of coals,* oil shales,* and source rocks). See also Epigenetic.*

TAI: See Thermal alteration index.*

TAR: Condensates from the thermal treatment (destructive distillation,* pyrolysis,* etc.) of organic materials; generally of liquid consistency; mainly soluble in carbon disulfide; comparatively volatile at high temperatures. Includes shale oil* and coal tar (see also Pitch*). Incorrectly used in term *tar sand.**

TAR SAND: Generally used but actually incorrect term for a sand or sandstone containing heavy crude oil or asphalt.

TCD: See Thermal conductivity detector.*

TELINITE: The coal maceral* of the vitrinite* group which shows clearly defined cell structure.

TERPANES: Ten-carbon hydrocarbons containing two isoprene* units.

TERPENOIDS: Naturally occurring compounds (both cyclic and noncyclic) constructed from two or more branched five-carbon isoprene units:

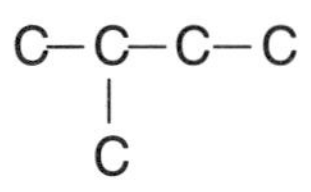

Terpenoids include hydrocarbons, alcohols, and acids. Among the common terpanes (hydrocarbons) are the hopanes* (pentacyclic triterpanes*).

TERTIARY MIGRATION: The movement of an oil or gas accumulation to a different part of the reservoir rock.

TEXTINITE: Consists of plant cell walls; both isolated occurrences of intact individual cells and also cell tissues which are ungelified. Variable form, sometimes torn or deformed. The primary structures of the cell walls (layers, pits, intercellular spaces, etc.) still clearly visible. Yellow-yellowish brown to reddish brown color in transmitted light.

THERMAL ALTERATION INDEX (TAI): A maturation color index for the particular organic matter of sedimentary rocks. The index indicates the degree of thermal alteration that the organic matter has undergone. As proposed by Staplin (1969), the index numbers 1 to 5 include color changes from yellow to brown to black, representing immature, mature, and metamorphosed facies of organic matter.

THERMAL CONDUCTIVITY: The material constant of a given rock defined as the amount of calorific units (cal) which passes in a unit time (sec) the square unit ($cm^{2)}$ of a body of unit thickness (cm). Together with the heat flow,* it defines the geothermal gradient.*

THERMAL CONDUCTIVITY DETECTOR (TCD): Instrument used to detect hydrocarbons as they emerge from a gas chromatograph.*

THERMAL CRACKING: The conversion of organic compounds to smaller molecules by purely thermal processes. In refineries, thermal cracking is used to convert heavy oils to lighter and more valuable products by means of heat alone (primarily by a free-radical process).

THERMOCLINE: See Pycnocline.*

THIN LAYER CHROMATOGRAPHY: Chromatography carried out on a thin layer of stationary phase spread out on a flat plate. The mixture to be separated is placed in a spot on one edge of the plate. The plate is then placed in a vertical position, with the spot at the bottom, in a shallow tray of solvent (the mobile phase). Upward movement of the solvent and mixture is by capillary action.

TIME–TEMPERATURE INDEX (TTI): Index of maturity calculated using Lopatin's method.*

TOC: Total organic carbon content, in weight percent

of dry rock or sediment.

TORBANITE: Synonym for boghead coal.*

TRACE ELEMENTS: Elements present in minor amount in the earth's crust, essentially all elements except the eight abundant rock-forming elements (O, Si, Al, Fe, Ca, Na, K, and Mg). In petroleum geochemistry, vanadium, nickel, and molybdenum are of special interest.

TRANS: Arrangement in which two groups attached to a molecule are on opposite sides. See also Cis.*

TRANSFORMATION OF OIL: Changes in the chemical composition of oil after expulsion from its source rock. These changes can occur by physical, thermal, and/or bacterial processes.

TRANSFORMATION RATIO: The ratio $S_1/(S_1 + S_2)$ derived from Rock-Eval* pyrolysis. High transformation ratios supposedly indicate either the occurrence of catagenesis* or contamination by migrated fluids or drilling additives. Low ratios indicate either immaturity or extreme overmaturity.

TRIS: Prefix meaning three.

TRITERPANES: Polycyclic (mostly pentacyclic) isoprenoid* hydrocarbons composed of three terpane* units. They are useful in maturity determinations and in interpretations of depositional environments as well as in oil typing and correlation.

TRITERPENOIDS: Biological precursors for triterpanes.*

TRUE LAYER DOM (TL DOM): See DOM.*

TTI: See Time–temperature index *

TTPF: Abbreviation for test tube pyrolysis-fluorescence. See Pyrolysis-fluorescence.*

ULMINITE: Partly or totally gelified plant cell walls in isolated occurrences of individual cells and cell tissues. Variable in size and shape. Cell lumens wholly or partly closed, cell walls swollen, shrinkage fissures typical, original cell walls no longer preserved. Yellow to reddish brown color in transmitted light, fluorescence weak.

ULTIMATE ANALYSIS: Elementary chemical analysis. In the case of coal,* the determination of carbon, hydrogen, oxygen, sulfur, nitrogen,, and ash.* See also Proximate analysis.*

UNSATURATED: A term applied to compounds in which two or more carbon atoms are held together by double bonds (i.e., aromatics* and olefins*).

UPWELLING: The vertical movement of subsurface marine or lake waters to the surface. Upwelling occurs where surface waters are moved aside by water and wind currents. Nutrients are brought into the photic zone by upwelling, and photosynthetic productivity is greatly enhanced.

UREA ADDUCTION: A method for separating *n*-alkanes from branched and cyclic hydrocarbons, similar in principle to molecular sieving.*

UV: Abbreviation for ultraviolet.

VAN KREVELEN DIAGRAM: A diagram developed by the coal scientist Van Krevelen in which the atomic H/C ratio of a coal is plotted against its atomic O/C ratio. The diagram distinguishes coals according to the combined effects of type of organic matter and rank (maturity*). Van Krevelen diagrams have also been adopted in modified form (hydrogen index* versus oxygen index) for interpreting pyrolysis data of kerogens.*

VISCOSITY INDEX (VI): A series of number ranging from 0 to 100 that indicate the rate of change of viscosity with temperature. A VI of 100 indicates a small change in viscosity between temperatures of 38° and 99°C (100° and 210°F), whereas a VI of 0 indicates a large change.

VITRAIN: The lithotype* of humic coal* designating the macroscopically recognizable, very bright bands or lenses, usually several millimeters thick.

VITRINITE: (1) Telecollinite, or normal vitrinite. The group of nonfluorescent coal macerals* derived from woody and cortical tissue, comprising collinite* and telinite* (and in low rank coals also gelinite* and detrinite*). For the petroleum geologist, this is the most important maceral group because its chemical analysis and reflectivity measurements are the best scales for the determination of coal rank,* DOM,* or maturity.* (2) Desmocollinite, or a hydrogen-rich variety of vitrinite, assumed to be a prescursor of gas and oil. Shows weak fluorescence. The hydrogen-rich character is caused by finely distributed, submicroscopic liptinites from synsedimentary bacterial transformation of organic matter. Should not be used for vitrinite reflectance* measurements.

VITRINITE: Synonym for microvitrain, the microlithotype* of humic coal* consisting principally (at least

95%) of vitrinite.* It is the most abundant constituent of humic coal. See also Coal structure.* In petroleum geology, it is of particular importance because the best scale for coal rank,* maturity,* or DOM* is based on it.

VITRINITE REFLECTANCE (VR): Measurement of the amount of light that is reflected by vitrinite (definition 1) used as maturity scale. See also R_a* and R_o.*

VOLATILE MATTER: Those products, exclusive of moisture, given off by a coal as gas and vapor, according to ASTM method D121-30.

VOLATILES: A colloquial term used to indicate the compounds boiling up to about 200°–250°C (392°–482°F), which occur in crude oils or in sediments.

VR/E: Equivalent vitrinite reflectance is the maturity measured by means other than vitrinite reflectance but expressed in vitrinite reflectance units.

WATER WASHING: The removal of the more soluble components (light hydrocarbons and aromatics*) of oil by dissolution in waters that are in contact with the oil. Water washing often occurs in conjunction with biodegradation,* and their effects are not easily distinguishable.

WAX: Plant or animal wax; a mixture of oxygen-containing molecules (esters) derived from long-chain fatty acids and alcohols with an even number of carbon atoms. See also Lipids* and Paraffin wax.*

WET GAS (HYDROCARBONS): Natural gas* that contains condensible (liquid) hydrocarbon constituents; also, in cuttings gas analysis, this term is used to describe gases that contain greater than about 5% ethane and heavier hydrocarbons; above 25% they are considered to be significantly wet and to indicate that any contiguous reservoirs should contain oil rather than gas. The natural gas liquid vapors amount to 4 L or more per 100 m^3 (≥0.3 gal per 1000 ft^3).

WURTZILITE: A massive, black or light brown elastic asphaltite* resisting the usual organic solvents, having brilliant luster, conchoidal fracture, 1.05–1.07 specific gravity, 2–25% fixed carbon, and light brown streak; is infusible and decomposes before melting.

ZOOPLANKTON: Tiny unicellular animals that feed on phytoplankton for their source of energy. They are not photosynthetic organisms.

References Cited

Abrahão, D., 1989, Well-log evaluation of lacustrine source rocks of the Lagoa Feia Formation, Lower Cretaceous, Campos Basin, offshore Brazil: SPWLA Twenty-sixth Annual Logging Symposium, Paper I.

Abrams, M. J., J. E. Conel, H. R. Lang, and H. N. Paley, 1985, The joint NASA\Geosat test case project: Final report: AAPG, Part 2, v. 2. p. 11-112.

Alexander, R., M. G. Strachan, R. I. Kagi, and W. van Bronswijk, 1986, Heating rate effects on aromatic maturity indicators: Organic Geochemistry, v. 10. p. 997-1003.

American Society for Testing and Materials (ASTM), 1976, Annual Book ASTM Standards: p. d 2798-72.

Anders, D. E., 1988, Geochemical correlations in petroleum systems, in L. B. Magoon, ed., Petroleum Systems of the United States: U.S. Geological Survey Bulletin 1870. p. 60-64.

Anders, D. E, and P. M. Gerrild, 1984, Hydrocarbon generation in lacustrine rocks of Tertiary age, Uinta Basin, Utah—Organic carbon, pyrolysis yield, and light hydrocarbons, in J. Woodward, F. R. Meissner, and J. L. Clayton, eds., Hydrocarbon source rocks of the greater Rocky Mountain region: Rocky Mountain Association of Geologists, p. 513-529.

Anders, D. E., and L. B. Magoon, 1986, Oil-source correlation study in northeastern Alaska, in J. Rullkötter, and D. Leythaeuser, eds., Advances in organic geochemistry 1985: v. 10, Pergamon Press, Oxford, p. 407-415.

Anders, D. E., L. B. Magoon, and C. M. Lubeck, 1987, Chapter 12. Geochemistry of surface oil shows and potential source rocks, in K. J. Bird, and L. B. Magoon, eds., Petroleum geology of the northern part of the Arctic National Wildlife Refuge, northeast Alaska: U.S. Geological Survey Bulletin 1778, p. 181-198.

Antonov, P. L., R. V. Bartachevich, L. M. Zorkin, G. A. Mogilevskiy, N. I. Macichenko, N. V. Prochneva, and V. A. Stroganov, 1971, The current and future state of developing geochemical methods for exploring for oil and gas, in The results of research study with direct geochemical prospecting for oil and gas: Ministry of geology of USSR, Niedra, Moscow, p. 3-16.

Antonov, P. L., R. V. Bartachevich, and L. M. Zorkin, 1978, Subsidence of landforms in Archean time: Noviska, v. 33, p. 100-108.

Arnosti C., and P. J. Muller, 1987, Pyrolysis-GC characterization of whole rock and kerogen-concentrate samples of immature Jurassic source rocks from NW-Germany: Organic Geochemistry, v. 6, p. 505-512.

Augustson, J. H., in press, A study of filling and drainage of some Barents Sea oil reservoirs—An evaluation based upon contents of heavy oil in core material.

Autric, A., and P. Dumesnil, 1985, Resistivity, radioactivity, and sonic transit time logs to evaluate the organic content of low permeability rocks: Log Analyst, May-June, p. 36-45.

Badley, M. E., T. Egeberg, and O. Nipen, 1984, Development of rift basins illustrated by the structural evolution of the Oseberg feature, Block 30/6, offshore Norway: Journal of the Geological Society of London, v. 141, p. 639-649.

Bailey, N. J. L., A. M. Jobson, and M. A.Rogers, 1973a, Bacterial degradation of crude oil: Comparison of field and experimental data. Chemical Geology, v. 11, 203-221.

Bailey, N. J. L., H. R. Krouse, C. R. Evans, and M. A. Rogers, 1973b, Alteration of crude oil by waters and bacteria—Evidence from geochemical and isotope studies: American Association Petroleum Geologists Bulletin, v. 57, p. 1276-1290.

Bailey, N. J. L., C. R. Evans, and C. W. D. Milner, 1974, Applying petroleum geochemistry to search for oil: examples from Western Canada Basin: American Association of Petroleum Geologists Bulletin, v. 58, p. 2284-2294.

Baker, E. W., and J. W. Louda, 1980, Organic geochemistry: highlights on the Deep Sea Drilling

Project, in A. G. Douglas, and J. R. Maxwell, eds., Advances in organic geochemistry 1979: Pergamon Press, Oxford, p. 295-319.

Barker, C., 1975, Oil source rock correlation aids drilling site selection: World Oil, Oct. 1975, p. 121-126.

Barker, C., 1980, Organic geochemistry in petroleum exploration: American Association of Petroleum Geologists Continuing Education Course Note Series #10, 159 p.

Barnard, P. C. and B. S. Cooper, 1981, Oils and Source Rocks of the North Sea Area, in C.V. Illing, and G.D. Hobson, eds., Petroleum Geology of the Continental Shelf North West Europe: Institute of Petroleum, London, p. 169-175.

Barnard, P. C., A. G. Collins, and B. S. Cooper, 1981, Identification and distribution of kerogen facies in a source rock horizon—Examples from the North Sea, in J. Brooks, ed., Organic Maturation Studies and Fossil Fuel Exploration: Academic Press, New York, pp. 271-282.

Barwise, A. J., and P. J. D. Park, 1983, Petroporphyrin fingerprinting as a geochemical marker, in Bjoröy, M. and others, eds., Advances in Organic Geochemistry, 1981: John Wiley, New York, p. 667-674.

Barwise, A. J., and I. Roberts, 1984, Diagenetic and catagenetic pathways for porphyrins in sediments: Organic Geochemistry, v. 6, p. 167-176.

Baskin, D. K., in press, Quantitative Estimates of Organic Matter Conversion and Expulsion from Source Rocks. American Association of Petroleum Geologists Bulletin.

Baskin, D. K., in press, Early Generation Characteristics of a Sulfur-Rich Monterey Kerogen: American Association of Petroleum Geologists Bulletin.

Bayliss, G. S., and L. B. Magoon, 1988, Organic facies and thermal maturity of sedimentary rocks in the National Petroleum Reserve in Alaska, in G. Grye, ed., Geology and exploration of the National Petroleum Reserve of Alaska 1974-1982: U.S. Geological Survey Professional Paper 1399, p. 489-518.

Berger, Z., 1985, Analysis of buried fault systems and related hydrocarbon plays using remote sensing data, in Proceedings of the International Symposium of Remote Sensing: Fourth Thematic Mapper Conference, San Francisco, p. 7.

Berner, R. A., and R. Raiswell, 1984, C/S method for distinguishing freshwater from marine sedimentary rocks: Geology, v.12, p.365-368.

Bethke, C. M., 1985, A numerical model of compaction driven flow and heat transfer and its application to the paleohydrology of intracratonic sedimentary basins: Journal of Geophysical Research, v. 90, p. 6817-6928.

Bethke, C. M., 1986, Inverse hydrologic analysis of the distribution and origin of Gulf Coast-type geopressured zones: Journal of Geophysical Research, v. 91, n. B6, p. 6535-6545.

Bjorøy, M., J. A. Williams, D. L. Dolcater, and J. C. Winters, 1988, Variation in hydrocarbon distribution in artificially matured oils, in L. Mattavelli, and L. Novelli, eds., Advances in Organic Geochemistry 1987: Organic Geochemistry, v. 13, p. 901-913.

Bonilla, J. V., 1986, Laboratory simulation of crude oil migration: Possible implications for oil exploration. Ph.D. thesis, Univ. of Oklahoma.

Bonilla, J. V., and M. H. Engel, 1986, Chemical and isotopic redistribution of hydrocarbons during migration: Laboratory simulation experiments, in D. Leythaeuser and J. Rullkîtter, eds., Advances in Organic Geochemistry 1985,: Pergamon Journals, Oxford. p. 181 190.

Bostick, N. H, 1979, Microscopic measurement of the level of catagenesis of solid organic matter in sedimentary rocks to aid exploration for petroleum and to determine former burial temperatures-a review: Society Economic Paleontologists and Mineralogists Spec. Pub. v. 26, p. 17-23.

Braun, R. L., and A. K. Burnham, 1987, Analysis of chemical reaction kinetics using a distribution of activation energies and simpler models: Energy and Fuels, v. 1, p. 153- 161.

Braun, R. L., and A. K. Burnham, 1989, A mathematical model of oil generation, degradation and expulsion: Energy and Fuels, v. 4, p. 132-146.

Bray, E. E., and E. D. Evans, 1965, Hydrocarbons in nonreservoir-rock source beds: Part 1: American Association of Petroleum Geologists Bulletin, v. 49, p. 248-257.

Brooks, J., and D. H. Welte, 1984, eds., Advances in Petroleum Geochemistry: Academic Press, London, v. 1.

Brooks, J., M. C. Kennicutt, and B. D. Grey, 1986, Offshore surface geochemical exploration: Oil and Gas Journal, v.84, n. 42, p. 66-72.

Brooks, J., C. Cornford, and R. Archer, 1987, The role of hydrocarbon source rocks in petroleum exploration, in J. Brooks, and A. J. Fleet, eds., Marine petroleum source rocks: Blackwell, Boston, p. 17-46.

Burkova, U. N., O. V. Ryadova, O. W. Serebrennikova, and V. I. Titov, 1980, Geoporphyrin composition as an indication of organic matter transformation: Geokhimiia, v. 9, p. 1417-1421.

Burnham, A. K., 1989, On the validity of the pristane formation index: Geochimica Cosmochimica Acta, v. 53, p. 1693-1697.

Burnham, A. K., and R. L. Braun, 1985, General kinetic model of oil shale pyrolysis: In Situ, v. 9, p. 1-23.

Burnham, A. K., and R. L. Braun, 1990, Development of a detailed kinetic and thermodynamic model of petroleum formation, destruction, and expulsion from lacustrine and marine source rocks: Advances in Organic Geochemistry 1989: Organic Geochemistry, v. 16, p. 27-40.

Burnham, A. K., R. L. Braun, H. R. Gregg, and A. M. Samoun, 1987, Comparison of methods for measuring kerogen pyrolysis rates and fitting kinetic parameters: Energy and Fuels, v. 1, p. 452-458.

Burnham, A. K., R. L. Braun, and A. Samoun, 1988, Further comparisons of methods for measuring kerogen pyrolysis rates and fitting kinetic parameters, in L. Mattavelli, and L. Novelli, eds., Advances in Organic Geochemistry 1987: Organic Geochemistry, v. 13, p. 839-845.

Burnham, A. K., M. S. Oh, R. W. Crawford, and A. M. Samoun, 1989, Pyrolysis of the Argonne premium coals—Activation energy distributions and related chemistry: Energy and Fuels, v. 3, p. 42-55.

Butenko, G and S.R. Østmo, in press, The Cenozoic along a profile from Oseberg to Hjartöy, 60°30′ N Northern Norwegian North Sea Basin.

Cain, R. B., 1980, Transformations of aromatic hydrocarbons, in Hydrocarbons in Biotechnology: Heyden and Son Ltd., London, p. 99-132.

Carlson, R. M. K., and D. E. Chamberlain, 1986, Steroid biomarker-clay mineral adsorption free energies: Implications to petroleum migration indices, in D. Leythaeuser, and J. Rullkötter, eds., Advances in Organic Geochemistry 1985: p. 163-180.

Carpentier, B., G. Bessereau, and A. Y. Huc, 1989, Diagraphies et roches-meres estimation des teneurs en carbone organique par la methode carbolog: Revue de l'Institut Français du Pétrole, v. 46, 699-719.

Castano, J. R., and D. M. Sparks, 1974, Interpretation of vitrinite reflectance measurements in sedimentary rocks and determination of burial history using vitrinite reflectance and authigenic minerals: Geological Society America Special Paper 153, p. 31-52.

Claypool, G. E., and I. R. Kaplan, 1974, The origin and distribution of methane in marine sediments, in I. R. Kaplan, ed., Natural Gases in Marine Sediments: Plenum Press, New York, p. 99-139.

Claypool, G. E., C. M. Lubeck, J. P. Baysinger, and T. G. Ging, 1977, Organic Geochemistry, in P. A. Scholle, ed., Geological Studies on the COST No B-2 well, U.S. Mid-Atlantic Outer Continental Shelf Area: U.S. Geological Survey Circular 750, p. 46-59.

Clayton, C. J., in press, Carbon isotope fractionation during natural gas generation from kerogen: Marine Petroleum Geology.

Clayton, J. L., 1988, Role of amount and type of organic matter in recognition of petroleum source rocks, in L. Magoon, ed. Petroleum systems of the United States: U.S. Geological Survey Bulletin 1870, p. 16-19.

Clayton, J. L., 1990, Role of kinetic modeling in petroleum exploration, in L. B. Magoon, ed., Petroleum Systems of the United States: U.S. Geological Survey Bulletin 1912, p. 43-45

Colombo, U., F. Gazzarrini, R. Gonfiantini, G. Sironi, and E. Tongiorgi, 1966, Measurements of 13C/12C-isotope ratios on Italian natural gases and their geochemical interpretations, in G. D. Hobson and M. C. Louis, eds., Advances in Organic Geochemistry 1964: Pergamon Press, Oxford, p. 279-292.

Connan, J., 1981, Un example de biodegradation preferentielle des hydrocarbonures aromatiques dans des asphaltes du bassin Sud-Aquitain (France): Bull. Centre Rech. Explor. Prod. Elf-Aquitaine, v. 5, n. 1, p. 151-171.

Connan, J., 1984, Biodegradation of crude oils in reservoirs, in J. Brooks and D. Welte, eds., Advances in Petroleum Geochemistry, v. 1: Academic Press, London. p. 300-330.

Connan, J., K. LeTran, and B. van der Weide, 1975,

Alteration of petroleum in reservoirs: 9th World Petroleum Congress, Tokyo, Proceedings, v. 2, p. 171-178.

Connan, J., A. Restle, and P. Albrecht, 1980, Biodegradation of crude oil in the Aquitaine basin, in A. A. Douglas and J. R. Maxwell, eds., Advances in Organic Geochemistry, 1979: Pergamon Press, Oxford, p. 1-17.

Cooles, G. P., A. S. Mackenzie, and T. M. Quigley, 1986, Calculation of petroleum masses generated and expelled from source rocks, in D. Leythaeuser and J. Rullkötter, eds., Advances in Organic Geochemistry 1985: Organic Geochemistry, v. 10, p. 235-245.

Cooper, B. S., and P. C. Barnard, 1984, Source Rocks and Oils from of the Central and Northern North Sea, in Demaison, G. and Murris, R. J., eds., Petroleum Geochemistry and Basin Evolution: American Association Petroleum Geologists Memoir 35, p. 303-314.

Cooper, B. S., S. H. Coleman, P. C. Barnard, and J. S. Butterworth, 1975, Paleotemperatures in the northern North Sea basin, in A. Woodland, ed., Petroleum and the Continental Shelf of Northwest Europe: Applied Science, v. 1, p. 487-492.

Cornford, C., 1986, Source Rocks and Hydrocarbons of the North Sea, in Introduction to the Petroleum Geology of the North Sea: Blackwell Scientific Publications, London, p. 197-236.

Cornford, C., J. A. Morrow, A. Turrington, J. A. Miles, and J. Brooks, 1983, Some geological controls on oil composition in the U.K. North Sea, in J. Brooks, ed., Petroleum Geochemistry and Exploration of Europe: Blackwell Scientific Publications, p. 175-194.

Cripps, R. E., and R. J. Watkinson, 1978, Polycyclic aromatic hydrocarbons: Metabolism aspects, in Developments in Biodegradation of Hydrocarbons, v. 1: Applied Science Publishers, London.

Crossey, L. J., E. S. Hagan, R. C. Surdam, and P. W. Lapointe, 1986, Correlation of organic parameters derived from elemental analysis and programmed pyrolysis of kerogen: Society of Economic Paleontologists and Mineralogists, p. 36-45

Dahl, B., 1987, Petroleum geochemistry and basin modelling, Oseberg Area, North Sea: PhD dissertation, Norwegian Institute of Technology, University of Trondheim, Norway.

Dahl, B., and G. C. Speers, 1985, Organic geochemistry of the Oseberg Field (I), in Thomas et al., eds., Petroleum Geochemistry in exploration of the Norwegian Shelf: Norwegian Petroleum Society, p. 185-195.

Dahl, B., and G. C. Speers, 1986, Geochemical charaterization of a tar mat in the Oseberg Field, Norwegian Sector, North Sea: Organic Geochemistry, v. 10, p. 547-558.

Dahl, B., E. Nysether, G. L. Speers, and M. A. Yükler, 1987, Oseberg area—Integrated basin modelling, in J. Brooks and K. Glennie, eds., Petroleum Geology of Northwest Europe: Graham and Trotman, p. 1029-1038.

Dahlberg, E. C., 1982, Applied Hydrodynamic in Petroleum Exploration: Springer-Verlag, New York

Daly, A. R., and J. D. Edman, 1987, Loss of organic carbon from source rocks during thermal maturation: American Association Petroleum Geologists Bulletin, v. 71, p. 546.

Dalziel, M. C., and T. J. Donovan, 1980, Biogeochemical evidence for subsurface hydrocarbon occurrence, Recluse oil field, Wyoming: Preliminary results: U.S. Geological Circular 837, 11p.

Davis, A., 1978, The Measurement of reflectance of coal macerals: Its automation and significance: Technical Report 10, Pennsylvania State University.

Davis, J. B., 1967, Petroleum Microbiology: Elsevier Press, Amsterdam, p. 604.

Davis, J. C., 1973, Statistics and Data Analysis in Geology: John Wiley, New York, p. 550.

Davis, R. W., 1987, Analysis of hydrodynamic factors in Petroleum Migration and Entrapment: American Association Petroleum Geologists Bulletin, v. 71, 643-649.

Deines, P., 1980, The isotopic composition of reduced organic carbon, in P. Fritz and C. H. Fontes, eds., Handbook of environmental isotope geochemistry: Elsevier, New York, v. 1, p. 329-406.

Dellenbach, J., J. Espitalié, and F. Lebreton, 1983, Source rock logging: Eighth European Formation Evaluation Symposium Transactions, SPWLA, London.

Demaison, G. J., and B. J. Huizinga, in press, Genetic classification of petroleum systems: American Association Petroleum Geologists Bulletin.

Demaison, G. J., and G. T. Moore, 1980, Anoxic environments and oil source bed genesis: American Association of Petroleum Geologists Bulletin, v. 64, p. 1179-1209.

Demaison, G. J., A. J. J. Holck, R. W. Jones, and G. T. Moore, 1984, Predictive source bed stratigraphy: A guide to regional petroleum occurrence, North Sea Basin and Eastern North American Continental Margin, Proceedings of the Eleventh World Petroleum Congress: John Wiley Ltd., London, p. 17-29.

Dembicki, H., Jr., and M. J. Anderson, 1989, Secondary migration of oil: Experiments supporting efficient movement of separate, buoyant oil phase along limited conduits: American Association Petroleum Geologists Bulletin, v. 73, p. 1018-1021.

Dembicki, H., B. Horsfield, and T. Y. Ho, 1983, Source rock evaluation by pyrolysis-gas chromatography: American Association of Petroleum Geologists Bulletin, v. 67, p. 1094-1103.

Didykeee, B. M., Y. A. Alturki, C. T. Pillinger, and G. Eglinton, 1975, Petroporphyrins as indicators of geothermal maturation: Nature, v. 256, p. 563-565.

Diecchio, R. J., 1985, Regional controls of gas accumulation in Oriskany Sandstone, Central Appalachian Basin: American Association Petroleum Geologists Bulletin, v. 69, p. 722-732.

Dolly, E. D., 1979, Geologic Techniques used in the discovery of Trap Springs field, in R. W. Newman and H. D. Goode, eds., 1979 Basin and Range Symposium: Rocky Mountain Association of Geologists, Denver, p. 455-467.

Dominé, F., 1989, Kinetics of hexane pyrolysis at very high pressures. 1. Experimental study: Energy and Fuels, v. 3, p. 69-96.

Donovan, T. J., 1974, Petroleum microseepage at Cement Field, Oklahoma. American Association Petroleum Geologists Bulletin, v. 58, p. 429-446.

Dore, A. G., J. Vollset, and G. P. Hamar, 1985, Correlation of the offshore sequences referred to the Kimmeridge Clay Formation—Relevance to the Norwegian sector, in, Thomas et al., eds., Petroleum Geochemistry in Exploration of the Norwegian Shelf: Norwegian Petroleum Society (Graham and Trotman), p. 27-37.

Dow, W. G., 1977, Kerogen studies and geological interpretations: Journal of Geochemical Exploration, v. 7, p. 79-99.

Dow, W. G., 1978, Petroleum source-beds on continental slopes and rises: American Association of Petroleum Geologists Bulletin, v. 62, p. 1584-1606.

Dow, W. G. and D.I. O'Connor, 1982, Kerogen maturity and type by reflected light microscopy, in How to Assess Maturation and Paleotemperatures: Society Economic Paleontologists and Mineralogists Short Course Number 7, p 133-158.

Duchscherer, W., 1980, Geochemical methods of prospecting for hydrocarbons: Oil and Gas Journal, December 1, p. 194-208.

Durand, B., 1980, Sedimentary organic matter and kerogen. Definition and quantitative importance of kerogen, in B. Durand, ed., Kerogen: Insoluble Organic Matter from Sedimentary Rock: Editions Technip, Paris, p. 13-34. Reproduced by permission in Geochemistry, Treatise of Petroleum Geology Reprint Series, No. 8, compiled by E. A. Beaumont and N. H. Foster, 1988, p. 81-102.

Durand, B., 1983, Present trends in organic geochemistry in research on migration of hydrocarbons, in M. Bjoröy et al., eds., Advances in Organic Geochemistry, 1981: John Wiley, Chichester, p. 117-128.

Durand, B., and J. L. Oudin, 1979, Exemple de migration des hydrocarbures dans use Sirie deltaoque: Le delta de la Mahakam, Indonisie: 10th World Petroleum Congress Proceedings. V. 2. p. 3-11.

Durand, B., and N. Paratte, 1983, Oil potential of coals: A geochemical approach, in J. Brooks, ed., Petroleum geochemistry and exploration of Europe: Blackwell, Boston, p. 255-265.

Dutta, N. C., 1988, Fluid flow in low permeable, porous media: Rev. de L'Institut Français du Pétrole, v. 43, n. 2, 165-179.

Eganhouse, R. P., and J. A. Calder, 1976, The solubility of medium molecular weight and aromatic hydrocarbons and the effects of hydrocarbon co-solutes and salinity: Geochimica et Cosmochimica Acta, v. 40, p. 556-561.

England, W. A., 1989, The organic geochemistry of petroleum reservoirs: 14th International Conference on Organic Geochemistry, Paris.

England, W. A., and A. S. Mackenzie, 1989, The

geochemistry of petroleum. Geologische Rundschau—Special Volume: Evolution of Sedimentary Basins, v. 78, n. l.

England, W. A., and I. Price, 1988, Geochemistry and condensate reservoirs, presented at The Development of Condensate Fields: IBC Technical Services, London.

England, W. A., A. S. Mackenzie, D. M. Mann, and T. M Quigley, 1987, The movement and entrapment of petroleum fluids in the subsurface: Journal of the Geological Society, v. 144, p. 327-347.

Ensminger, A., A. van Dorsselaer, C. Spyckerelle, P. Albrecht, and G. Ourisson, 1974, Pentacyclic triterpanes of the hopane type as ubiquitous geochemical markers: Origin and significance, in B. Tissot, and F. Bienner, eds., Advances in Organic Geochemistry, 1973: Pergamon Press, Oxford, p. 245-260.

Epstein, A. G., J. B. Epstein, and L. D.Harris, 1977, Conodont color alteration—An index to organic metamorphism: U.S. Geological Survey Professional Paper 995, 27 p.

Erdman, J. G., and D. A. Morris, 1974, Geochemical correlation of petroleum: American Association of Petroleum Geologists Bulletin, v. 58, p. 2326-2337.

Espitalié J., M. Madec, and B. Tissot, 1977a, Source rock characterization method for petroleum exploration. 9th Annual Offshore Technology Conference, Paper 2935.

Espitalié, J., J. L. Laporte, M. Madec, F. Marquis, P. Leplat, J. Paulet, and A. Boutefeu, 1977b, Méthode rapide de caractérisation des roches méres, de leur potentiel pétrolier et de leur degré d'évolution: Revue de l'Institut Français du Pétrole, v. 32. p. 23-42.

Espitalié, J., M. Madec, and B. Tissot, 1980, Role of mineral matrix in kerogen pyrolysis: Influence on petroleum generation and migration: American Association of Petroleum Geologists Bulletin, v. 64, p. 59-66.

Espitalié, J., F. Marquis, and I. Barsony, 1984, Geochemical logging, in K. Voorhees, ed., Analytical Pyrolysis Techniques and Applications: Butterworths, London, p. 276-304.

Espitalié, J., G. Deroo, and F. Marquis, 1985, Rock-Eval pyrolysis and its applications (preprint): Revue de l'Institut Français du Pétrole, Rueil, France.

Espitalié, J. P., P. Ungerer, I. Irwin, and F. Marquis, 1988, Primary cracking of kerogens. Experimenting and modeling C1, C2-C5, C6-C15, and C15+ classes of hydrocarbons formed, in L. Mattavelli, and L. Novelli, eds., Advances in Organic Geochemistry, 1987: Organic Geochemistry, v. 13, p. 893-899.

Evans, C. R., and Staplin, F. R., 1971, Regional facies metamorphism in geochemical exploration, in 3rd International Geochemical Exploration Symposium Proceedings: Canadian Institute of Mining and Metallurgy, p. 517-520.

Fertl, W. H.,and G. V. Chilingar, 1988, Total organic carbon content determined from well logs: Society of Petroleum Engineers, SPE Formation Evaluation, p.407-419.

Field, J. D. ,1985, Organic geochemistry in exploration of the Northern North Sea, in Thomas et al., eds., Petroleum Geochemistry in Exploration of the Norwegian Shelf: Norwegian Petroleum Society (Graham and Trotman), p. 39-57.

Foster, N. H., 1979, Geomorphic exploration utilized in the discovery of Trap Springs field, Nye County, Nevada, in R. W. Newman and H. D. Goode, eds., 1979 Basin and Range Symposium: Rocky Mountain Association of Geologists, Denver, p. 477-486.

Fuex, A. N., 1977, The use of stable carbon isotopes in hydrocarbon exploration: Journal of Geochemical Exploration, v. 7, p. 155-188.

Fuex, A. N., 1980, Experimental evidence against an appreciable isotopic fractionation of methane during migration, in A. G. Douglas and J. R. Maxwell, eds., Advances in Organic Geochemistry: Pergamon Press, Oxford, p. 725-732.

Furguson, J. D., 1975, Surface alteration and mineralization of Permian redbeds overlying several oil fields in southern Oklahoma: MS thesis, Oklahoma State University, 95p.

Galimov, E. ,1973, Carbon Isotopes in Oil and Gas Geology: National Aeronautics and Space Administration, Washington, D.C. (Translation, 1974, from Russian)

Galimov, E., 1974, Organic geochemistry of carbon isotopes, in B. P. Tissot, and R. Bienner, eds., Advances in Organic Geochemistry 1973: Editions Technip, Paris. p. 439-452.

Galimov, E., 1980, 13C/12C in kerogen, in. B. Durand, ed., Kerogen: Editions Technip, Paris, p. 271-299.

Gehman, H. M., 1962, Organic matter in limestones: Geochimica et Cosmochimica Acta, v. 26, p. 885-894.

Gibson, D. T. ,1976, Microbial metabolism of polycyclic aromatic hydrocarbons: American Chemical Society, Division of Petroleum Chemistry Preprints, p. 409.

Glasø, O., 1980, Generalized pressure-volume temperature correlations. Journal Petroleum Technology, v. 32, p. 785-795.

Glennie, K. W. ,1984, The structural framework and the pre-Permian history of the North Sea area, in K. W. Glennie, ed., Introduction to the Petroleum Geology of the North Sea: Blackwell, London, p. 1739.

Goff, J. C. ,1983, Hydrocarbon generation and migration from Jurassic rocks in the East Shetland Basin and Viking Graben of the northern North Sea: Journal Geological Society of London, v. 140, p. 445-474.

Goodwin, N. S., P. J. D. Park, and A. P. Rawlinson, 1983, Crude oil biodegradation under simulated and natural conditions, in M. Bjoröy, et al., eds, Advances in Organic Geochemistry 1981: John Wiley, Chichester, p. 650-658.

Goossens, H., A. Due, J. W. de Leeuw, B. Van de Graff, and P. A. Schenck, 1988, The Pristane Formation Index, a new molecular maturity parameter: A simple method to assess maturity by pyrolysis/evaporation-gas chromatography of unextracted samples: Geochimica et Cosmochimica Acta, v. 52, p. 1189-1193.

Gransch, J. A., and E. Eisman, 1970, Geochemical aspects of the occurrence of porphyrins in West Venezuelan mineral oils and rocks, in G. D. Hobson and G. C. Speers, eds., Advances in organic geochemistry 1966: Pergamon Press, Oxford, p. 69-86.

Haq, B. U., J. Hardenbohl, and P. K. Vail, 1988, Mesozoic and Cenozoic chronostratigraphy and cycles of sea-level Change, in C. K. Wilgus et al., eds., Sea-Level Changes: An Integrated Approach, SEPM Special Publication 42, p. 71-108

Hatch, J. R., S. R. Jacobson, B. J. Witzke, J. B. Risatti, D. E. Anders, W. L. Watney, K. D. Newell, and A. K. Vuletich, 1987, Possible late Middle Ordovician carbon isotope excursion: Evidence from Ordovician oils and hydrocarbon source rocks, mid-continent and east-central United States: American Association Petroleum Geologists Bulletin, v. 71, p. 1342-1354.

Heacock, R. L., and A. Hood, 1970, Process for measuring the live carbon content of organic samples: U.S. Patent 3508877, April 28, 1970.

Henderson, J. H., and L. Weber, 1965, Physical upgrading of heavy crude oils by the application of heat: Journal of Canadian Petroleum Technology, v. 4, p. 206-212.

Héritier, F. E., 1979, Frigg field: Large submarine fan trap in lower Eocene rocks of North Sea Viking Graben: American Association of Petroleum Geologists Bulletin, v. 63, p. 1999-2020.

Heroux, Y., A. Chagnon, and R. Bertrand, 1979, Compilation and correlation of major thermal maturation indicators: American Association Petroleum Geologists Bulletin, v. 63, p. 2128-2144.

Herron, M. M., and S. L. Herron, 1990, Geological applications of geochemical well logging, in A. Hurst, M.A. Lovell, and A.C. Morton, eds., Geological applications of wireline logs, Geological Society Special Publication No. 48: The Geological Society, London, p. 165-175.

Herron, S. L., 1986, A total organic carbon log for source rock evaluation: Transactions of Twenty-Seventh Annual Society of Professional Well Log Analysts Annual Meeting, Houston, Paper HH: Log Analyst, v. 28, p. 520-527.

Herron, S. L., and L. Le Tendre, 1990, Wireline source rock evaluation in the Paris Basin, in A. Y. Huc, ed., Deposition of Organic Facies, AAPG Studies in Geology #30, American Association Petroleum Geologists, p. 57-71.

Hertzog, R, L., L. Colson, B. Seeman, M. O'Brien, H. Scott, D. McKeon, P. Wraight, J. Grau, D. Ellis, J. Schweitzer, and M. Herron, 1989, Geochemical logging with spectrometry tools: Society of Petroleum Engineers, SPE Formation Evaluation, v.4, p 153-162.

Heum, O. R., A. Dalland, and K. K. Meisingset, 1986, Habitat of hydrocarbons at Haltenbanken, PVT modelling as a predictive tool in hydrocarbon exploration, in A. M. Spencer, ed., Habitat of the Hydrocarbons of the Norwegian Shelf: Graham and Trotman, p. 259-274.

Higgins, I. J., and P. D. Gilbert, 1978, The biodegradation of hydrocarbons, in K. W. A. Chater, and H. J. Somerville, eds., The Oil Industry and Microbial Ecosystems: Heyden and Son, London, p. 80-117.

Hirschberg, A., 1988, Role of asphaltenes in compositional grading of a reservoir's fluid column: Journal of Petroleum Technology, v. 89.

Hite, R. J., and D. E. Anders, 1991, Petroleum and evaporites, in J. L. Melvin, ed., Development of Sedimentology Series: Elsevier Science Publishers, Amsterdam, B.V.

Hitzman, D. O., 1961, Comparison of geomicrobiological prospecting methods used by various investigators: Developments in Industrial Microbiology, v. 2, p. 33-43.

Ho, T. T. Y., 1977, Geological and geochemical factors controlling electron spin resonance signals in kerogen: paper presented at ASCOPE/CCOP Seminar on Generation and Maturation of Hydrocarbons in Sedimentary Basins, Manila.

Hoffman, C. F., A. S. Mackenzie, C. A. Lewis, J. R. Maxwell, and A. Hirschberg, 1988, Role of asphaltenes in compositional grading of a reservoir's fluid column: Journal of Petroleum Technology, v. 89.

Holt, T., E. Lindberg, and S. K. Ratkje, 1983, The effects of gravity and temperature gradients on methane distribution in oil reservoirs: Society of Petroleum Engineers, unsolicited paper 11761.

Horowitz, A., D. Gutnick, and E. Rosenberg, 1975, Sequential growth of bacteria on crude oil: Applied Microbiology, v. 30, p. 10-19.

Horsfield, B., and A. G. Douglas, 1980, The influence of minerals on the pyrolysis of kerogens: Geochimica et Cosmochimica Acta, v. 44, p. 1110-1131.

Horsfield, B., H. Dembicki, Jr., and T. T. Y. Ho, 1983, Some potential applications of pyrolysis to basin studies: Geological Society of London, v. 140, p. 431-443.

Horvitz, L., 1939, On geochemical prospecting: Geophysics, v.4, p. 210-225.

Horvitz, L., 1954, Near-surface hydrocarbons and petroleum accumulation at depth: Mining Engineering, p. 1205-1209.

Horvitz, L., 1972, Vegetation and geochemical prospecting for petroleum. American Association Petroleum Geologists Bulletin, v. 56, 925-940.

Horvitz, L., 1985a, Geochemical exploration for petroleum: Science, v. 229, 821-827.

Horvitz, L., 1985b, Near-surface hydrocarbons and nonhydrocarbon gases in petroleum exploration: Paper presented at, Association of Petroleum Geochemical Explorationists-AAPG Rocky Mountain Section, Denver, June 2, 1985, p. D1-D52.

Howell, V. J., J. Connan, and A. K. Aldridge, 1984, Tentative identification of demethylated tricyclic terpanes in nonbiodegraded and slightly biodegraded crude oils from the Los Llanos Basin, Columbia: Organic Geochemistry, v. 6, p. 83-92.

Hubbert, M. K., 1953, Entrapment of petroleum under hydrodynamic conditions: American Association Petroleum Geologists Bulletin, v. 37, 1954-2026.

Huizinga, B. J., Z. A. Aizenshtat, and K. E. Peters, 1988, Programmed pyrolysis-gas chromatography of artificially matured Green River shale kerogen: Energy and Fuels, v. 2, p. 74-81.

Hunt, J. M, 1975, Origin of gasoline range alkanes in the deep sea: Nature, v. 254, p. 411-413.

Hunt, J. M., 1979, Petroleum geochemistry and geology: W. H. Freeman, San Francisco, 617 p.

Hunt, J. M., 1990, Generation and migration of petroleum from abnormally pressured fluid compartments: American Association Petroleum Geologists Bulletin, v. 74, p. 1-12.

Hunt, J. M., and G. W. Jaieson, 1956, Oil and organic matter in source rocks of petroleum: American Association of Petroleum Geologists Bulletin, v. 40, p. 477-488.

Hunt, J. M., Miller, R. S., and Whelan, J. K., 1980, Formation of C4-C7 hydrocarbons from bacterial degradation of naturally occurring terpenoids: Nature, v. 288, 577-588.

Hutton, A. C., and A. C. Cook, 1980, Influence of alginite on the reflectance of vitrinite from Joadja, NSW, and some other coals and oil shales containing alginite: FUEL, v. 59, p. 711-714.

Hutton, A. C., A.J. Kantsler, A.J. Cook, and D.M. McKindy, 1980, Organic matter in oil shales: A.P.E.A., v. 20, n. 1, p. 44.

Itzan, H., 1985, Geochemical correlations between crude oils from the Misener formation and potential source rocks in central and north-central Oklahoma: MS thesis, University of Tulsa, Tulsa, Oklahoma.

Jacobson, S. R., J. R. Hatch, S. C. Teerman, and R. A.

Askin, 1988, Geologic note: Middle Ordovician organic matter assemblages and their eeffect on Ordovician-derived oils: American Association Petroleum Geologists Bulletin, v. 72, p. 1090-1100.

James, A. T., 1983, Correlation of natural gas using the carbon isotopic distribution between hydrocarbon components: American Association Petroleum Geologists Bulletin, v. 67, 1176-1191.

James, A. T., 1990, Correlation of reservoired gases using the carbon isotopic compositions of wet gas components: American Association Petroleum Geologists Bulletin, v. 74, p. 1441-1458.

Jarvie, D. M., in press, Variations in empirically derived Rock-Eval kinetic parameters: Chemical Geology.

Jobson, A. M., F. D. Cook, and D. W. S. Westlake,1979, Interaction of aerobic and anaerobic bacteria in petroleum biodegradation: Chemical Geology, v. 24, p. 355-365.

Jones, R. W., 1981, Some mass balance and geological constraints on migration mechanisms: American Association Petroleum Geologists Bulletin, v. 65, p. 102-122.

Jones, R. W., 1984, Comparison of carbonate and shale source rocks, in J. G. Palacas, ed., Petroleum Geochemistry and Source Rock Potential of Carbonate Rocks: American Association Petroleum Geologists Studies in Geology No. 18.

Jones, R. W., 1987, Organic Facies, in J. Brooks and D. H. Welte, eds., Advances in Petroleum Geochemistry, v. 2, Academic Press, London, p. 1-90.

Jones, R. W., and G. J. Demaison, 1982, in A. Saldiva-Sali, ed., Proceedings of the Second ASCOPE Conference and Exhibition, Manila: p. 51-68.

Jones, R. W., and T. Edison, 1979, Microscopic observations of kerogen related to geochemical parameters with emphasis on thermal maturation, in Olitz, D. F., ed., Low temperature metamorphism of kerogen and clay minerals: Society Economic Paleontologists and Mineralogists, Tulsa, Oklahoma, p. 1-12.

Jones, V. T., and R. J. Drozd, 1983, Predictions of oil and gas potential by near-surface geochemistry: American Association of Petroleum Geologists Bulletin, v. 67, p. 932-952.

Jones, V. T., and H. W. Thune, 1982, Surface detection of retort gases from an underground coal gasification reactor in steeply dipping beds near Rawlins, Wyoming: Society of Petroleum Engineers, SPE No. 11050, 24p.

Jones, V. T., S. G. Burtell, R. A. Hodgson, T. Whelan, C. Milan, T. Ando, K. Okada, T. Agtsuma, and O. Takono, 1985, Remote sensing and surface geochemical study of Railroad Valley, Nye County, Nevada: Paper presented at the Fourth Thematic Mapper Conference, Remote Sensing for Exploration Geology, San Francisco, California, April 1-4, 1985.

Juhasz, I, 1981, Normalized Qv—The key to shaly sand evaluation using the Waxman-Smits equation in the absence of core data: Transactions of the SPWLA Twenty-Second Annual Logging Symposium, 1981, Paper Z.

Kalkreuth, W. D, 1982, Rank and petrographic composition of selected Jurassic-Lower Cretaceous coals of British Columbia, Canada: Canadian Petroleum Geologists Bulletin, v. 30, p. 112-139.

Katz, B. J., 1983, Limitations of Rock Eval pyrolysis for typing organic matter: Organic Geochemistry, v. 4, p. 195-199.

Keith, L. A., and J. D. Rimstidt, 1985, A Numerical Compaction Model of Overpressuring in Shales: Mathematical Geology, v. 17, n. 2, p. 115-135.

King, J. D., and G. E. Claypool, 1983, Biological marker compounds and implications for generation and migration of petroleum in rocks of the Point Conception deep stratigraphic test well, OCS-GAL 78-164 No. 1 offshore California, in J. D. King, G. E. Claypool, C. M. Isaacs, and R. E. Garrison, eds., Petroleum Generation and Occurrence in the Miocene Monterey Formation, California: Society Economic Paleontologists and Mineralogists, Pacific Section, p. 191-200.

Klusman, R. W., and K. S. Voorhees, 1983, A new development in petroleum exploration technology: Mines Magazine, Colorado School of Mines, v. 73.

Koch, G. S., and R. F. Link, 1970, Statistical Analysis of Geological Data, v. 1 and 2: Dover Publications, New York, p. 438.

Koons, C. B., J. G. Bond, and F. L. Peirce, 1974, Effects of depositional environment and postdepositional history on chemical composition of Lower Tuscaloosa oils: American Association of Petroleum Geologists Bulletin, v. 58, p. 1271-1280.

Kroob, B., Leythaeuser, D., and Schaefer, R. G., 1986,

Experimental determinationof diffusion parameters for light hydrocarbons in water saturated rocks: Some selected results, in D. Leythaeuser and J. Rullkötter, eds., Advances in Organic Geochemistry 1985: Pergamon Journals Ltd., Germany, p. 291-298.

Lafargue, E., and C. Barker, 1988, Effect of water washing on crude oil compositions. American Association Petroleum Geologists Bulletin, v. 72, p. 263-276.

Larter, S. R., 1988, The molecular characterization of kerogen—Applications to primary and secondary migration studies and to maturation modelling: 7th World Palynological Congress, August 1988.

Landrum, J. H., D. M. Richers, L. E. Maxwell, and W. Fallgatter, 1989, A comparative soilgas of the Flathead Valley and Townsend Valley, Montana: Journal of Geochemical Exploration, v. 34, p. 303-335.

Lane, E. C., and E. L. Garton, 1935, "Base" of a crude oil: U.S. Bureau of Mines, Report of Investigations 3279.

Langford, F. F., and M. M. Blanc-Valleron, 1990, Interpreting Rock-Eval data using graphs of pyrolizable hydrocarbons vs. total organic carbon: American Association of Petroleum Geologists Bulletin, v. 74, p. 799-804.

LaPlante, R. E., 1974, Hydrocarbon generation in Gulf Coast Tertiary sediments: American Association of Petroleum Geologists Bulletin, v. 58, p. 1281-1289.

Larsen, V. B., S. M. Aasheim, and J. M. Marset, 1981, Alpha structure—A field case study in the Silver Block, in Norwegian Symposium on Exploration, 1981: Norwegian Petroleum Society, Bergen, p. 14.1-14.36.

Larter, S. R., 1984, Application of analytical pyrolysis techniques to kerogen characterization and fossil fuel exploration exploitation , in K. Voorhees, ed., Analytical Pyrolysis Techniques and Applications: Butterworth, London, p. 212-275.

Larter, S. R., 1985, Improved kerogen typing for petroleum source rock evaluation: Nature, v. 318, p. 277-280.

Larter, S. R., and A. G. Douglas, 1980, A pyrolysis-gas chromatography method for kerogen typing, in A. G. Douglas, and J. R. Maxwell, eds., Advances in organic geochemistry 1979: Pergamon Press, Oxford, p. 579-584.

Larter, S. R., and J. T. Sentfle, 1985, Improved kerogen typing for petroleum source rock analysis: Nature, v. 318, p. 277-280.

Laubmeyer, G., 1933, A new geophysical prospecting method, especially for deposits of hydrocarbons: Petroleum, v. 29, p. 1-4.

Lawrence, T. D., S. Ball, and M. Harris, 1984, Continuous carbon/oxygen and neutron lifetime log proposed interpretation for organic and/or shaly depositional environments, SPWLA Twenty-Fifth Annual Logging Symposium, Paper QQ.

Lehner, F. K., D. Marsal, L. Hermans, and A. Van Kugk, 1988, A model of secondary hydrocarbon migration as a buoyancy-driven separate phase flow: Migration of Hydrocarbons, in B. Doligez, ed., Sedimentary Basins: Editions Technip, Paris, p. 457-472.

LeTran, K., 1971, Geochemical study of hydrogen sulfide sorbed in sediments, in H. R. von Gaertner and H. Wehner, eds., Advances in Organic Geochemistry 1971: Pergamon Press, Oxford, p. 717-726.

Lewan, M. D., 1983, Effects of thermal maturation on stable organic carbon isotopes as determined by hydrous pyrolysis of Woodford Shale: Geochimica et Cosmochimica Acta, v. 47, p. 1471-1479.

Lewan, M. D., 1985, Evaluation of petroleum generation by hydrous pyrolysis experimentation: Philosophical Transactions of the Royal Society of London, v. A 315, p. 124-134.

Leythaeuser, D., Schaefer R. G., Corford, C., and Weiner, B., 1979, Generation andmigration of light hydrocarbons (C2-C7) in sedimentary basins: Organic Geochemistry, v. 1, p. 191-204.

Leythaeuser, D., Schaefer, R. G., and Yükler, 1980, Diffusion of light hydrocarbons through near-surface rocks: Nature, v. 284, p. 522-525.

Leythaeuser, D., Schaefer, R. G., and Yükler, 1982, Role of diffusion in primary migration of hydrocarbons: American Association Petroleum Geologists Bulletin, v. 66, p. 408-429.

Leythaeuser, D., Schaefer, R. G., and Pooch, H., 1983a, Diffusion of light hydrocarbons in subsurface sedimentary rocks: American Association Petroleum Geologists Bulletin, v. 67, p. 889-895.

Leythaeuser, D., M. Bjoröy, A. S. Mackenzie, A. G. Schaefer, and F. J. Altebaumer, 1983b, Recognition of migration and its effects within two coreholes in shale/sandstone sequences from Svalbard, Norway, in M. Bjoröy, et al., eds., Advances in Organic Geochemistry 1981: John Wiley, Chichester, p. 136-146.

Leythaeuser, D., Mackenzie, A. S., Schaefer, R. G.,and Bjoröy, M., 1984, A novel approach for recognition and quantitation of hydrocarbon migration effects in shale/sandstone sequences: American Association Petroleum Geologists Bulletin, v. 68, p. 196-219.

Leythaeuser, D., R. G. Schaefer, and M. Radke, 1987, Geochemical effects of primary migration of petroleum in Kimmeridge source rocks from Brae field area, North Sea. I. Gross composition of C15+-soluble organic matter and molecular composition of C15+-saturated hydrocarbons: Geochimica Cosmochimica Acta, v. 52, p. 701-713.

Leythaeuser, D., R. Littke, H. S. Poelchau, M. Radke, and R. G. Schaefer, 1989, Differences in expulsion mechanisms and effects depending on kerogen quality of source rocks (abs.): 14th International Meeting on Organic Geochemistry, Sept. 18-22, 1989, Paris , France, Abstract no. 139.

Lopatin, N. V., 1971, Temperature and geologic time factors in coalification (in Russian): Akademiia Nauk SSSR, Izvestiia Seriia Geologicheskalaya, v. 3, p. 95-106.

Lopatin, N. V., 1976, Historical-genetic analysis of petroleum generation: application of a model of uniform continuous subsidence of the oil-source bed (in Russian): Akademica Nauk SSSR, Izvestiya Seriya Geologicheskaya, v. 8, p. 93-101.

Louis, M., 1967, Cours de Geochimie du Pétrole: Societe des Editions Technip. et Institut Français du Pétrole, p. 244-260.

Loutit, T. S., J. Hardenbohl, P. R. Vail, and G. R. Baum, 1988, Condensed sections: The key to age determinations and correlation of continental margin sequences, in C. K. Wilgus et al., eds, Sea-Level Changes: An Integrated Approach: SEPM Special Publication 42, p. 183-213.

McAuliffe, C. D. ,1966, Solubility in water of paraffin, cycloparaffin, olefin, acetylene, cycloolefin, and aromatic hydrocarbons: Journal Physical Chemistry, v. 70, p. 1267-1275.

McAuliffe, C. D., 1979, Oil and gas migration—Chemical and physical contraints: American Association Petroleum Geologists Bulletin, v. 63, p. 761-781.

McAuliffe, C. D. ,1980, The multiple gas-phase equilibrium method and its application to environmental studies, in L. Petrakis and F. T. Weiss, eds., Petroleum in the Marine Environment: American Chemical Society, p. 193-218.

McCartney, J. T., and M. Teichmüller, 1972, Classification of coals according to degree of coalification by reflectance of the vitrinite component: FUEL, v. 51, p. 64-68.

McCaslin, 1984, AAPG looks at Utah, Nevada basins: Oil and Gas Journal, October 8, p. 107-108.

McCrossan, R. G., N. L. Ball, and L. R. Snowdon, 1971, An evaluation of surface geochemical prospecting for petroleum, Olds-Caroline area, Alberta: Geological Survey of Canada Paper 71-31, 101p.

Macgregor, D. S., and A. S. Mackenzie, 1987, Quantification of oil generation in the Malacca Straits region: Proceedings of 15th Annual Convention of Indonesian Petroleum Association, Jakarta, 305-319.

McIver, R. D., 1962, The crude oils of Wyoming—Product of depositional environment and alteration, in Symposium on Early Cretaceous Rocks of Wyoming and Adjacent Areas: Wyoming Geological Association 17th Annual Field Conference, p. 248-251.

McKenna, E. J. ,1972, Microbial metabolism of normal and branched chain alkanes, in Degradation of Synthetic Organic Molecules in the Biosphere: National Academy of Sciences. p. 73-97.

Mackenzie, A. S., 1984, Application of biological markers in petrogeochemistry, in J. Brooks and D. Welte, eds., Advances in Petroleum Geochemistry, v. 1.: Academic Press, London, p. 115-214.

Mackenzie, A. S., and J. R. Maxwell, 1981, Assessment of thermal maturation in sedimentary rocks by molecular measurements, in J. Brooks, ed., Organic maturation studies and fossil fuel exploration: Academic Press, London, p. 239-254.

Mackenzie, A. S., and D. H. McKenzie, 1983, Isomerization and aromatization of hydrocarbons in sedimentary basins formed by extension: Geological Magazine, v. 120, n. 5, p. 417-528.

Mackenzie, A. S., and Quigley, T. M., 1988, Principles of Geochemical Prospect Appraisal: American Association Petroleum Geologists Bulletin, v. 72, n. 4, p. 399-415.

Mackenzie, A. S., R. L. Patience, J. R. Maxwell, M. Vandenbroucke, and B. Durand, 1980a, Molecular parameters of maturation in the Toarcian shales, Paris Basin, France. I. Changes in the configurations of acyclic isoprenoid alkanes, steranes and triterpanes: Geochimica et Cosmochimica Acta, v. 44, p. 1709-1721.

Mackenzie, A. S., J. M. E. Quirke, and J. R. Maxwell, 1980b, Molecular parameters of maturation in the Toarcian shales, Paris Basin, France. II. Evolution of Metallo-porphyrins, in A. G. Douglas, and J. R. Maxwell, eds., Advances in Organic Geochemistry 1979: Pergamon Press, Oxford, p. 239-248.

Mackenzie, A. S., C. F. Hoffmann, and J. R. Maxwell, 1981a, Molecular parameters of maturtion in the Toarcian shales, Paris Basin, France. III. Changes in aromatic steroid hydrocarbons: Geochimica et Cosmochimica Acta, v. 45, p. 1345-1355.

Mackenzie, A. S., C. A. Lewis, and J. R. Maxwell, 1981b, Molecular parameters of maturation in the Toarcian shales, Paris Basin, France. IV. Laboratory thermal alteration studies: Geochimica et Cosmochimica Acta, v. 45, p. 2369-2376.

Mackenzie, A. S., S. C. Brassell, G. Eglington, and J. R. Maxwell, 1982, Chemical fossils: The geological fate of steroids. Science: v. 217, p. 491-504.

Mackenzie, A. S., J. R. Maxwell, M. L. Coleman, and C. E. Deegan, 1983a, Biological marker and isotope studies of North Sea crude oils and sediments: Proceedings 11th World Petroleum Congress 1983, Heyden, v. 2, p. 17-29.

Mackenzie, A. S., G. A. Wolff, and J. R. Maxwell, 1983b, Fatty acids in some biodegraded petroleums: Possible origins and significance, in M. Bjoröy et al., eds, Advances in Organic Geochemistry 1981: John Wiley, Chichester, p. 637-649.

Mackenzie, A. S., D. Leythaeuser, R. G. Schaefer, and M. Bjoröy, 1983c, Expulsion of petroleum hydrocarbons from shale source rocks: Nature, London, v. 301, p. 506-509.

Mackenzie, A. S., D. Leythaeuser, P. Muller, T. M. Quigley, and M. Radke, 1988, The movement of hydrocarbons in shales: Nature, v. 331, p. 63-63.

McKenzie, D. P., 1984, The generation and compaction of partial melts: Journal of Petrology, v. 25, p. 713-765.

McLimans, R. K., 1987, The application of fluid inclusions to migration of oil and diagenesis in petroleum reservoirs: Applied Geochemistry, v. 2, p. 585-603.

Magoon, L. B., and G. E. Claypool, 1985, Alaska North Slope oil/rock correlation study: American Association of Petroleum Geologists Studies in Geology, n. 20, 682p.

Mann, D. M., and A. S. Mackenzie, 1990, The prediction of pore fluid pressures in sedimentary basins: Marine and Petroleum Geology, v. 7, p. 55-65.

Mattavelli, L., T. Ricchiuto, D. Grignani, and M. Schoell, 1983, Geochemistry and habitat of natural gases in Po Basin, Northern Italy: American Association of Petroleum Geologists Bulletin, v. 67, p. 2239-2254.

Matthews, M. D., 1985, Effects of hydrocarbon leakage on earth surface materials: Paper presented at Symposium IV, Unconventional Methods in Exploration for Petroleum and Natural Gas, Southern Methodist University. Dallas. May 1-2, p. 27-44.

Matthews, M. D., V. T. Jones, and D. M. Richers, 1984, Remote sensing and hydrocarbon leakage, in Proceedings of the 3rd International Symposium on Remote Sensing of the Environment: ERIM, Colorado Springs, April, 1984, v. 2, p. 663-670.

May, R., W. Freund, E. P. Muller, and K. P. Dostal, 1968, Modellversuche uber Isotopenfraktionierung von Erdgaskomonenten wahrend der Migration: Z. angew. Geol, v. 14, p 376-380.

May, W. E., S. P. Wasik, and D. H. Freeman, 1978a, Determination of the aqueous solubility of polynuclear aromatic hydrocarbons by a coupled column liquid chromatographic technique: Analytical Chemistry, v. 50, p. 175-179.

May, W. E., S. P. Wasik, and D. H. Freeman,1978b, Determination of the solubility behavior of some polycyclic aromatic hydrocarbons in water: Analytical Chemistry, v. 50, p. 997-1000.

Meissner, F. F., 1978, Petroleum geology of the Bakken Formation, Williston Basin, North Dakota and Montana, in Williston Basin Symposium: Montana Geological Society, 24th Annual Conference, Billings, Montana, p. 207-227.

Mello, M. R., P. O. Gaglianone, S. C. Brassell, and J. R. Maxwell, 1988, Geochemical and biological marker assessment of depositional environments using Brazilian offshore oils: Marine and Petroleum Geology, v. 5, p. 205-223.

Mendelson, J. D., and M. N. Toksoz, 1985, Source rock characterization using multivariate analysis of log data: SPWLA Twenty-sixth Annual Logging Symposium.

Menendez, R., 1973, Composition isotopique du carbone dans les gaz provenant des sondages d'Aquitaine: Bulletin du Centre de Recherches de Pau, v. 7, p. 69-81.

Meyer, B. L., M. H. Nederlof, 1984, Identification of source rocks on wireline logs by density/resistivity and sonic transit time/resistivity crossplots: American Association Petroleum Geologists Bulletin, v. 68, n. 2, p 121-129.

Milner, C. W. D., M. A. Rogers, and C. R. Evans, 1977, Petroleum transformations in reservoirs: Journal of Geochemical Exploration, v. 7. p. 101-153.

Mogilevskii, G. A., 1938, Microbiological investigation in connection with gas surveying: Rasvedka Nedr, v. 8, p. 59-68.

Moldowan, J. M., and W. K. Seifert, 1980, First discovery of botryococcane in petroleum: Chemical Communications, v. 34, p. 912-914.

Moldowan, J. M., W. K. Seifert, E. Arnold, and J. Clardy, 1984, Structure proof and significance of stereoisomeric 28,30-bisnorhopanes in petroleum and petroleum source rocks: Geochemica et Cosmochimica Acta., v. 4, p. 1651-1661.

Moldowan, J. M., W. K. Seifert, and E. J. Gallegos, 1985, Relationship between petroleum composition and depositional environment of petroleum source rocks: American Association of Petroleum Geologists Bulletin, v. 69, p. 1255-1268.

Momper, J. A., 1978, Oil migration limitations suggested by geological and geochemical considerations, in W. H. Roberts, III, and R. J. Cordell, eds., Physical and chemical controls on petroleum migration: American Association of Petroleum Geologists Continuing Education Course Note Series 8, p. B1-B60.

Momper, J. A., and J. A. Williams, 1984, Geochemical exploration in the Powder River basin, in G. Demaison and R. J. Murris, eds., Petroleum Geochemistry and Basin Evaluation: American Association of Petroleum Geologists Memoir 35, p. 181-191.

Montel, F., and P. L. Gouel, 1985, Prediction of compositional grading in a reservoir column: Society of Petroleum Engineers, SPE No. 14410, Las Vegas.

Monthioux, M., P. Landais, and J. C. Monin, 1985, Comparison between natural and artificial maturation of humic coals from the Mahakam delta, Indonesia: Organic Geochemistry, v. 8, p. 275-292.

Monthioux, M., P. Landais, and B. Durand, 1986, Comparison between extracts from natural and artificial maturation series of Mahakam delta coals, Organic Geochemistry, v. 10, p. 299-311.

Moser, R. D., 1986, Mass transfer by thermal convection and diffusion in porous media, in Proceedings of the 10th U.S. National Congress of Applied Mechanics, Austin, Texas: American Society of Mechanical Engineers, New York, p. 89-96.

Newman, J., and N. A. Newman, 1982, Reflectance anomalies in Pike River coals: Evidence of variability in vitrinite type, with implications for maturation studies and "Suggate rank": New Zealand Journal of Geology and Geophysics, v. 25, 233-243.

Nie, N. H., C. H. Hull, J. G. Jenkins, K. Steinbrenner, and D. H. Bent, 1975, Statistical Package for the Social Sciences (2nd edition): McGraw-Hill, New York.

Oehler, D. Z., and B. K. Sternberg, 1984, Seepage-induced anomalies, "false" anomalies, and implications for electrical prospecting: American Association of Petroleum Geologists Bulletin, v. 68, p. 1121-1145.

Orr, W. L., 1983, Comments on pyrolitic hydrocarbon yields in source-rock evaluation, in M. Bjoröy et al., eds., Advances in Organic Geochemistry 1981, p. 775-787

Orr, W. L., 1986, Kerogen/asphaltene/sulfur relationships in sulfur-rich Monterey oils: Organic Geochemistry, v. 10, p. 499-516.

Orr, W. L., and J. S. Sinninghe-Damste, 1990, Geochemistry of sulfur in petroleum systems, in W. L. Orr and J. S. Sinninghe-Damste, eds., Geochemistry of sulfur in fossil fuels: American Chemical Society, Washington, D. C., American Chemical Society Symposium Series No. 429, p 2-29.

Oudin, J. L., B. Durand, and M. Vandenbroucke, 1984, A biological marker study of coals, shales and oils from the Mahakam delta, Kalimantan, Indonesia: Chemical Geology, v. 42, pp 1-23.

Palacas, J. G., 1978, Preliminary assessment of organic carbon content and petroleum source rock potential of Cretaceous and lower Tertiary carbonates, South Florida Basin: Gulf Coast Association of Geological Societies Transactions, v. 28, p. 357-381.

Palacas, J. G. 1988, Characteristics of carbonate source rocks of petroleum, in L. B. Magoon, ed., Petroleum Systems of the United States: U.S. Geological Survey Bulletin 1870, p. 20-25.

Palacas, J. G., D. E. Anders, and J. D. King, 1984, South Florida Basin—Prime example of carbonate source rocks of petroleum, in J. G. Palacas, ed., Petroleum geochemistry and source rock potential of carbonate rocks: American Association of Petroleum Geologists Studies in Geology, n. 18, p. 71-96.

Palacas, J. G., D. Monopolis, C. A. Nicolaou, and D. E. Anders, 1986, Geochemical correlation of surface and subsurface oils, Western Greece, in D. Leythaeuser and J. Rullkötter, eds., Advances in Organic Geochemistry 1985: Pergamon Press, Oxford, p. 417-423.

Palacas, J. G., D. E. Anders, and J. D. King, 1988, Use of biological markers in determining thermal maturity of biodegraded heavy oils and solid bitumens: Preprints 4th International Conference on Heavy Crude and Tar Sands, Aug. 7-12, 1988, Edmonton, Alberta, 43 p.

Palciauskas, V. V., and P. A. Domenico, 1980, Microfracture development in compacting sediments: Relation to hydrocarbon-maturation kinetics: American Association Petroleum Geologists Bulletin, v. 64, n. 6.

Palmer, S. E., 1984, Effect of water washing on C15+ hydrocarbon fraction of crude oils from northwest Palawan, Philippines: American Association of Petroleum Bulletin, v 68, p. 137-149.

Passey, Q. R., S. Creaney, J. B. Kulla, F. J. Moretti, and J. D. Stroud, 1989, Well log evaluation of organic-rich rocks (abs.): 14th International Meeting on Organic Geochemistry, Paris, September 18-22, Paper 75.

Passey, Q. R., S. Creaney, J. B. Kulla, F. J. Moretti, and J. D. Stroud, 1990, A practical model for organic richness from porosity and resistivity logs: American Association Petroleum Geologists Bulletin, v. 74, p. 1777-1794.

Patton, K. H., and M. Manwaring, 1984, Evaluation of a Landsat derived tonal anomaly for hydrocarbon microseeps, southwest Kansas: Proceedings of the Third International Symposium on Remote Sensing of the Environment, Colorado Springs, Colorado, April 1984, v. 1, p. 441-450.

Peters, K. E., 1986, Guidedlines for evaluating petroleum source rocks using programmed pyrolysis: American Association Petroleum Geologists Bulletin, v. 70, p. 318-329.

Peters, K. E., and B. R. T. Simoneit, 1982, Rock-Eval pyrolysis of Quaternary sediments from Leg 64, Sites 479 and 480, Gulf of California, in J. R. Curray et al., eds., Initial Reports of the Deep Sea Drilling Project, v. 64, p. 925-931.

Peters, K. E., J. M. Moldowan, M. Scholle, and W. B. Hempkins, 1986, Petroleum isotopic and biomarker composition related to source rock organic matter and depositional environment, in D. Leythaeuser, and J. Rullkötter, eds., Advances in Organic Geochemistry 1985: v. 10, p. 17-27.

Philp, R. P., 1985a, Fossil fuel biomarkers: Applications and spectra: Methods in Geochemistry and Geophysics, Elsevier, New York, v. 23, p. 294.

Philp, R. P., 1985b, Biological markers in fossil fuel production: Mass Spectrometry Review, v 4, 1-54.

Philp, R. P., and P. T. Crisp, 1982, Surface geochemical methods used for oil and gas prospecting—A review: Journal of Geochemical Exploration, v. 17, p. 1-34.

Philp, R. P., and T. D. Gilbert, 1982, Unusual distribution of biological markers in Australian crude oil: Nature, London, v. 299, p. 245-247.

Philp, R. P., and T. D. Gilbert, 1985a, Biomarker distributions in Australian oils predominantly derived from terrigenous source material, in D. Leythaeuser and J. Rullkötter, eds., Advances in Organic Geochemistry 1984: Pergamon Journals, Oxford, p. 73-84.

Philp, R. P., and T. D. Gilbert, 1985b, Source rock and asphaltene biomarker characterization by pyrolysis-gas chromatography-mass spectrometry-multiple ion detection: Geochimica et Cosmochimica Acta, v. 49, p. 1421-1432.

Philp, R. P., and C. A. Lewis, 1987, Organic geochemistry of biomarkers: Annual Review of Earth and Planetary Science, v. 15, p. 363-395.

Picard, M. D., 1960, On the origin of oil, Eagle Springs field, Nye County, Nevada, in Guidebook to the Geology of East-Central Nevada, InterMountain Association of Petroleum Geologists and Eastern Nevada Geological Society, 11th Annual Field Conference. Salt Lake City, Utah, p. 237-244.

Powell, T. G., 1984, Some aspects of the hydrocarbon geochemistry of a Middle Devonian barrier reef complex, western Canada, in J. G. Palacas, ed., Petroleum geochemistry and source rock potential of carbonate rocks: American Association of Petroleum Geologists Studies in Geology, n. 18, p. 45-61.

Powell, T. G., and D. M. McKirdy, 1975, Geological factors controlling crude oil composition in Australia and Papua, New Guinea: American Association of Petroleum Geologists Bulletin, v. 59, p. 1176-1197.

Prauss, M., and W. Riegel, 1989, Evidence from phytoplankton associations for causes of black shale formation in epicontinental seas: Neues Jahrbuch fur Geologie und Palaontologie, Monatshefte, v. 11, p. 671-685.

Price, L. C., 1976, Aqueous solubility of petroleum as applied to its origin and primary migration: American Association of Petroleum Geologists Bulletin, v. 60, p. 213-244.

Price, L. C., 1979, Aqueous solubility of methane at elevated pressures and temperatures: American Association Petroleum Geologists Bulletin, v. 63, p. 1527.

Price, L. C., 1983, Solubility of crude oil in methane as a function of pressure and temperature: Organic Geochemistry, v. 4, p. 201-221.

Price, L. C., 1986, A critical review and proposed working model of surface geochemical exploration, in M. J. Davidson, ed., Unconventional Methods in Exploration for Petroleum and Natural Gas. Symposium IV: Dallas, Southern Methodist University, p. 245-303.

Price, L. C., and Barker, C. E., 1985, Suppression of vitrinite reflectance in amorphous rich kerogen—A major unrecognized problem: Journal Petroleum Geology, v. 8, p 59-84.

Price, L. C., L. M. Wegner, T. Ging, and C. W. Blount, 1983, Solubility of crude oil in methane as a function of pressure and temperature: Organic Geochemistry, v. 4, p. 201-221.

Primmer, T. J., S. A. Kerr, and K. J. Myers, 1990, Feasibility of in situ elemental analysis in mudrock evaluation, in A. Hurst, M.A. Lovell and A.C. Morton, eds., Geological applications of wireline logs, Geological Society Special Publication No. 48: The Geological Society, London, p. 203-210.

Pusey, W. C. 1973, How to evaluate potential gas and oil source rocks: World Oil, April 1973, p. 71-75.

Radke, M., 1987, Organic geochemistry of aromatic hydrocarbons, in J. Brooks and D. Welte, eds., Advances in Petroleum Geochemistry, v. 2: Academic Press, London, p. 141-207.

Radke, M., and D. H. Welte, 1983, The methylphenanthrene index (MPI): A maturity parameter based on aromatic hydrocarbons, in M. Bjoröy et al., eds., Advances in Organic Geochemistry 1981: John Wiley, Chichester.

Radke, M., D. H. Welte, and H. Willsch, 1982, Geochemical study on a well in the Western Canada Basin: Relation of the aromatic distribution pattern to maturity of organic matter: Geochimica et Cosmochimica Acta, v. 46, p. 1-10.

Radke, M., D. H. Welte, and H. Willsch, 1986, Maturity parameters based on aromatic hydrocarbons: Influence of the organic matter type, in D. Leythaeuser and J. Rullkötter, eds., Advances in Organic Geochemistry 1985, v. 10, p. 51-63.

Raiswell, R., and R. A. Berner, 1987, Organic carbon losses during burial and thermal maturation of normal marine shales: Geology, v. 15, p. 853-856.

Reed, J. D., H. A. Illich, and B. Horsfield, 1986, Biochemical evolutionary significance of Ordovician oils an their source: Organic Geochemistry, v. 10, p. 347-358.

Reitsema, R. H., F. A. Lindberg, and A. J. Kaltenback, 1978, Light hydrocarbons in Gulf of Mexico water: Sources and relation to structural highs: Journal of Geochemical Exploration, v. 10. p. 139-151.

Rejebian, V. A., A. G. Harris, and J. S. Huebner, 1987, Conodont color and textural alteration: An index to regional metamorphism, contact metamorphism, and hydrothermal alteration: Geological Society America Bulletin, v. 99, p. 471-479.

Rice, D. D., and G. E. Claypool, 1981, Generation accumulation, and resource potential of biogenic gas: American Association of Petroleum Geologists Bulletin, v. 65, p. 5-25.

Richardson, M., M. A. Arthur, J. S. Quinn, J. K. Whelan, and B. J. Katz, 1988, Depositional setting and hydrocarbon source potential of Miocene Gulf of Suez synrift evaporites: American Association of Petroleum Geologists Bulletin, v. 72, p. 1010-1021.

Richers, D. M., R. J. Reed, K. C. Horstman, G. D. Michels, R. N. Baker, L. Lundell, and R. W. Marrs, 1982, Landsat-soilgas geochemical survey of Patrick Draw oil field, Sweetwater County, Wyoming: American Association Petroleum Geologists Bulletin, v. 66, p. 903-922.

Richers, D. M., V. T. Jones, M. D. Matthews, J. Maciolek, R. J. Pirkle, and W. C. Sidle, 1986, A Landsat-soilgas geochemical survey of the Patrick Draw area, Sweetwater County, Wyoming, 1983 survey: American Association of Petroleum Geologists Bulletin, v. 70, p. 869-887.

Rigby, D., and J. W. Smith, 1982, A reassessment of stable carbon isotopes in hydrocarbon exploration: Erdoll Kohle Erdgas Petrochem., v. 35, p. 415-417.

Robert, P., 1980, The optical evolution of kerogen and geothermal histories applied to oil and gas exploration, in B. Durand, ed., Kerogen: Insoluble Organic Matter from Sedimentary Rocks: Editions Technip., Paris, p. 385-414.

Roberts, W. H., and R. J. Cordell, 1980, Problems of Petroleum Migration: American Association of Petroleum Geologists Studies in Geology, n. 10, p. 273.

Ronov, A. B. 1958, Organic carbon in sedimentary rocks (in relation to presence of petroleum): Translation in Geochemistry, n. 5, p. 510-536.

Ronov, A. B., and A. A. Yaroshevsky, 1969, Chemical composition of the earth's crust, in J. M. Hunt, ed., The earth's crust and upper mantle: American Geophysical Union Geophysical Monograph No. 13, p. 37-57.

Rosaire, E. E., 1940, Geochemical prospecting for petroleum: American Association Petroleum Geologists Bulletin, v. 24, p. 1401-1433.

Rossini, F. D., and B. J. Mair, 1959, The work of the API research project 6 on the composition of petroleum: 5th World Petroleum Congress, v. 5, p. 223-245.

Rouchet, J. du, 1980, Une explication paleoclimatique des grands gisements superficiels de gaz sec du nord de la Siberie occidentale: Bull. des Centres de Reserches Elf Aquitane, v. 4, n. 1, p. 119-142.

Rouchet, J. du, 1981, Stress fields: A key to oil migration: American Association Petroleum Geologists Bulletin, v. 65, p. 74-85.

Royden, L., F. Horvath, A Nagymarosy, and L. Stegena, 1983, Evolution of the Pannonian basin system. 2. Subsidence and thermal history: Tectonics, v. 2, p. 91-137.

Rubinstein, I., O. P. Strausz, C. Spyckerelle, R. J. Crawford, and D. W. S. Westlake,1977, The origin of the oil sand bitumens of Alberta: A chemical and microbial simulation study: Geochimica et Cosmochimica Acta, v. 41, p. 1341-1353.

Rullkötter, J., and D. Wendish,1982, Microbial alteration of 17a(H)-hopanes in Madagascar asphalts: Removal of C-10 methyl group and ring opening: Geochimica et Cosmochimica Acta, v. 46, p. 1545-1553.

Runge, A., 1980, Kohlenstoff and Wasserstoff-isotopenvariationen in organischen Sedimenten und Gasen: Chemie der Erde, v. 39, p. 52-62.

Sackett, W. M., 1977, Use of hydrocarbon sniffing in offshore exploration: Journal of Geochemical Exploration, v. 7. p. 243-245.

Saenz, G., 1984, Geochemical prospecting in Mexico, in P. A. Schenck, J. W. de Leeuw, and G. W. M. Lijmbach, eds. Advances in Organic Geochemistry 1983: Pergamon Press, Oxford, p. 715-725.

Saenz, G., 1987, Geochemical exploration for petroleum in a marshy area: examination and statistical analysis of C1-C7 hydrocarbons in near-surface samples: MS thesis, University of Texas at El Paso, El Paso, 130p.

Saenz, G., and Pingitore, N., 1989, Surface organicgeochemical prospecting for hydrocarbons: Multivariate analysis: Journal Geochemical Exploration, v. 34, p. 337-349.

Sage, B. H., and W. N. Lacey, 1939, Gravitational concentration gradients in static columns of hydrocarbon fluids: Transactions AIME, 132 120.

Sajgo, C., 1984, Organic geochemistry of crude oils from SE Hungary, in P. A. Schenck, J. W. de Leeuw, and G. W. M. Lijmbach, eds., Advances in Organic Geochemistry 1983: Pergamon Press, Oxford, p. 569-578.

Scalan, R. S., and T. D. Morgan, 1970, Isotope ratio mass spectrometer instrumentation and application to organic matter contained in recent sediments: International Journal of Mass Spectrometry Ion

Physics, v. 4, p. 267-281.

Schaefer, R. G., D. Leythaeuser, and H. van der Dick, 1983, Generation and migration of low molecular weight hydrocarbons in sediments from Site 511 of DSDP/IPOD Leg 71, Falkland Plateau, South Atlantic, in M. Bjoröy, et al., eds., Advances in Organic Geochemistry 1981: John Wiley, Chichester, p. 355-363.

Schmoker, J. W., 1981, Determination of organic-matter content of Appalachian Devonian shales from gamma-ray logs: American Association Petroleum Geologists Bulletin, v. 65, p. 1285-1298.

Schmoker, J. W., and T. C. Hester, 1983, Organic carbon in Bakken Formation, United States portion of Williston Basin: American Association Petroleum Geologists Bulletin, v. 67, p. 2165-2174.

Schmoker, J. W., and T. C. Hester, 1990, Formation resistivity as an indicator of oil generation—Bakken Formation of North Dakota and Woodford Shale of Oklahoma : The Log Analyst, Jan.-Feb., p. 1-9

Schoell, M., 1977, Die Erdgase der süddeutschen Molasse: Erdoel Erdgas Zeitschr, v. 93, p. 311-322.

Schoell, M., 1980, The hydrogen and carbon isotopic composition of methane from natural gases of various regions: Geochimica et Cosmochimica Acta, v. 44, p. 649-661.

Schoell, M., 1983a, Genetic characterization of natural gases: American Association Petroleum Geologists Bulletin, v. 67, 2225-2238.

Schoell, M., 1983b, Wasserstoff and kohlenstoff isotopenvariationen in organischen substanzen, Erdulen und Erdgasen: Geologisches Jahrbuch, Reihe D., v. 67, 164p.

Schoell, M., 1984, Recent advances in petroleum isotope geochemistry, in P. A. Schenck, J. W. de Leeuw, and G. W. M. Lijmbach, eds., Advances in Organic Geochemistry 1983: Pergamon Press, Oxford, p. 645-663.

Schoell, M., M. Teschner, H. Wehner, B. Durand, and J. L. Oudin, 1983, Maturity related biomarker and stable isotope variations and their application to oil/source rock correlation in the Mahakam delta. Kalimantan, in M. Bjoröy et al., eds., Advances in Organic Geochemistry 1981: John Wiley, Chichester, p. 156-163.

Schowalter, T. T., 1979, Mechanics of secondary hydrocarbon migration and entrapment: American Association Petroleum Geologists Bulletin, v. 63, p. 723-760.

Schulte, A. M., 1980, Compositional variations within a hydrocarbon column due to gravity: Society of Petroleum Engineers, SPE No. 9235, Dallas.

Sclater, J. G., and P. A. F. Christie, 1980, Continental stretching—An explanation of the post mid-Cretaceous subsidence of the central North Sea basin: Journal of Geophysical Research, v. 85, p. 3711-3739.

Segal, D. V., M. D. Ruth, I. S. Merin, H. Watanebe, K. Soda, O. Takano, and M. Sano, 1985, Correlation of remotely detected mineralogy with hydrocarbon production, Lisbon Valley, Utah, in technical papers of the American Society of Photogrammetry Annual Meeting, v. 51, p. 115-127.

Seifert, W. K., 1978, Steranes and terpanes in kerogen pyrolysis for correlation of oils and source rocks: Geochimica et Cosmochimica Acta, v. 42, p. 473-484.

Seifert, W. K., and J. M. Moldowan, 1978, Application of steranes, terpanes and monoaromatics to the maturation, migration, and source of crude oils: Geochimica et Cosmochimica Acta, v. 42, p. 77-95.

Seifert, W. K., and J. M. Moldowan, 1979, The effect of biodegradation of steranes and terpanes in crude oils: Geochimica et Cosmochimica Acta, v. 43, p. 111-126.

Seifert, W. K., and J. M. Moldowan, 1980, The effect of thermal stress on source-rock quality as measured by hopane stereochemistry, in A. G. Douglas and J. R. Maxwell, eds., Advances in Organic Geochemistry 1979: Pergamon Press, Oxford, p. 229-237.

Seifert, W. K., and J. M. Moldowan, 1981, Paleoreconstruction by biological markers: Geochimica et Cosmochimica Acta, v. 45, p. 783-794.

Seifert, W. K., and J. M. Moldowan, 1986, Use of biological markers in petroleum exploration, in R. B. Johns, ed., Biological Markers in the Sedimentary Record: Methods in Geochemistry and Geophysics 24: Amsterdam, Elsevier, p. 261-290.

Seifert, W. K., J. M. Moldowan, and R. W. Jones, 1980, Application of biological marker chemistry to petroleum exploration: 10th World Petroleum Congress Proceedings, v. 2, p. 425-438.

Seifert, W. K., J. M. Moldowan, and G. J. Demaison, 1984, Source correlation of biodegraded oils, in P. A. Schenck, , J. W. de Leeuw, and G. W. M. Lijmbach, eds., Advances in Organic Geochemistry 1983: Pergamon Press, Oxford, p. 633-643.

Shi, Jin-Yang, A. S. Mackenzie, R. Alexander, G. Eglinton, A. P. Gowar, G. A. Wolff, and J. R. Maxwell, 1982, A biological marker investigation of petroleums and shales from the Shengli oil field, the People's Republic of China: Chemical Geology, v. 35, p. 1-31.

Sidle, W. C., and D. M. Richers, 1985, Geochemical reconnaissance of Cretaceous inliers in north-central Oregon: American Association Petroleum Geologists Bulletin, v. 69, p. 412-421.

Siegel, F. R., 1974, Geochemical prospecting for hydrocarbons: Applied Geochemistry: Wiley-Interscience, New York, p. 228-252.

Silverman, S. R., 1965, Migration and segregation of oil and gas, in A. Young and J. E. Galley, eds., Fluids in Subsurface Environments: American Association Petroleum Geologists Memoir 4, p. 53-65.

Silverman, S. R., and S. Epstein, 1958, Carbon isotopic composition of petroleums and other sedimentary organic materials: American Association of Petroleum Geologists Bulletin, v. 42, p. 998-1012.

Sitler, J. A., 1979, Effects of source material on vitrinite reflectance: MS thesis, West Virginia University, Morgantown, Virginia, 108 p.

Smagala, T. M., C. A. Brown, and G. L. Nydegger, 1984, Log-derived indicator of thermal maturity, Niobrara formation, Denver basin, Colorado, Nebraska, Wyoming, in J. Woodward, F. F. Meissner, and J. L. Clayton, eds., Hydrocarbon Source Rocks of the Greater Rocky Mountain Region: Rocky Mountain Association of Geologists, Denver, p. 355-363.

Smith, H. M., 1968, Qualitative and quantitative aspects of crude oil composition: U.S. Bureau of Mines Bulletin 642, 136 p.

Smith, J. E., J. G. Erdman, and D. A. Morris, 1971, Migration, accumulation, and retention of petroleum in the earth: Proceedings of the 8th World Petroleum Congress, Moscow, U.S.S.R., p. 13-26.

Smith, J. F., 1971, The dynamics of shale compaction and the evaluation of pore fluid pressures: Journal Mathematical Geology, v. 3, p. 239-263.

Smith, J. W., D. Rigby, K. W. Gould, G. Hart, and A. J. Hargraves, 1985, An isotopic study of hydrocarbon generation processes: Organic Geochemistry, v. 8, p. 341-347.

Smith, N. A. C., 1927, The interpretation of crude oil analyses: U.S. Bureau of Mines, Report of Investigations No. 2806.

Snowdon, L. R., and K. J. Roy, 1975, Regional organic metamorphism in the Mesozoic strata of the Sverdrup Basin: Bulletin of Canadian Petroleum Geology, v. 23, p. 131-148.

Sofer, Z., 1984, Stable carbon isotope composition of crude oils: Application to source depositional environments and petroleum alteration: American Association of Petroleum Geologists Bulletin, v. 68, p. 31-49.

Sofer, Z., J. E. Zumberge, and V. Lay, 1986, Stable carbon isotopes and biomarkers as tools in understanding genetic relationship, maturation, biodegradation, and migration of crude oils in the Northern Peruvian Oriente (Maranon) basin, in Advances in Organic Geochemistry 1985: Organic Geochemistry, v. 10, p. 377-389.

Sokolov, V. A., 1933, The gas survey as a method of propecting for oil and gasformations: Technika, v. 1.

Sokolov, V. A., 1970, The theoretical foundations of geochemical prospecting for petroleum and natural gas and the tendencies of its development: Canadian Institute Mining and Metallurgy, v. 11, p. 544-549.

Spiro, B., D. H. Welte, J. Rullkötter, and R. G. Schaeffer, 1983, Asphalts, oils, and bituminous rocks from the Dead Sea area—A geochemical correlation study: American Association of Petroleum Geologists Bulletin, v. 67, p. 1163-1175.

Stach, E., D. Chandra, M. Mackowsky, G. Taylor, M. Teichmüller, and R. Teichmüller, 1982, Stach's Textbook of Coal Petrology: Gebruder Borntraeger, Berlin, 535p.

Stahl, W. J., 1973, Carbon isotope ratios of German natural gases in comparisons with isotope data of gaseous hydrocarbons from other parts of the world, in B. Tissot and F. Bienner, eds., Advances in Organic Geochemistry 1973: Editions Technip., Paris, p. 453-461.

Stahl, W. J., 1975, Kohlenstoff-Isotopenverhältnisse von Erdgasen: Erdoel und Kohle, Erdgas, Petrochemie, v. 28, p. 188-191.

Stahl, W. J., 1977, Carbon and nitrogen isotopes in hydrocarbon research and exploration: Chemistry, v. 20, p. 121-149.

Stahl, W. J., 1978, Reifeabhangigkeit der Kohlenstoff-Isotopenverhältnisse des Methans von Erdölgasen aus Norddeutschland: Erdoel und Kohle, Erdgas, Petrochemie, v. 31, p. 515-517.

Stahl, W. J., 1979, Carbon isotopes in petroleum geochemistry, in E. Jäger and J. C. Hunziken, eds., Lectures in Isotope Geology: Springer-Verlag, Berlin.

Stahl, W. J., 1980, Compositional changes and 13C/12C fractionations during the degradation of hydrocarbons by bacteria: Geochimica et Cosmochimica Acta, v. 44, p. 1903-1907.

Stahl, W. J., and B. D. Carey, Jr., 1975, Source-rock identification by isotope analyses of natural gases from fields in the Val Verde and Delaware Basins, West Texas: Chemical Geology, v. 16, p. 257-267.

Stainforth, J. G., and J. E. A. Reinders, 1990, Primary migration of hydrocarbons by diffusion through organic matter networks and its effect on oil and gas generation, in B. Durand, and F. Behar, eds., Advances in Organic Geochemistry 1989: Organic Geochemistry, v. 16, p. 61-74.

Staplin, F. L., 1969, Sedimentary organic matter, organic metamorphism, and oil and gas occurrence: Bulletin of Canadian Petroleum Geology, v. 17, p. 47-66.

Stocks, A. E., and S. R. Lawrence, 1990, Identification of source rocks from wireline logs, in A. Hurst, M.A. Lovell, and A.C. Morton, eds., Geological Applications of Wireline Logs, Geological Society Special Publication No. 48: The Geological Society, London, p. 241-252.

Stroganov, V. A., 1969, Some recommendations of the employment of geochemical prospecting: Geophysics, v. 7.

Sundararaman, P., W. R. Biggs, J. G. Reynolds, and J. C. Fetzer, 1988, Vanadylporphyrins, indicators of kerogen breakdown and generation of petroleum: Geochimica et Cosmochemica Acta, v. 52, p. 2337-2341

Supernaw, I. R., D. M. Arnold, and A. J. Link, 1978, Method for in situ evaluation of the source rock potential of earth formations: U.S. Patent 4,071,755.

Sweeney, J. J., and A. K. Burnham, 1990, Applications of a simple model of vitrinite reflectance based on chemical kinetics: American Association Petroleum Geologists Bulletin, v. 74, p. 1559-1570; errata v. 75, p. 848 (1991).

Sweeney, J. J., A. K. Burnham, and R. L. Braun, 1987, A model of hydrocarbon generation from Type I kerogen—Application to the Uinta basin, Utah: American Association Petroleum Geologists Bulletin, v. 71, p. 967-985.

Talukdar, S., O. Gallango, C. Vallejos, and A. Ruggiero, 1987, Observations on the primary migration of oil in the La Luna source rocks of the Maracaibo basin, Venezuela, in B. Doligez, ed., Migration of Hydrocarbons in Sedimentary Basins: Editions Technip, Paris, France, p. 59-78.

Tarafa, M. E., J. K. Whelan, and J. W. Farrington, 1988, Investigation on the effects of organic solvent extraction on whole-rock pyrolysis: Multiple-lobed and symmetrical P2 peaks: Organic Geochemistry, v. 12, p. 137-149.

Teichmüller, M., and B. Durand, 1983, Fluorescence microscopical rank studies on liptinites and vitrinites in peat and coals, and comparison with the results of the Rock-Eval pyrolysis: International Journal of Coal Geology, v. 2, p. 197-230.

Telnæs, N., and B. Dahl, 1986, Oil-oil correlation using multivariate techniques: Organic Geochemistry, v. 10, p. 425-432.

ten Haven, H. L., J. W. de Leeuw, T. M. Peakman, and J. R. Maxwell, 1986, Anomalies in steroid and hopanoid maturity indices: Geochimica Cosmochimica Acta, v. 50, p. 853-855.

Thode, H. G., 1981, Sulfur isotope ratios in petroleum research and exploration—Williston Basin: American Association of Petroleum Geologists Bulletin, v. 65, p. 1527-1535.

Thomas, B. M., P. Möller-Pedersen, M. F. Whitaker, and N. D. Shaw, 1985, Organic facies and hydrocarbon distributions in the Norwegian North Sea, in Thomas et al., eds., Petroleum Geochemistry in Exploration of the Norwegian Shelf: Norwegian Petroleum Society, Graham and Trotman, London, p. 3-26.

Thompson, K. F. M., 1988, Gas-condensate migration and oil fractionation in deltaic systems: Marine and Petroleum Geology, v. 5, p. 237-246.

Thompson S., 1985, Oil-generating coals, in B. M. Thomas et al., eds. Petroleum Geochemistry in

Exploration of the Norwegian Shelf: Graham and Trotman, London, p. 59-73.

Thompson, S., Cooper, B. S., Morley, R. J., and Barnard, P. C., and Smith, J. F., 1971, The dynamics of shale compaction and the evaluation of pore fluid pressures: Journal Mathematical Geology, v. 3, p. 239-263.

Ting, F. T. C., 1977, Microscopial investigation of the transformation (diagenesis) from peat to lignite: Journal of Microscopy, v. 109, 75-83.

Tissot, B. P., 1987, Migration of hydrocarbons in sedimentary basins: A geological, geochemical, and historical perspective, in Migration of Hydrocarbons in Sedimentary Basins: 2nd IFP Exploration Research Conference, Editions Technip, Paris, p. 1-20.

Tissot, B. P., and J. Espitalié, 1975, L'evolution thermique de la matiére organique des sédiments: Applications d'une simulation mathématique: Revue de l'Institut Français du Pétrole, v. 30, p. 734-777.

Tissot, B. P., and M. Vandenbroucke, 1983, Geochemistry and pyrolysis of oil shales, in Geochemistry and Chemistry of Oil Shales, Washington: ACS Symposium Series No. 230.

Tissot, B. P., and D. H. Welte, 1984, Petroleum formation and occurrence: A new approach to oil and gas exploration: Springer-Verlag, Berlin, 538p.

Tissot, B. P., B. Durand, J. Espitalié, and A. Combaz, 1974, Influence of nature and diagenesis of organic matter in formation of petroleum: American Association of Petroleum Geologists Bulletin, v. 58, p. 499-506.

Tissot, B. P., G. Deroo, and A. Hood, 1978, Geochemical study of the Uinta basin—Formation of petroleum from the Green River Formation: Geochimica Cosmochimica Acta, v. 42, p. 1469-1485.

Tissot, B. P., G. J. Demaison, P. Masson, J. R. Delteil, and A. Combaz, 1980, Paleoenvironment and petroleum potential of Middle Cretaceous black shales in Atlantic Basins: American Association Petroleum Geologists Bulletin, v. 64, p. 2051-2063.

Tissot, B. P., R. Pelet, and P. Ungerer, 1987, Thermal history of sedimentary basins: Maturation indices, and kinetics of oil and gas generation: American Association Petroleum Geologists Bulletin: v. 71, p. 1145-1466.

Toth, J., 1987, Petroleum hydrogeology: A potential application of groundwater science: Journal Geological Society of India, v. 29, n. 1, p. 172-179.

Toxopeus, Buiskool-, J. M. A., 1982, Selection criteria for the use of vitrinite reflectance as a maturity tool in petroleum exploration: Journal Petroleum Geology, v. 5, p. 215.

Ulmishek, G. F., and H. D. Klemme, 1990, Depositional controls, distribution and effectiveness of world's petroleum source rock—U.S: Geological Survey Bulletin, v. 1931, 59p.

Ungerer, P., and R. Pelet, 1987, Extrapolation of kinetics of oil and gas formation from laboratory experments to sedimentary basin: Nature, v. 327, p. 52-54.

Ungerer, P., A. Chiarelli, and J. L. Oudin, 1985, Modelling of petroleum genesis and migration with a biodimensional computer model in the Frigg sector, Viking Graben, in Thomas et al., eds., Petroleum Geochemistry of the Norwegian Shelf: Norwegian Petroleum Society, Graham and Trotman London, p. 121-130.

Ungerer. P., F. Behar, M. Villalba, O. D. Heum, and A. Audibert, 1988a, Kinetic modeling of oil cracking, in L. Mattavelli, and L. Novelli, eds., Advances in Organic Geochemistry 1987: Organic Geochemistry, v. 13, p. 857-868.

Ungerer, P. J. Espitalié, F. Behar, and S. Eggen, 1988b, Mathematical modelling of the interations between thermal cracking and migration in the formation of oil and gas: Compte Rendu Academie Science, Paris, v. 307, Series II, p. 927-934.

Ungerer, P., B. Doligez, P. Y. Chenet, J. Burrus, F. Bessis, E. Lafargue, G. Giroir, O. Heum, and S. Eggen, 1988c, A 2-D model of basin scale petroleum migration by two-phase fluid flow: Application to some case studies, in B. Doligez, ed., Migration of Hydrocarbons in Sedimentary Basins: Editions Technip, France, p. 415-456.

Vail, P. K., R. M. Mitchum, R. G. Todd, J. M. Widmier, S. Thompson, J. B. Sangree, J. N. Bubb, and W. G. Hatlelid, 1977, Seismic stratigraphy and global changes of sea level, in C. E. Payton, ed., Seismic Stratigraphy—Applications to Hydrocarbon Exploration: American Association Petroleum Geologists Memoir 26, p. 49-212.

Vail, P. K., F. Audemard, S. A. Bowman, P. N. Eisner, and G. Perez-Cruz, in press, The stratigraphic signatures of tectonics, eustasy and sedimentation, in

A. Seilacher and G. Eisner, eds., Cycles and Events in Stratigraphy: Springer-Verlag, Berlin/New York.

Vandenbroucke, M., B. Durand, and J. L. Oudin, 1983, Detection of migration phenomena in a geological series by means of C1-C35 hydrocarbon amounts and distributions, in M. Bjoröy et al., eds., Advances in Organic Geochemistry 1981: John Wiley, Chichester, p. 147-155.

van Eggelpoel, A., 1964, Comparison entre les porphyrins extraites de la rocke reservoir (argiles silicifices) d'Ozouri (Gabon) et celles de l'huile: Proceedings 2nd International Geochemical Congress, Paris.

van Grass, G., de Leeuw, J. W., Schenck, P. A., and Haverkamp, J., 1981, Kerogen of Toarcian shales of the Paris Basin: A study of its maturation by flash pyrolysis techniques: Geochimica et Cosmochimica Acta, v. 45, p. 2465-2474.

Van Krevelen, D. W., 1961, Coal: Amsterdam, Elsevier, 514p.

Vincent, P. W., I. R. Mortimore, and D. M. McKirdy, 1985, Hydrocarbon generation, migration and entrapment in the Jackson-Naccowlah area, ATP 259 P, Southwestern Queensland: APEA Journal, v. 25, p. 62-84.

Volkman, J. K., R. Alexander, R. I. Kagi, and G. W. Woodhouse, 1983, Demethylated hopanes in crude oils and their applications in petroleum geochemistry: Geochimica et Cosmochimica Acta, v. 47, p. 1033-1040.

Volkman, J. K., Alexander, R., Kagi, R. I., Rowland, S. J., and Sheppard, P. N., 1984, Biodegradation of aromatic hydrocarbons in crude oils from the Barrow Subbasin of western Australia: Organic Geochemistry, v. 6, p. 619-632.

Van Wagoner, J. C., H. W. Posamentier, R. M. Mitchum, P. R. Vail, J. F. Sarg, T. S. Loutit, and J. Hardenbohl, 1988, An overview of the fundamentals of sequence stratigraphy and key definitions, in Wilgus, C., Hastings, B., Ross, C. Posamentier, H., Van Wagoner, J., and Kendall, C. G. St. C, eds., Sea-Level Changes: An Integrated Approach: Society of Economic Paleontologists and Mineralogists Special Publication 42, p. 39-45.

Waples, D. W., 1980, Time and temperature in petroleum formation: Application of Lopatin's method to petroleum exploration: American Association Petroleum Geologists Bulletin, v. 64, p. 916-926.

Waples, D. W., 1985, Geochemistry in petroleum exploration: International Human Resources Development Corporation, Boston, 232p.

Wardroper, A. M. K., C. F. Hoffmann, J. R. Maxwell, A. J. G. Barwise, N. S. Goodwin, and P. J. D. Park, 1984, Crude oil biodegradation under simulated and natural conditions—II. Aromatic steroid hydrocarbons, in P. A. Schenck, J. W. de Leeuw, and G. W. M. Lijmbach, eds., Advances in Organic Geochemistry 1983: Pergamon Press, Oxford, p. 605-618.

Watts, N. L., 1987, Theoretical aspects of cap-rock and fault seals for single- and two-phase hydrocarbon columns: Marine and Petroleum Geology, v. 4, p. 274-307.

Weart, R. C., and G. Heimberg, 1981, Exploration radiometrics postsurvey drilling results, in Symposium II, Unconventional Methods in Exploration for Petroleum and Natural Gas: Southern Methodist University, Dallas, 1980, p. 116-123.

Webster, R. L., 1984, Petroleum source rocks and stratigraphy of the Bakken Formation in North Dakota, in J. Woodward et al., eds., Hydrocarbon Source Rocks of the Greater Rocky Mountain Region: Rocky Mountain Association of Geologists, Denver, p. 57-81.

Weisman T. J., 1980, Developments in geochemistry and their contributions to hydrocarbon exploration: Proceedings of 10th World Petroleum Congress, v. 2. p. 369-386.

Welte, D. H., 1988, Migration of hydrocarbons: Facts and Theory, in B. Doligez, ed., Migration of Hydrocarbons in Sedimentary Basins: Editions Technip, Paris, p. 393.

Welte, D. H., and M. A. Yükler, 1981, Petroleum origin and accumulation in basin evolution—A quantitative mode: American Association Petroleum Geologists Bulletin, v. 65, p. 1387-1396.

Welte, D. H., and M. A. Yükler, 1984, Petroleum origin and accumulation in basin evolution—A quantitative model, in G. Demaison, and R. J. Murris, eds., Petroleum Geochemistry and Basin Evaluation: American Association of Petroleum Geologists Memoir 35.

Welte, D. H., H. Kratochvil, J. Rullkötter, H. Ladwein, and R. G. Schaefer, 1982, Organic geochemistry of crude oils from the Vienna Basin and an assessment of their origin: Chemical Geology, v. 35, p. 33-68.

Whelan, J. K., J. M. Hunt, J. Jasper, and A. Huc, 1984, Migration of C1-C8 hydrocarbons in marine sediments, in P. A. Schenck, J. de Leeuw, and G. W. M. Lijmbach, eds., Advances in Organic Geochemistry 1983: Pergamon Press, Oxford, p. 683-694.

Williams, J. A., 1974, Application of oil correlation and source rock data to exploration in the Williston basin: American Association of Petroleum Geologists Bulletin, v. 58, p. 1243-1252.

Williams, J. A., M. Bjoröy, D. L. Dolcater, and J. C. Winters, 1986, Biodegradation in South Texas Eocene oils—Effects on aromatics and biomarkers, in Advances in Organic Geochemistry 1985: Organic Geochemistry, v. 10, p. 451-461.

Winter, A., 1987, Percolative aspects of petroleum migration, in B. Doligez, ed., Migration of Hydrocarbons in Sedimentary Basins: Editions Technip, Paris, p. 237-256.

Winters, J. C., and J. A. Williams, , 1969, Microbiological alteration of crude oil in the reservoir, in Symposium on Petroleum Transformation in the Geological Environment: American Chemical Society, Division of Petroleum Chemistry, New York, Sept. 7-12, Preprints 14, v. 4, p. E22-E31.

Winters, J. C., J. A. Williams, and M. D Lewan., 1983, A laboratory study of petroleum generation by hydrous pyrolysis, in M. Bjoröy, ed., Advances in Organic Geochemistry 1981: John Wiley, London, p. 524-533.

Wold, S., 1976, Pattern recognition by means of disjoint principal models: Pattern Recognition, v. 8, p. 127-139.

Wood, A. R. O., 1988, Reservoir simulation applied to the development of the Magnus field: Society of Petroleum Engineers, SPE No. 18353.

Wood, D. A., 1988, Relationships between thermal maturity indices calculated using Arrhenius equation and Lopatin method: Implications for petroleum exploration: American Association Petroleum Geologists Bulletin, v. 72, p. 115-134.

Yalcin, M. N., and D. H. Welte, 1987, 3-D computer-aided basin modelling of the Cambay Basin: A case history of hydrocarbon generation: Proceedings of the First Conference of Petroleum Geochemistry and Exploration, Balkema, Nederlands.

Yeh, H. W., and S. Epstein, 1981, Hydrogen and carbon isotopes of petroleum and related organic matter: Geochimica et Cosmochimica Acta, v. 45, p. 753-762.

Yükler, M. A., 1987, How essential is quantitative basin modelling in petroleum exploration?, in Proceedings of the 7th Biannual Petroleum Congress of Turkey: Turkish Association of Petroleum Geologists, p. 392-404.

Yükler, M. A., and D. H. Welte, 1980, A 3-dimensional deterministic dynamic model to determine geologic history and hydrocarbon generation and accumulation in a sedimentary basin, in Fossil Fuels, Hydrocarbons C2, 26th International Geological Congress,: Editions Technics, Paris, p. 271-285.

Zumberge, J. E., 1987, Prediction of source rock characteristics based on terpane biomarkers in crude oils: A multivariate statistical approach: Geochimica et Cosmochimica Acta, v. 51, p. 1625-1637.

Zumberge, J. E., S. E. Palmer, and C. J. Schiefelbein, 1984, Kinetics of sterane and hopane epimerization: Simulation by hydrous pyrolysis (abs.): Geological Society of America, 97th Annual Meeting Abstracts with Programs, v. 16, p. 706.

Index